FOLDING circle TETRAHEDRA

TRUTH IN THE GEOMETRY OF WHOLEMOVEMENT

8/2/05
To Bob —

Bradford Hansen-Smith

Wholemovement Publications
Chicago, IL 2005

Dedicated to those that have come before, and to all students and teachers I have had the pleasure to work with during this extraordinary exploration. And to Grounder who provides balance and direction... BH-S

ISBN # 0-9766773-0-X

First Edition

Cover images, front and back: Spiral, 23 folded circles, another view on page 277, Fig.2.

Published/distributed by:
Wholemovement Publications
4606 N. Elston # 3
Chicago IL, 60630
(773) 794-9763
bradhs@interaccess.com
www.wholemovement.com

Other books by Bradford Hansen-Smith:

The Geometry of Wholemovement; folding the circle for information, Wholemovement Publications. Chicago, IL. 1999
The Hands-on Marvelous Ball Book; Books for Young Readers, Scientific American, W.H. Freeman & Co./New York, 1995. Distributed by Wholemovement Publications.

Distributed by:
Wholemovement Publications; www.wholemovement.com
FroebelUSA; A division of American Traditional Toys. 1-888-774-2046
http://www.froebelusa.com

"Truth often becomes confusing and even misleading when it is dismembered, segregated, isolated, and too much analyzed. Living truth teaches the truth seeker aright only when it is embraced in wholeness and as a living spiritual reality, not as a fact of material science or as inspiration of intervening art."

The Urantia Book, Urantia Foundation, Chicago IL. 1955 p.2075

TABLE OF CONTENTS

PREFACE

When drawing a circle without first having folded the circle, we do not know what the image represents that we are drawing. Folding circles is experiential and generates information in ways not found in the symbol. This book offers a direct and simple approach to understanding the inclusive nature of the circle through folding the tetrahedron as a primary function of the circle. I have tried to stay to with what is fundamental to the nature of pattern and the formation process observed in folding the circle, keeping to a minimum information that is easily accessible in other books. To recognize patterned information crosses all educational and cultural boundaries and is crucial to all fields of understanding. This is particularly important in the development of young children. Through observations in folding and joining circles there is clarity about pattern of movement and the transforming process that occurs in what must be considered the only shape that is inherently and inclusively Whole.

Wholemovement is a comprehensive understanding of the word geometry. It claims self-referencing movement of the Whole. There are levels of truth about our spatial universe that are inherently demonstrated in the Wholeness of the circle form. Were this not so, the circle could not do what it does nor could we construct what we do when we draw circles. A distinction is made between the circle as Whole and the circle as part, which is not possible with any other shape or form. The economics of the circle shows no waste, all parts are multi-functional and necessary for full expression of ongoing generation and continual reformation.

This book is limited to using and exploring the nine creased lines in the circle that come from folding the tetrahedron. The pictures and step-by-step instructions show a process that provides a wide range of possible directions and connections to be explored. Everything in this book can be made by anyone who is able to fold these 9 lines. Folding the circle in half forms a tetrahedron pattern that holds information to then systematically fold the regular formed tetrahedron, which appears fundamental to all spatial formation. By folding multiple reformed tetrahedral units and joining in various systems, a great variety can be developed of which many are known and surprisingly there are many more that are unknown.

When folding, we observe and reflect on what is generated within the circle, bringing out geometry functions, mathematical thinking, and awareness that helps us facilitate clarity in many areas of understanding. All the information that is accumulated with each proportional fold is always principled to the first movement and unity of the circle. Folding circles demonstrates an inclusive, comprehensive, consistent, principled, hands-on process of in-formation generation that is simple, direct, and accessible to anyone who can fold a circle in half.

Originally this book was to be about folding only the tetrahedron to demonstrate the fundamentals of geometry for primary grade level teachers. With continued exploration, the tetrahedron opened far beyond my expectations. I hope it still serves that initial intention, and at the same time offers the means to explore further into the richness of the circle and the primacy of the tetrahedron. Folding the circle is a low tech, high yield process that requires we get in touch with our physical universe, that we respond to mind function, and we open to the divine creative spirit.

BH-S Jan. 2005

INTRODUCTION

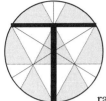raditionally we all draw pictures of circles.

Draw a picture of a circle.

Cut out the image from the paper on which it was drawn.

You are now holding the circle in your hand.

The dynamic information contained in this circle was not predictable by simply looking at the drawing. The nature and functioning of the circle remain unexplored, except as a symbolic image used in the development and teaching of mathematics. Pictures are used to construct other pictures in geometry, to substantiate and prove the value of abstract thinking and concepts formed in the mind. Take a moment and hold the circle, feel the movement, play with it, and wonder.

The circle is our most inclusive and effective manipulative teaching tool. This book explores the movement of the circle and the information generated through simple proportional folding. The only requirement is the ability to fold a circle in half and that makes this process experientially accessible to young children and to people without an educational background.

Folding circles is about revealing Wholeness in the form of the circle, not possible with any other shape. It is not about the "whole" used to describe single digit numbers, squares or ideas to which we wish to give emphasis and importance. The word "whole" is used as a relative indicator meaning coherent parts rather than anything *comprehensive, inclusive,* or *absolute.* What we call "whole" usually refer to sub sets of generalized parts. To differentiate between the all-inclusive Whole and the partial whole I will use a capital "W". The concept of Wholeness is infinitely beyond our human experience. Still it is important to acknowledge Wholeness and the inclusive nature of a single source; origin to existence that extends far beyond the limits of our imagination. The Whole is the source of life and where we live.

Folding and joining circles is working from a single patterned source that generates endless points, lines, angles, shapes and forms in countless combinations all through self-referencing divisional movements. It is impossible to put into one book all of the information and reconfigurations possible within a single circle. I have therefore limited this book to nine of an endless number of creases that can be folded to form the triangle/hexagon matrix of the circle. These nine creases are the channels of movement that form the tetrahedron. We will use these creased lines to explore the principles, the patterns, and the forming process that is foundational to understanding geometry functions that are basic to mathematical understandings.

The first fold of the circle generates a tetrahedron relationship. There is no discernable tetrahedron form, only the pattern of movement recognizable in folding two points together, which generates two more points, or four points in space each triangulated to each other. All subsequent folding is systematically and consistently principled to that first fold. The following eight folds give form to the first fold as pattern. By opening the tetrahedron to various angulations and joining in multiples, the "Five Platonic Solids" can be formed. We will go through the process fold-by-fold, discussing the information generated and the fundamentals of the regular and semi-regular polyhedra as they are demonstrated. Some extraordinary reformations are generated that do not appear in other forms of model making.

The most important observation about the circle is that it functions as both Whole and as a part simultaneously in a way no other shape or form can do. Our thinking changes when we experience the circle as origin to both 2-D and 3-D information. As we observe the interrelationships of all parts in context to each other within circle unity, the limitations of trying to understand each part in separation vanish. The multifunctional and interrelated nature of all parts to the Whole is reflected in the abstract mathematical formulations. The circle carries the same patterned information regardless of scale, number, or reconfiguration. Replication in multiples can be joined in many ways, forming an endless variety of complex systems. This fractal process shows how each individual unit circle, reconfigured differently, can join to form diverse individual systems that are all interconnected. There is no separation of parts from the whole circle to the Wholeness of the circle. The circle is the only formed expression that can inclusively demonstrate the concept of Wholeness.

The purpose is not to simply reproduce the circle-folded objects in this book, but rather to present a process by which people can discover the things that are not in this book. The fullness of the circle is revealed one fold at a time for anyone who will take the time to fold the circle and to observe the interactions formed between various combinations of parts. Through this principled and patterned process one can discover what the circle will do. Through play we become sensitive to the movement of the circle, the quality of folding, feeling what it will and will not do. We experience the symmetry--and out of symmetry--shifting that occurs between reconfigurations and the subtle variations of movement from one form to another. The body feels the movement, the mind sees the changes, and the spirit delights in the process.

It is only through exploring the Whole of the circle that we see demonstration of a "big picture". If we start with less, we get less. The Wholeness of the circle is instructive to the multiplication of individualized parts and relationships of arrangements through division without causing separation. This adherence to unity in the circle reveals a greater diversity than what can be constructed combining separated parts. What is constructed from pieces will always fall apart. What is formed from Wholeness will change and reform but never will it be less than Whole. There is no unity in assembling pieces; at best there is reflection of unity that belongs exclusively to the Whole.

Discrete geometry is used in many practical ways in our day-to-day living, but we give little thought to the larger implications as it applies to our personal lives. The study of geometry reveals generalizations about the mechanics of the physical universe. The geometry of Wholemovement reveals patterns of formation that universally work, making it clear what does not. Morals and ethics as concepts of "right and wrong" are also about what works and what does not work. The circle demonstrates an ethics providing the greatest potential for interaction between all parts that supports individual sustainability, and collective expression that allows full potential of the circle. Putting it another way, the Whole provides the greatest benefit and well being for an endless number of individualized parts over the greatest length of time necessary for each part towards maximum individual expression of both parts and Whole. The Whole is origin, support, potential, and ultimate expression. There is nowhere to go other than into what is yet unrevealed. Folding the circle is the only form of modeling I know that demonstrates an ethical context, where structural pattern forms to an ongoing stability through endless generations of reformation. Mathematics is an abstracted expression of the same patterns of universal organization.

Each individualized part is informational and contributes to the evolving movement of the Whole. Ethics is the workability of creation. Morality is how we individually relate to those ethics. When we separate ourselves, acting without regard to other parts, ethics breaks down and inhibits individual moral development. In the circle each part is individually essential to the movement of the Whole; otherwise it would not be there. Sometimes parts of the circle are folded to the inside, becoming invisible, and sometimes those parts move out into visibility, but always interacting to the forming, reforming, transforming, and informing. No part moves without affecting all parts of the circle. Passive and active are reciprocal functions of mechanical interactions. The human mind has capacity to recognize the geometric inter-workings of our physical reality, making mental connections to the unseen parts of this universe, and to conceptualize beyond the physical. We are developing moral beings in an ethical universe.

Folding paper circles is more than just learning how to make interesting objects. It is about understanding the whole of the circle shape and that it can be used to model the conceptual Whole. In talking about everything being in the circle, and considering the question of how to get a only a few things of everything that is in the circle out, I asked a first grade class what we needed to do first. One 6-year-old student said, "We need to make space". His solution to making space was to fold the circle in half. This is what we did. Space is defined through the edge-traced spherical movement by folding the circle in half in both directions. This movement folds a duality of inside and outside surfaces, a reciprocal repositioning of the circle, a spherical pattern. The creased line, axis/diameter/bisector, in the circle plane is not separate from the circle. The relationship between diameter and circumference set up a triangulated proportional division generating straight and curved edges, angles, areas, ratios, quadrilaterals, congruencies, and other more abstract information. Division occurs as a natural function of movement creating multiplicities of parts, all interrelated in the circle without separation.

Rarely do we consider the circle Whole. We have defined the image to be empty, zero, meaning nothing. The opposite of the nothing symbol is everything. The zero point on a number line, between positive and negative, is not "nothing". It is origin to endless positive and negative numbers in all directions, reciprocal functions in opposite sets. This location point is a small circle; concentric to any measured scale. Therefore all positive and negative functions lie within the Wholeness of the circle that is balanced to the first diameter folded line. The circle is both origin and container providing the greatest meaning for everything. When we start with nothing, we get nothing. Starting with the Whole we get everything together with the unidentified nothing.

The circle is origin to the number sequence 0-9. The circle comes first. Nine whole numbers are necessary to express primary number relationships within the circle. Folding the circle into the tetrahedron is nine moves, nine creases. From 0 to 9 are countless numbers of inexhaustible supply. The truth of this is in the experience of folding the circle, observing what happens and counting the parts as they are generated.

Drawing a straight line between two points generates nothing, it simply makes visible the distance that already exists between the points. The line is a symbol that represents a connection that is not there. When two points on any plane touch each other a connection is made and a creased line is formed half way between the two points and at right angles to the direction of movement. This line forms two new points of intersection with the boundary of that plane. This is what happens when any two points on the circumference are touched together. This right angle spatial self-referencing movement of the circle is principle to all geometry and mathematics.

The first fold of the circle shows a symmetrical dual of tetrahedra, a positive and negative function that is reciprocal one to the other. The tetrahedron is the only polyhedron that generates its own dual, which gives primacy to the tetrahedron over all other regular and irregular polyhedral formations. A single tetrahedron by itself, without the self-referencing dual function, is a generalized concept, it is missing all contextual information.

A sphere is the only form that inherently demonstrates the concept of absolute unity, an unqualified totality reflected in the order and the organization of our presently evolving universe. Within the divisions of the sphere are all patterns, symmetries, and geometries of associative organization that we have discovered. Within the sphere we find our own biological origin and individual development forming to the same universe principles and pattern reflected in the circle. When we cut a sphere in two parts, two separate circle planes are revealed, and the unity of the sphere has been destroyed. When we compress a sphere to a flat plane a circle is generated showing two individual circle planes without separation, without destroying Unity. Nothing of spherical information is lost in compression; it is simply compressed. Through folding the circle, spherical information is decompressed. Similar to a compact disk, distortion is accommodated and replication of the original is formed. The folded lines show 2-D compression and 3-D is in the spatial reconfigurations of the circle. These two forms, traditionally separated, are in fact the same pattern and process of common spherical origin.

Through individual experience we learn to differentiate and understand combinations of relationships that enlarge our context for understanding. We learn by observing spatial complexities and making conceptual connections to physical interactions. We play with what we learn, modeling in some way, to see if we have it "right". Modeling as a form of play is an important part of the self-learning process that is taken away from students when we remove direct experience. It is through direct involvement and interaction that meaningful problem solving occurs. The self-learning process, already in place at birth, observable in all animals, is denied the student as we teach down at them. Reciprocally this violates the teacher as well.

Folding the circle is a natural feed-back activity which continues this self-learning process. It provides the informational context necessary to identify and extend skills towards greater abstract mathematical thinking, as well as in all other areas where pattern recognition is essential. Hands-on experience is particularly important today where many children are without development in their motor skills, and have little connection with nature that has not been modified in some way. Most hands-on materials do not demonstrate comprehensive principles and have minimum information with which to make connections. The toys and educational materials we give children come from a reductionist way of thinking, lacking any inherent sense of a greater order. The largest context we provide for children is always to our own limitations.

Periodic house cleaning is necessary. It is essential to update words as our understanding of concepts evolves. The word "geometry" long defined as "earth measure", itself reveals its own greater meaning. "Geo" means earth. The earth is spherical. The sphere is the only form to inherently demonstrate the idea of the *Whole*. The suffix "metry" means measure. Measurement is about *movement,* keeping tract from one place to another. The word "geometry" comprehensively means *Wholemovement*. This redefining embodies the movement of spherical compression, and embraces folding the circle without denying any earth measuring system. This is movement of the Whole to itself.

When starting from the Whole and looking at the relationship of part to Whole we begin to understand the nature of connection between parts. How we understand our position relative to the Whole sets habits of reactions we have to everything around us. The relative cohesiveness of parts is always reforming and redefining the limits of any given combinations of systems. How we relate to the largest imaginable context outside of ourselves becomes the guiding morality that determines how we interact with the ethical conscript of what is around us. The largest context we find acceptable is what gives meaning and brings value to our lives. What we value determines the choices we make. From the world condition we can see that our collective choices have not been highly ethical. It is becoming increasingly important to educate ourselves and our children to the concept of the Whole. Wholeness, which the circle experiential represents, can serve as a model towards ethical behavior inherent to the interrelated nature of all things. There is an equalization of all parts in the circle, precisely because every one is individually and uniquely positioned. It is in the individualized differences of each part that give meaning to the interactions between parts. The order of pattern is what gives value, allowing it all to work.

The most comprehensive and *individual* understanding of the concept about the Whole is embodied in the word *God*. This is not about anybody's religion. There have been, are, and will be, many religions of widely diverse social construction. It is a natural impulse to give personality to origin and source creator of this greater universe environment. A name makes it easier to understand a relationship to the inexplicable Whole that is neither purely conceptual nor mechanical, but personal, as we ourselves have been personally created to individually experience. It is through the individualized spirit and mind that personality takes on meaning beyond our physical bodies. Physical points, lines, areas, and angles without context have no existence. The circle provides the meaning where each part is understood relative to the context of all others. The individual Whole-to-part relationship is fixed and directive to the changing relationships between parts-to-parts and parts-to-Whole. This triangulated pattern is observed in the first fold of the circle and is true for all human beings. By counting the points and the lines of relationship in this first fold of the circle, the number ten becomes apparent. This is the number pattern of the tetrahedron and will be discussed later as we make our first fold into the circle. One diameter in one circle is binary, the symbol ten without separation of parts.

To find what we don't see we have to look closer, look longer, perhaps look faster, quickly, with intention, and with great curiosity at what appears not to be there. Unformed relationships do not mean they are not there, only that we do not see them. The word "invisible" tell us that what is not visible is with *in* what is *visible.* To expand understanding we must transcend boundaries with vision towards developing sight to unseen relationships. This is what science, art, and religion are about; a process of the human mind trying to understand as much as possible about what we see, and to observe with greater clarity what we cannot see. These three basic components, this triangulation of human expression; the physical, the mental and the spiritual together provide understanding into the marvelous, mysterious, and extraordinary things that happen everywhere, all the time, in both the seen and unseen space of this universe. Folding circles is not just to teach what we already know about geometry and mathematics or to reproduce given forms, although that is there to do. It is about discovering Wholeness and being able to give attention to things usually considered outside the boundaries of academic and social, and often personal concerns. The circle gives demonstration to these things.

The inclusive nature of the biggest picture, the principles of what happens first, and the ethics of interactions are the primary reasons we should be studying geometry. Only to the degree that we are willing to comprehensively approach the inclusive nature of pattern—that which is principle, and structural generation—will we understand that it becomes essential in how we live our lives. Pattern is not the reality we experience; it is the living and evolving truth that leads us forward towards the source of all things. We have the capability to make choices based on intelligence mind function that bridges the physical and the spiritual. Collectively we make few ethical choices of a structural nature because it is not in our education to do so. Decisions made in the largest context, over the greatest period of time, for the greatest good for all people, will yield the greatest educational and evolutionary benefits.

As we approach education more comprehensively we will find there is an accumulated knowledge base that supports the development towards a more spiritually based ethical interaction that is available to all who are willing to make that choice. This is not a religious issue. It is an educational issue that incorporates balance of the spiritual, mental, and physical reality. Introducing ideas about structural patterns of universe formation and sustainability within the largest imaginable context is educational health. While human activities may not appear to be geometry, the movement of the circle clearly demonstrates a principled, organizational approach that gives balance and symmetry to the structural generation of all parts within and multiple extensions outward. The developing geometry of interaction is spiritual symmetry formed to physical expression. This must include human existence.

Higher mathematics has become abstract and separated from the experience of spatial phenomena that it describes and thus remains silent on the ethics inherent in the subject. Proportional relationships of interaction between parts have become number differentials. Students are asked to visualize and understand those things they have not experienced. In the same way it is unfair to ask teachers to teach that with which they have little or no experience. Students and teachers cannot be expected to make ethical choices when we do not have educational models or materials that extend beyond a complex of philosophical constructions. Folding circles is a principled activity providing a simple and common hands-on experience that is fundamental to the physical, supports conceptual understanding, and touches the spirit. The Wholeness of the circle gives meaning to the unknown factor in the balance of human equations.

One only needs to fold a tetrahedron to understand the principles, the ordering of relationships, and pattern necessary for ongoing development and exploration of the circle. This takes time; it is not a quick fix for inadequate education, but it does recalibrate intention. A fifth grade student asked why we needed to fold all these lines just to make a triangle. The answer of course is that we are not just making a triangle. When we do not start from the beginning we are missing information, which later on, because it is not there, will disrupt further development. We find the ability to balance in learning to walk. Without all nine lines we have missing information, thereby having to make assumptions based on guessing. This leads to faulty construction and misconceptions. The more information we have, the more we can do, the greater is our potential for accuracy and for being able to make meaningful connections and changes. One diameter gives the information for two more diameters. Three diameters give the information for six more lines. Nine lines form the tetrahedron. The tetrahedron is a seed pattern.

Folding the circle is an easy and engaging way to give students an understanding of traditional Euclidean geometry and at the same time the fundamentals for all other kinds of geometries. They are all the same pattern, albeit different forms that have been separated from each other. The circle functions as a tool and container for conceptualizing abstract mathematical information. Everything is experienced in the context of everything else. This process provides a comprehensive field, a free range for making diverse connections which gives a greater understanding of why what works, works. It is a healing process to educationally move towards Wholeness, based on moral choices made with understanding towards appropriate ethical behavior.

While some readers may disagree with the understanding that comes from my observations, I present them with a sincerity that reflects fifteen years of folding paper plate circles. These observations are simply what I am able to see and understand of what is happening. Anyone can fold the circle for themselves and through their own understanding find meaning. The process is the same for everyone with everyone getting something a bit different from it. Each teacher uses folding circles differently depending on what they bring to it, the connections they make, and how deep and wide they are willing to go.

Everything in this book represents only a few of the many reconfigurations and combinations possible by folding the circle and reconfiguring the nine lines used to form the tetrahedron. This is a first step in beginning to understand the nature of the circle. It is a process that replicates the forming of spatial pattern sequentially through time revealing much of what we observe as we look around us on this planet.

Our galaxy is a great breathing circular disk, a cosmic creation forming within a greater context. There are laws that govern all physical interactions. There are spiritual principles that are directive towards mind that develops a functioning balance and order between the seen and the unseen. There is no explaining the unseen truth in the divine nature of the Whole. It is to be experienced, even as reflected in the folding of the circle, or the opening of a flower, or the turning of the earth to reveal the effects of the sun every morning.

Section I

General information

USING THIS BOOK

This book is written as a guide for teachers of all grade levels, and for all those interested in geometry, math, art, origami, modeling and building things in general; and those who are just plain curious about the circle. This book will guide you towards some of what can be done folding circles. The circle, like any tool, does nothing if you only look at it, it must be used. Folding circles is an information-generating process. From the knowledgeable to the novice, all that use this tool will gain from it. To what purpose is for the user to decide. As with all tools, everybody uses them differently to individual intention.

The book is designed in three sections:
1) General information.
2) Mechanical tips for folding and joining.
3) Folding circles. This is divided in two sections intermixed between the folding activities and discussions about the information observed in what is folded.

The order of how things are presented and the connections I make have a logic that is not necessarily standard, but more freely explores what is in the circle. Repetition of information cannot be helped, given the multifunctional nature of interconnected parts. When things reappear there is usually a changed focus and perspective showing different connections. As you begin to fold you will discover reconfigurations of the circle not shown in this book; you will form your own connections, repeating in different ways what becomes important for you.

Section one

General information includes philosophical concepts that have surfaced from many years of folding circles and my experience working with students and teachers at all grade levels. The practical reasons for studying geometry are in the connections we make in our quest to understand the universe, its origin, and our collective and individual purpose. Religious connections come up as people reflect upon folding the circle. Students often identify in the folds of the circle various symbols such as those called "sacred" or religious, cult and gang markings, crop circles, letters, numbers, and signs of both ancient and contemporary usage. This makes studying the circle relevant to issues beyond traditional geometry. Wholemovement stimulates many aspects of our experiences which brings forth a broad perspective of information you will not get in any geometry book or mathematics curriculum. For that reason it is important for everyone to read this section. There are reflections on education as it relates to the folding process, particularly for primary grades. I present observations and connections to numbers, drawn images, and algebra, as descriptions to better understand the relationships within the circle pattern. Some of this information you will be familiar with, some of it you will not.

Section two

The *mechanical and technical* aspects of folding and joining circles are simply suggestions that will make the process easier. Accurate folding, strong creasing, and different ways of joining circles are covered. Not attending to these, some aspects can cause problems. I suggest you go over this information before working with students; you can help avoid unnecessary frustrations that arise from sloppy work habits and inadequate techniques. This is particularly important for young students and those with little experience working with their hands. Mechanical problem solving is simply part of doing things and not to be underestimated. A few students of all ages will have some initial trouble in this area, nothing unsermountable.

Section three

Folding circles is the last and largest section. Following many of the folding activities is a short discussion of general information and observations. The folding activities are on two-column pages with instructional text on the left and notated illustrations on the right. *Discussions* are spread across the full-page. This different formatting helps distinguish between folding and discussion, mostly, making each easily accessible without the other. Keep in mind it is one process. Interspersed are photographs of models providing another kind of information to go with the drawings.

In the classroom, discussion can be handled as it comes up fold-by-fold, or after much folding by going back and picking up that information. <u>Discussing what you fold is important.</u> During observations, reflection, and discussion, connections are often made between the abstractions of mathematics and the experience. This can happen individually, in groups, or with the entire class. Folding circles without the information limits it to just another folding activity. This information comes up in discussion, but it can only be a discussion if you participate and add your own observation and reflections. This is a do-it-yourself with everybody else process.

It is important to observe, describe, and explore the circle. Observations reveal basic directional information. Every circle starts with folding one diameter and develops sequentially to form the tetrahedron. The Five Platonic Solids are then formed and the process opens up to boundless exploration. Many of the activities are simply demonstrations of possible directions to explore and not things to be made as an end in themselves, though many are beautiful objects and instructive to have around for deeper reflection and contemplation.

Inside/outside folding, transformational systems, soft folding and pocket folding methods are treated individually as subdivisions of the basic circle folding activities.

Different approaches

Consistency is important. Each fold develops information based on previous folds. Without observation about what is being generated we do not know what options are available and are then working with an unnecessary handicap. Intuition is also an important part of this process, since we will never be aware of all the information that is available at any given time.

One teacher related how she proposed to her third and fourth grade classes that they could make a 3-D form from a 2-D circle without using pencils, rulers or scissors. They did not believe her. She walked them through their own discovery process, letting them figure it out. When they had folded the tetrahedron, it amazed them to do that. Her next step, without showing them anything, was for them to use four tetrahedra and find ways to join them. They were all delighted with the static and kinetic sculptures they made. It was from their intuitive approach, with guidance, that they had naturally begun to talk about what they had done. Introducing new terms and ideas was then easy. <u>They were now making connections from the solid experience of self-discovery that started with working intuitively to explore what they did not believe possible.</u>

After learning to fold the tetrahedron it is not necessary to follow the book step-by-step. Teachers sometimes learn along with their students. Sometimes more can be learned this way than by reading about it first. There is no one "right" approach. Whatever you do with it will be instructive if you pay attention. Then the book becomes simply a useful guide.

The general procedure I often use in classrooms is described in the beginning of the folding process. My approach varies depending on the needs and attitude of the class, grade level, size, focus, and my own interest at the moment. With limited time in any one classroom my demands are different than that of an ongoing classroom teacher. You will find the best way to introduce the circle to your students is through the connections that you think are important for them to make. One way does not work for all classes. Often I ask for observations and discussions about the circle before we even consider folding. This opens their minds to make connections they would not usually make. It can initiate questions about things we do not yet know. We talk about what we know only when it supports making connections to expand understanding. Discussions as a way to review often leads to making connections. Most important is the students' own discovery experience, and allowing them to give that expression through talking about it and to collectively share in the process.

Philosophical reflections

In *section 1* are philosophical reflections that come from my experience in folding circles. By folding you will make your own sense of it, individually integrating what becomes meaningful to you. Philosophical understanding is a natural synthesizing that comes from reflection and discussion about the process that gives value. Even young children have developed a personal understanding about their life and can talk about the generalizations and connections they experience. Unifying our experience through reflecting on the physical, mental and spiritual parts of ourselves is what we inherently do, and this becomes apparent in talking about the first fold in the circle. How you choose to integrate these three aspects determines what and how you present geometry to your students. This first approach is crucial to their sense-making experience. It can open them to, or limit future interest, both in the classroom and beyond.

Introductory activity

Folding circles is a totally hands-on process that can be used for introducing geometry and algebra at the same time. It provides an experiential pattern base to understand geometry that is inclusive of mathematics, art, biology, chemistry and all things we observe in the natural world. It is not necessary to talk about all of this at one time. It is important for students to question the origin of these shapes, forms, and ideas and to experience the interrelatedness and integration from a comprehensive approach. Do not underestimate five and six year-olds ability to understand and coordinate complex patterns; they have been doing so from birth. Now they need to talk about what they have experienced. They will, with guidance, talk about their ideas and what is meaningful in their young lives. Reflection is not age-determined; it is a feed back system we all have from birth. They will wonder in their own ways. This begins self-learning. Basic folding of the tetrahedron, forming the Platonic Solids, and exploring expanded systems are easily within the range of primary grade children. Algebraic thinking is a wonderful way for young minds to play with the abstractness of the information they folding. It encourages and develops the mind to find multiple relationships between parts that are not obvious.

Folding promotes discussion as an effective review for older students in both 2 and 3-D geometry. At the same time the circle gives them a comprehensive place from which to make connections between the abstract formulas, theorems, and functions of analytical thinking. This is as important for accelerated and gifted math students as it is for average or disadvantaged students. It gives concrete experience to the abstract expressions in the mind. It engages the creative spirit in the physical activities through mental calculations. Most importantly this process teaches us to feel and observe, and think about what we do.

The circle empowers people to explore on their own. The process and materials are not regulated by someone else's experience or understanding. There are no rules to discovery, no black line templates to follow, nobody else's limited understanding to get in the way. No special training is necessary, only the ability to fold a paper circle in half and giving attention to what is generated. The rest is observation, exploration, and problem solving, which seems particularly difficult for most students in our public school culture. The comprehensive nature of the folding-circle process allows students to go as far as they chose to. This activity provides comprehensive exploration on many levels of interest for all levels of abilities. It is as easy to introduce as an art activity and talk about the patterns of the folds as directive to designing both 2 and 3-D projects, as it is to talk about the mathematics of the same relationships. Students are always interested in showing others what they have learned, spontaneously showing others what they can do when they are excited about it. Sharing with someone is a natural response that goes along with self-learning. That is a crucial step in assimilating understanding.

Things to pay attention to:
- Paying attention.
- Group reflection and individual discoveries, sharing experiences.
- Understanding what we have done.
- Use common language, introduce specific words to add clarity.
- Look for things you cannot see.
- Consider the largest thing before making conjectures about little things.
- Rediscovering old things makes them new.
- Visualization developed through experience.
- The mind is a tool for problem solving the unknown.
- What comes first is principle, everything else changes.
- The difference of pattern is in the form.
- Feel the movement of the circle; let the circle feel you.
- - - -

You might make your own classroom list where students can add thoughts they feel deserve attention. Discover your own directives congruent to family, friends, community, and the entire human population. That is a good place from which to explore many areas of concern.

Purpose and Use

Basic vocabulary words are underlined when used for the first time. They are important for accurately describing geometric functions. Students will pick up new words in talking about what they do. The meaning usually comes out of the context if the connections are clear. Clarity with words helps students to better express what they know and helps them discover what they do not know. Students will sometimes go to their geometry books, looking for the definition of words we use in our discussions about what they have folded. Reading words is different than using them to describe our own activities.

This book can serve as a visual and conceptual stimulus for folding and joining circles as well as drawing and designing. My intention is not to go deeply into the details of analytical geometry, algebra, or the complexities of 2-D designs, proofs and construction methods; many reference books already exist in those areas. My attempt is to provide in one place a common activity for people to experience something of the total interconnections and co-ordination of the order, patterns and process reflected in all these separated and diverse areas of study.

Students must be allowed to make their own discoveries, step by step, guiding them in a process of how to look for information. Some teachers use my books to explore with the students. Others make the books available only after a certain amount of folding is experienced. A few teachers give it to students who learn for themselves and in turn teach the rest of the class. Some teachers use folding circles for extra curricular projects, sometimes for art activities, or geometry, or for math review, at other times to demonstrate similarities of functions in the sciences, or as after school and family activities. Folding circles are often used as a supplement for enrichment activities. Some find it useful for slower learners who have trouble making connections to abstract information while others use it primarily for folding with children who need experience doing things with their hands. How you use it will depend on your interest and needs. Some teachers think it looks too complicated and don't think they can use it at all, until they have folded their first tetrahedron. Then they begin to use it for many things on all grade levels.

16

PRINCIPLES

The word "principle" is indiscriminately used to give importance to our ideas. Ask people what the principles of geometry, of art, of math or of any subject are, and you will get many different answers as to what they think is most important. Having been reduced to an amplifier, the word principle has little meaning. To simplify and eliminate posturing I use the term "principle" to mean that which happens first. Comprehensively what is principle is the fundamental truth that brings into existence essential qualities of creation; of which we are part. What is principle is principle through out, without exception. Everything else changes. The principle fold in the circle distributes those qualities to everything that can be folded within the circle and to all combinations of multiple reformations.

The sphere is principle to the circle. The compression of the sphere is the first self-referencing movement that is reflected in the first movement of folding the circle in half. Listed are qualities that I have observed in describing the first fold of the circle; what are principle.

> Wholeness
> movement
> division
> duality
> triangulation
> consistency
> inner-dependency

Wholeness initiates *movement,* creating *division* that establishes *duality* in the pattern of *triangulation.* There is *consistency* of all parts to the movement, where each part is *inner-dependent* to the Whole. These seven principle qualities of the first fold are directives for continuation of all formative development. Notice the first five are generally about function and the last two are about quality of relationship.

These appear to be universal qualities, principle to all expressions of spatial formation. I find nothing that is more fundamental. These qualities are reflected everywhere in the natural world and are instructive to all aspects of folding and joining circles. These principle qualities describe a process that is far greater than the physical demonstration.

Folding the circle is a logical place to initiate understanding these fundamental qualities. There is no other form of modeling that as clearly and comprehensively reveal the progressive development extending from the first single movement into unexpected complexities. The circle reveals the beauty and truth about the process of formation. If these seven qualities were applied to our own lives we might manage the social interactions on this planet in a more responsible way. Seeing how we treat each other clearly shows there is no agreement among us about what we believe is principle. Without agreement on what is most important, there is little chance of agreement about those things of less importance.

This process is principled to the first movement, which causes division, forming a multiplicity of parts that are then added and subtracted, creating innumerable, individualized, interrelated and coordinated systems. Teaching adding and subtracting first, and then multiplication and division is backwards to comprehensive logical thinking. Where do the numbers come from that we first add and subtract? In our attempts to simplify understanding, we reduce things to meaning-less, and then we lose understanding of what is principle. We have not been taught to think comprehensively. We have developed a system of constructed logic based on the progressive difficulty of digitized information, rather than anything that reflects a principled process.

Folding the circle [O] in half creates a line of *division* [1] that forms a *duality* [2] of area. This is *triangulation* [3]; a relationship of 0+1+2=3. These number symbols; 0, 1, 2, 3, add up to the number 6. Counting the parts there are two points touched together forming one line with two points of intersection creating two individual areas (p.36) The circle has not been broken, only reformed showing each part to be *inner-depended* upon the *movement* of the *Whole* and *consistent* to the circle/sphere. The total number of parts in this movement is 7, a principled reflection.

Before all else, these seven qualities stand as principle information generated about the relationship of parts in this one fold of the circle in half. They are qualities that describe the same movement event. They also need to be considered individually and therefore it is useful to elaborate on the individual aspects of these qualities by using different words descriptive of general observations. You can add your own observations to this list.

Wholeness............origin, absolute, inclusive, context, unity, singular, source...
movement.............conscious, change, aware, other, expression, continuous...
division................individualization, unique, difference, multiply, growth...
duality......... ....symmetry, balance, reciprocal, opposite, relationship, interaction...
triangulation..........pattern, structure, strength, stability, trinity, three, primary...
consistency............generation, continuation, progression, growth, system, development...
inner-dependency... supply, spirit, connection, reliance, meaning, purpose, resource...

These qualities are relevant to geometry and need to be discussed as much as any theorem or abstract ideas about what we agree is going on in this universe. Geometry is a way to model life's movements of organization to understand how things work together. These principle qualities help us to comprehend; they serve as guides, as tools for balancing experience with understanding. This is reason enough for why it is important to study geometry—not just for the bits and pieces of abstract formulations, but for greater meanings about the larger questions.

Often the center is thought to be principle to the circle. Euclid defines the circle by the center point, radial line, and the circumference line of movement; another triangulated definition. We accept a point moving in a curved path restricted by the distance from a center point as the definition about the circle, yet this is simply a description of how we draw a circle using the compass. By appointing one end of the divider/compass as center, and setting a distance to the other point, we draw an image. The idea of the center of the circle is a concession to the tool used to draw it. The principle of duality shows two points per line segment, three parts. Each end point of the radius is both a center and circumference point, interchangeable to the other. All line segments are radial with two intersecting circles. The center is an important and useful concept when referring to what is absolute; otherwise "center" is a relative term of location.

The circle is infinitely large and in the same way it is infinitely small. It is impossible for the circle to have one outermost boundary or one center point. No matter how far out or in, the boundary is always relative to infinite extremes. The circle is its own center. The Whole is the encompassing center, which suggests origin without location, scale, or measure. Every point in the circle can be demonstrated to be a center point relative to all others. Each multiple of the circle is a replicated gravity location of centeredness. The unity of the point/circle/sphere is the same Whole at different scales of partial and relative observation. In determining what is principle to any function it is necessary to view it comprehensively within the greatest context. If we do not do that we error towards the what is less.

18

WHOLENESS

"Whole" is another word used to give emphasis and importance to favored ideas. It generally describes what is coherent and is often used to project a sense of knowing. Much like the word "principle", it has been corrupted and is rarely used comprehensively. Whole has become a word that refers to some generalized condition of parts. There is even greater confusion when we introduce the word 'holistic', which indicates healthy and natural associations of interrelated systems. To keep it simple I capitalize the word 'Whole' when it is used comprehensively; meaning everything we know and can ever know, and all that we will never know. Whole is capitalized in the same way we capitalize the word 'God' to make a distinction between the idea of many gods, and the one infinite, creator source, God.

Whole is a word I use because I know of no other word in the English language that embodies a larger, more inclusive concept. God as creator is Wholeness with personality. This personal naming of "God", provides the means for individual relationship to what otherwise is of an inconceivable magnitude and non-personal Whole. Personally we can understand that each created part is inner-dependent to the source we call God. The two words in their capitalized forms are different functions of the same infinite source, absolute center, and eternal creator/container. One aspect is personally identified and knowable and the other is collectively evolving to what is unimaginable, non-ending, and unknowable.

In reference to what is less than Whole a lower case 'w' is used, such as the whole world. The sphere functions as a whole form reflecting the nature of the Whole absolute. The Whole is origin and destination sustaining all created parts within. This seems paradoxical when we assume separation, yet there is only division of individual recognition. "Throughout the ages, the wise have taught that every person can experience the whole because all beings are of this selfsame whole."[1] The truth of this statement is evident regardless of the size of the "w".

We can see that individual parts are not the Whole of the circle. And it is quit clear that the circle is all revealed parts plus the rest that remains unformed. Parts are never separate from the Whole even as we adhere to the illusion of separation. Each individually formed part is intimately dependent upon the movement of the circle, which determines the continually changing relationships between parts. There is no separation between parts and the whole circle and spherical Wholeness that gives form to it all.

Of the seven principles, Wholeness is origin, it is first and the following six are attributes that are born of this one self-referenced movement. Without the circle shape no polygons would exist. The circle is all that is represented in this book as it is reformed through a self-referenced folding that produces nine lines. The sphere is preexistent to the circle plane and source for the drawn image in our minds. The mind is positioned at the intersection between the physical and spiritual. Our interactions with all that we know, and all unknowns outside of ourselves is in direct expression to how we understand our inner-dependency to the Whole, our personal

---

[1] Alex Gerber Jr. *Wholeness On Education, Buckminster Fuller, and Tao*, (Gerber Educational Resources, Kirkland, WA, 2001 p.99)

relationship to God. How we place ourselves in relationship to what we conceive the Whole to be is how we live our lives, determining the direction of choices we make. The inner and outer, the formed and unformed —the reciprocal function of the first fold of the circle— are interlocked one to the other, expressing a unity that is experiential only to the Whole. We can never directly see more than half of the surface of a circle at any one time, no matter how it is folded. That is a reflection of spherical reality.

It is often said, "the whole is more than the sum of its parts". If there is a sum of parts then we are not talking about the Whole, but rather what is less; a sub-total of larger and smaller wholes. Experiential reality reminds us that everything functions within the context of everything else. This evolving universe appears to continually reveal an endless outpouring of parts on all perceptible and conceptual levels. This means there is no "sum total". Within our own bodies are boundaries of individual systems that allow continuous flow and interaction between widely diverse scales of parts and functions, allowing connections to an "outside" universe of parts that are both seen and unseen to even greater scale. The mind derives meaning from the endless interactions with a multitude of other bodies, other parts, other systems, finding value within the larger context that extends beyond perceivable reality. Whole is a concept that is beyond our local understanding of individual and social, so faith is required if the unknown is to be acknowledged in our lives. Much of what is commonplace today, that was unseen yesterday, has been revealed by faith that something was there. It is the same faith in the unknown that will tomorrow reveal what is unimaginable today. Without faith there is no value in education. One cannot prove faith; one can only live through the truth of it. As our faith in the unseen grows, more of what is beyond our knowing and greater than our imagining can then be revealed.

The circle clearly demonstrates a process of Wholeness that is revealed in the form of the circle. This is not observed in any other shape. This process becomes accessible to anyone who will spend time folding the circle and reflecting on what they are doing. It takes faith that we are not wasting our time, that there is greater meaning and value to be gained form folding paper plates. These philosophical concepts have been developed out of folding circles and observing and thinking about the inherent patterns that formulate individualized parts, that in coordination create systems of interwoven, ordered, and principled complexities of movement. The experience of folding circles offers a new interactive approach to old and static concepts of traditional understanding. Faith impels us towards the unknown, introducing the unexpected into our lives. Faith is the motivation behind all advancing discoveries, works of art, and so many little things that add quality to our individual day-to-day living. The unknown (the x factor) that appears in mathematical equations is there on faith that there is an unknown. And of course the greatest unknown is the Wholeness of where, who, and what we are.

BIGGEST PICTURE

"And while the growth of the whole is thus a totalizing of the
collective growth of the parts, it equally follows that the
evolution of the parts is a segmented reflection of the purposive
growth of the whole."[2]

If you capitalize the word Whole in the sentence above you have the BIG picture. The big picture is the Whole, but the Whole is not the big picture, it is more. In education we talk about presenting the biggest picture, but rarely do we extend beyond the current educational frame. We fall short talking about little pictures. It gets confusing when little pictures are called big to give them importance: whole language, whole math, whole learning, whole numbers, whole schools, etc. We have all seen little pictures put into big frames to give importance. Pictures are static and represent isolated past remembering. To get away from pictures we must look to the changing moment, an evolving universe of information. We cannot considered anything that is not inclusive to all aspects to be the big picture.

This universe is a time place within Wholeness, revealed through the energy creation and evolution of matter. Everything in this immensity of space is integrated and organized to an order of magnitude that is so far inconceivable to human intelligence. The geometry in this universe is not the static concepts that we teach; it is of a higher order of intelligent interaction of life. One day we will discover this to be so and we will teach with great interest towards that truth.

There is inconsistency to the idea that we are created as personal beings from a non-personal, even non-existent source. Human beings are personal; we give names to things and concepts outside of ourselves. We have an individualized, recognizable personality by which we are identified even as our bodies change. Personality source is God, the name given to what is creator beyond our understanding. God is Whole and cannot be framed in any mind-making sense except through individualized personal experience. So we give it a name. Each part of the circle is unique and individual with a specific relationship to the whole circle that cannot be shared by any other part of the circle; we give each part a name. Knowing there are many names for the Whole, and knowing there are diverse interpretations of the name God, I also understand that it will bring up a wide spectrum of emotional reactions in many readers. I rely on your individual experiences to bring meaning and value to the words and concepts of what can only be understood on a personal experiential level. I do not intend to offend, diminish, or take away anyone's understanding about this universe or their individual belief in, or not in, God, or to demean in any way ones personal values in life. This is not about ones religion, or individual metaphysics or philosophy, beliefs or favorite theory. It is simply a story about the big picture served up on a paper plate circle. This story is about the geometry of life, a story about understanding what is

---

[2] The Urantia Book p.1275

21

principle and acknowledging the three-fold nature of our being in *body*, with *mind,* and *spirit.* All three must be fed for balanced healthy growth. But then most public schools do not give children healthy food for their bodies, let alone what is healthy for their minds. And the spirit is starved for lack of attention.

The three-fold question; 'where did we come from, where we are, and where we are going' assumes an overriding intelligence designed into the system that is creating and sustaining this space-in-time physical universe with an obvious sense of purpose. If we are to make progress in understanding the biggest picture we must include the question of purpose on a cosmic scale. This leads to the question, what are we suppose to be doing, what is the plan?

Our mind synthesizes understanding about human evolving experience. Religion has been a vital part of social development around the natural curiosity of "where we are" questions. The creative spiritual component has been largely separated from social education in favor of production rituals. By doing so we become unbalanced without knowledge of self-alignment. We barely educate two of our three-parts of being human. Structural balance is critical for the continuation of this evolving story of human participation in a universe of cosmic creation.

The past combined with the present generates the future, each moment. The Fibonacci number series is a good mathematical analogy to this growth process (p.46). This ratio of evolving numbers is based on a consistent linear progression of infinite number parts from the Whole of the circle. The quotation at the top of this section accurately describes the function reflected in the Fibonacci series. The triangular pattern moving forward through these numbers is also reflected in the principles observed in the folded circle. The equilateral triangle in geometry and the trinity forms in religions are congruent as manifest structural pattern. We do not know how to talk about the spiritual divinity of God without getting caught up in each other's personal/social religious concepts. Nor do we know how to talk about geometry in any way that is at all comprehensive, or personal. They have both become fragmented concepts describing the same pattern of a single, ongoing creation process. Talking about these concepts in school is <u>not</u> teaching somebody's religion. It is simply addressing the biggest picture possible, to get the greatest yield from what we already do in a very fragmented way. Whether we are *body* with *mind* and *spirit*, or *spirit* that has a *body* and *mind*, or *mind* that is imagining both *body* and *spirit*, this tri-unity must be acknowledged and fully engaged to function towards full capacity. By not engaging the triangulated nature of being we grow up confused, collectively in conflict, and individually undernourished.

It takes faith to believe in what we cannot see and do not understand. Faith is a higher function of the human mind combined with spirit. Faith accounts for most things that happen in our lives. Faith is truth-based knowing that goes beyond provable facts. It is tied to intuitive thinking; spontaneously to know in a moment. Discoveries and innovation are brought forth by an individual's faith in something of a greater unseen nature. Often faith is sparked by intuitive knowing. Scientific and mathematical concepts, inspired works of art, all discovered through faith; faith in numbers, in process, in unseen connections, in greater visions of truth, and in beauty and in the goodness of creation. Every creative act of the human mind is an act of faith in the creative spirit, in the unseen and pre-existence. Faith is what pulls the reality of the unknown down into the physical realm of the knowable. Faith is based in trust, in knowing there is truth within a greater context beyond that of ourselves.

It is through faith that we participate in the ongoing creation of this universe. Faith in what sustains us, ultimately guides us towards being better stewards of this planet. Awareness of that faith component must eventually become part of the educational dialogue in order to responsibly integrate all three aspects of human function into this creation process. We are part of this divine creation and active participation is a choice we have to make. We are accountable to the mind, the body, and the spirit that we use and that we are.

There is need to agree on the biggest picture without educational, religious, or other special interest framing. We must go openly into the ethical nature of sustainability, to find

22

spiritual connectives, giving presents to mind, and to participate with and to enjoy the physical beauty of the progressive universe.

Let's look at some triangulated connections between concepts symbolized in some familiar "big picture" words:

absolute – pattern - creation	words referring to <u>plan</u>
whole – movement – formation	words referring to <u>process</u>
truth – beauty - goodness	words referring to <u>value</u>
father – mother - child	words referring to <u>personality</u>
God	This word refers to personal origin including all concepts above.

These words are arranged to show the triangulation of pattern in various expressions through qualities of progressive relationships. Unity is only with God, singular, unknowable, beyond description and symbolized in the form of the circle/sphere. Division within unity creates duality expressed in triangulation; a structural pattern reflected throughout.

EDUCATION

Twenty six hundred years ago Lao-tse declared, "Unity arises out of the Absolute Tao, and from Unity there appears cosmic Duality, and from such Duality, Trinity springs forth into existence, and Trinity is the primal source of all reality."[3] What I have come to realize through folding the circle was understood centuries ago by Lao-tse, as well as others. This story is our origin and is crucial to all forward movement. This is science and mathematics as much as it is religion and philosophy, but it has been effectively removed from basic education curriculum. Lao-tse is informing us about a process, about patterns of formation on the grandest imaginable cosmic scale. He is describing the origins of reality, about the geometry of the organization and forming of all the things in life that we study in school, and all of the things that remain unstudied. He is talking about the triangulation of structural generative pattern. Lao-tse addresses a process that is principle to all human experience, social, political, economic, and personal. This statement accurately describes what is reflected in the observations of the first folding of the circle in half. It is not beyond the mind of a five or six-year-old to talk about this while folding circles. If young children are given the opportunity to discuss what they have done, they will in their own words discover this truth appropriate for themselves. They will find their own stories by observing what happens with the circle. This process cannot be demonstrated with a square or any other shape. Unity is a Whole function and not the function of wholes. In this regard it would show extreme neglect on my part to not mention God, or as Lao-tse would say, "Absolute Tao".

Faith is the corner stone of understanding geometry as an ongoing pursuit to find order in the mechanics of a non-mechanical creation, and find evidence of some purposeful plan to an intelligence that is unimaginably greater than our own. Early attempts in geometry were to understand creation and glorify God, giving praise to what was not understood. It is in this faith of something greater that mathematics and science truly lives. Mathematics has opened views of reality greater than what we would have known otherwise. By incorporating the body/mind/spirit trinity into our educational process, greater truth awaits to be open to us and for our children. Folding circles opens a process that otherwise remains unexplored in the mere drawing of geometric symbols.

The circle shows that what works with principle does not work when principle is lacking. Without principles there is no direction. Educationally this means to give faith to the greater

---

[3] The Urantia Book p.1033

rather than reducing life to what can be "proved". Facts are about the perceived past; faith is towards the future. The mind functions balancing the physical with the spiritual giving attention to the greatest meaning. Parts become less confusing and the working relationships become more clear by expanding the context. The integration of the tri-unification of being human is an important part of the educational big picture.

Children come with questions that do not have simple answers. It is important to explore these questions with them, allowing them, and us, to ponder the answers. How these questions are handled is crucial and will directly impact decisions they make in their lives. They are important questions that get to issues about our individual and social existence. The quick answers often given are usually confused mythologies, accepted facts, imagined futures, superstitions, and theories based on old concepts. The 'God' questions are usually sidelined, separated out as religious, used to control, to promote fear, favor, and special interest. We do not address these questions in school because we fear what they may bring to surface in ourselves. The fear goes deep, because the answers might bring greater clarity, require changing how we treat each other, and place greater responsibility upon our individual actions. That would mean a major house cleaning. We all resist what must eventually be done. Inherent to the question of education is why we educate our children in the first place, to what purposeful end? Once that question is well answered we will then know how.

At preschool level children are given balls to play with. They exhibit a natural fascination with movement of the spherical form. There is no better demonstration of Wholeness that we can give to a child. The sphere reveals unity as does no other form. Inherent within the spherical form of the ball is found divisional order to the organization and arrangements of spatial patterns yielding information about the planet we live on, the universe we live in. We have developed from a very small sphere moving out into an enormous amalgamation of spherical order. The ball is the most comprehensive and informational tool we have; yet this spherical object is thrown and kicked around. We use it to separate sides, encouraging a competitive, aggressive attitude in our children towards each other. These qualities are used to create conflict for our own amusement. In placing the sphere in the hands of children we give them the universe, and unthinkingly teach them to abuse it without regard to the information inherent in what they hold. The sphere can be used to demonstrate order and the principles of formation where everything is inherently multifunctional, structurally patterned, and completely interconnected. If children can learn rules of "ball" games they are certainly capable of learning other things about the ball.

The foundation of geometry is held within the divisions of the sphere. Mathematical functions have developed from the geometry of relationships between parts within the sphere, to define basic universal laws. The sphere is the unrecognized seed that holds the information we would hope our children will learn. Yet little is done with the sphere that enlightens and supports their mind or their creativeness of spirit. We provide little beyond play on the physical level. By separating God from play, from learning, and from most social interactions, we have successfully eliminated the biggest picture in educating our children. They learn parts, they construct with parts, they learn to take apart, and eventually to fall apart. We give them little to have faith in beyond the frame of our own belief systems. There is nothing structural or sustainable in using only two parts of a three-part system.

In paying attention to observable information generalized patterns will eventually emerge to be discovered. There is nothing hidden in nature, all is revealed in the details of larger and smaller parts. There are no secrets: creation is endlessly self-revealing of the Whole. Only human minds conditioned to competition have developed that attitude of withholding. While grounded in survival, keeping information from others has become a corruption, creating separation, fear and conflict that prevent any greater understanding or real socialization. We teach young people to fear life by what we do educationally, economically, politically, and religiously. What we say has little effect, what we do is keenly observed. We model with our behavior. Our children observe and they replicate the model to see, if they have it right. Children usually reform our corruption to their own expression. This is perpetuated in families, communities, and in larger cultures. Our behavior shows where we place our faith. If we change our behavior our children will change theirs. Our direction is the direction our children take to make their own path. To change the direction of our children we must first redirect ourselves.

Children need principled hands-on activities early in their development that reflect the comprehensive nature of the big picture. This is crucial for early education; otherwise they have little pieces to build on. Children are small, so everything to them is big. They will do extraordinary things if we can support their experience of living in an enormous and extraordinary universe. Unfortunately most of the educational tools we give children are not principled to universal considerations, nor is the divine nature of anything ever inferred. Tools and toys we give children do not embody the grandeur of life processes; rather they support relic experiences, outdated concepts, and imagined fantasies, all a denial of experiential health.

Teaching Wholeness as the big picture is the one thing we have not tried in our ongoing attempts to find a meaningful solution to the problems that surround the education of our children. As long as we talk about God as the property of religious beliefs and ideas that are formed to philosophical church dogma, the big picture will never be allowed in schools. If we continue to eliminate comprehensive thinking, excluding principled information from our schools, we will never become an educated people.

TRUTH

It is true that a picture is only an image of the thing it represents, and it is true we are drawn to images with fascination. It is also true that there is much more to be learned by folding circles than drawing pictures of them. The truth of this statement can only be known through the experience of folding the circle. It is also true that how we understand life is through living the experience. When we draw a circle we experience making the image. We do not experience the circle. When we fold the circle we experience something directly about the nature of the circle. It is then our pleasure to find out what that is.

When you discuss observations from folding the circle in half, stay with what happens. Reflect on your experience with the circle, not the concepts you have learned from looking at pictures. Look for what just happened, for the things you have not seen before. Put away what you know, so that it will not take attention away from what you do not know. This is hard to do. Take in the experience, and reflect on it. Fold the circle again and notice the difference. Look for the unseen relationships in the movement. Our discrete, time/space reality is imbedded in this singular Wholemovement. It is all *information*, everything is a process of *in-forming* us to the experience of creation as a continuous, ongoing event.

Where is the truth in the circle? It is in the whole of its shape and the Wholeness of spherical movement. To settle for less than the truth is shortsighted, complacent and arrogant. It is not our job to sort through and eliminate what we think is not important; rather it is to see appropriately the action necessary in the choices available.

To spend time folding circles demonstrates a faith in something more that what is seen. When we touch together any two locations on the circumference of the circle, it affects the entire circle. Everything the circle can subsequently do is inherent in that first fold. Touching is always a personal thing. It is how we are designed to interact as individual, personal, physical, human beings. Mental and spiritual touching is a higher order. When touching we are connecting with other than ourselves, as with any point on the circle. But were it not a part of our awareness we could not touch it. Currently there is little regard about whom we touch or how we touch. So much touching has become a corrupt activity producing unhealthy relationships. The most tragic touching of each other is always done without any sense of personal connection. The strategy of fear and of warring action can only happen when we have lost touch with each other. A line connecting two points is not the relationship of touching; it only shows the distance apart. Relationship is revealed in the right angle action of two points touching. This goes for all relationships. Touching is a triangulated action. The one touching is also the one being touched and that generates interaction that was not there before. The truth in touching is the interaction between parts within the Whole. With that connection is when healing can happen, and information is generated. Withdrawing from touching creates separation, loss of connection.

An often talked about aspect of geometry is *love*. Yes, people seem to love it or hate it, and a few just plain don't understand it. How we are first introduced to anything is how we are touched by it, and touching determines how we feel and to what degree of openness we will have about it. People who love geometry are often those who have been in some way touched by love. Clarity and openness usually go with love. There is a feeling of connection to other. The greatest love must be with God, for love is first movement. Peter Rogers in his Quest series has stated this another way; "Until you learn to Love the Whole before you Love the part, Your Love is but a vanity, you have not made a start"[2]

There is a profound mystery to the simplicity of the circle and touching two points together, and of love. Everything that you fold has a familiar touch to it, yet each shape and configuration and movement will be different. The folding of circles is space related in time, measuring differences between what has been folded, what is presently configured, and what has yet to be formed. While the properties of reformations keep changing the nature of the circle itself remains consistent and unchanged. What's not to love about that?

ETHICS

Ethics is about relative "right and wrong" conduct as it applies to human interactions. The separation between self and other creates ethical conflict. As we fold and reform the circle we get another view about ethics. In the first fold, principles are identified that regulate the relationships between all parts of the circle through endless reconfigurations and reformations. All parts work in an organized, integrated way within the whole circle. Moving the circle continually changes the relationships and interactions between parts. Everything works to the patterns of movement in the circle. This workability is a direct function of the individual part-to-Whole relationship that is expressed as the interrelationships between parts-to-parts. An astounding number of arrangements and systems are possible when all parts are individually working in coordination with all other parts. Every part supports the ongoing expression of the entire circle, which completely functions to the generation and support of all parts. There are no favored parts. This reciprocal circle/parts function gives definition to the potential of both, that would otherwise exist for neither. There is a necessary balance to mutual interaction of giving and receiving through movement into and out from, with a continual changing focus that is always towards the greater expression of Whole-to-part, and part-to-Whole. The highest order of this interaction is love.

The absolute functionality of Whole-to-part is necessary for the sustainability of all parts. The individual responsibility of part-to-Whole takes the form of specific relationships in parts-to-parts. The workability of these relationships largely depends on the triangulated function and level of touching. There is certain predictability about interactions in regards to the degree of consciousness. Free will is not absolute; it is always relative to the context already in place and our level of awareness. With that in mind we do have responsibility to support and/or change that context in a progressive manner that benefits all that has been, what is here now, and all that yet is to come. Ethical considerations are social workabilities and must accompany all creative action in order to protect each individual boundary against violation. The only insurance is adherence to the principles first established within Wholeness, particularly those two of relationship (p.18).

---

[2] Peter Rogers *The Quest*, (La Mancha Press, Roswell, NM, 1971, p. 6

Anything less will eventually lead to conflict. These seven principles—qualities of the first fold in the circle—are foundational for all ethical interaction. Ethics is a living geometry where appropriate interactions sustain progress and benefits all occurring situations. Again, the biggest picture provides understanding towards the greater benefit.

Attention to accuracy in touching points together will fold creases exactly where the lines need to be in order to function appropriately as they need to with what is already there and with creases yet to be folded. With alignment comes proportionally correct information, expanded options, ease of movement, and secure joining. The unfolded "empty" circle shape is the context for giving form to all the lines, planes, shapes, points of intersection, angles, proportions and properties of all spatial reformations possible. By being observant to what happens first, we will know what is to follow. We need faith, in this order, to explore, to let go of expectations and allow the circle to direct our actions. We must place our mind into the Wholeness of the movement of the circle that we hold in our hands, and learn to feel the nature of what it is in order to understand the information that is held.

Ethics is about sustainability of individual expression in the expanding progression of intelligence and spiritual attention. It can be viewed as the workings of an evolving balance of individual human interaction towards greater truth. In making ethical choices we become more aware that nothing is ours, we are transient caretakers with responsibility beholden to what has been, what is, and what is to be. The fractal nature of the Whole is self-referential, evidenced in the largest and in the least of parts. The expression of that self replicating pattern is in the individual forming that is unique at each level of development.

Only intelligence that far exceeds our understanding could possibly generate, keep track of, and maintain all individualized parts, potential and manifestations of all possible combinations of endlessly changing, organizing, and ordering of matter. Genetic and cosmic forming is an interactive process of unimaginable intelligent weaving of Creation. We have only to look around, marveling at the *truth*, the extraordinary *beauty* and the inherent *goodness* of the creative spirit, to feel touched by the immeasurability of divine existence.

This brings us back to the Urantia Book quote at the beginning of this chapter:

> "And while the growth of the whole is thus a totalizing of the collective growth of the parts, it equally follows that the evolution of the parts is a segmented reflection of the purposive growth of the whole".

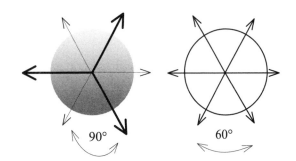

SPHERE AND CIRCLE

90°   60°

We live in three dimensions; up-down, front-back, and side-to-side. These three sets of two opposite directions are evident everywhere, covering a full range of spherical movement, appropriately call 3-D. The six directions, three axial divisions, are based on right angle relationships between each other. Compressed into the circle these right-angled axial divisions are the hexagon pattern of three folded diameters.

<u>Above</u>. These two images represent the same division. One shows the sphere with three lines of division equally at 90°, the other represents the compressed sphere of the same lines of division equally at 60°. The 60° angles between the three axial diameters in the circle (the hexagon) are the same as the 90° angles in the three axial divisions of the sphere (the octahedron). The difference of 30° is the compressive distortion between 2-D and 3-D. It is in the right angle compressing action that we understand the connection of one to the other.

The uniqueness of the folded circle is that it gives demonstration to the static 2-D information and the spatial dynamics of 3-D. They are without separation. The image of the circle, even drawn with great accuracy, generates nothing. Cut the circle image from the plane and it becomes a self-referencing, dynamic, 3-D generator of endless information.

Over two thousand years ago Democritus of Abdera "is said to have regarded a sphere as being really a polyhedron with imperceptibly small faces. This may have created a division in thinking about the sphere; an actual sphere would be made up of small atoms and the mathematical idea of a sphere would have a smooth continuous surface. This probably carried over to the circle being a polygon of infinite sides and the ideal circle "with a perfectly smooth curved surface". [1] One wonders about the distortions necessary for compressing an "ideal" sphere into an experiential sphere of atomic composition, or of mathematical description. We may not know the difference, for this down-stepping is unimaginable except from the position of descending. The circle plane can always be a polygon shape but the polygon can never be the circle.

The idea of the circle being an infinitely sided polygon is misleading and confusing when compared to atomic arrangements in space. A polygon is only one part of a circle. The fractal process demonstrates a circle cannot be compared to a digitized boundary of discrete small faces that appear continuous. An infinite-sided polygon or polyhedra is always less in area or volume than the circle/sphere. The first fold of the circle shows only one of an unlimited number of diameters representing a specific relationship between two opposite and furthest points on the circumference. Unlimited means there is always, without end, room for more. A straight line between unlimited numbers of points will never be a circle or a sphere.

Folding the circle will never generate the ideal we expect from mathematics, but then nor will numbers ever generate that ideal. The ideal is towards the experience not the image. What the circle can do is to model the conditional push and pull and the mechanical transformation of pattern into specific designs of life-like formations that we observe everywhere in nature. The circle demonstrates divisional reformation of a compressed spherical surface by generating individualized parts that can be reconfigured without separation. The reality of the sphere-circle reformation is based in faith that the inevitability of pattern is the non-provable ideal and can be demonstrated by folding the circle. The drawings in this book are representations of circles. The

---

[1] *Polyhedra* , Peter Cromwell

circle has a self-referential boundary that moves. The movement of the circle is contained in folding and the reconfiguration of the surface, and is reflected in the abstract generalizations of mathematical description about the movement relationships between parts as they are formed. The movement of the circle is 3-dimensional. The compressed 2-dimensional information is in the creased lines. Folding both 2 and 3-dimensional information at the same time makes the circle a generalized 5-dimensional reformation of the sphere.

Folding the circle in half, with movement going in two directions, traces a spherical pattern in space reflecting the spherical origin of the circle. The primary relationship of straight line to the sphere is the axis/diameter/bisector. The straight line is a function of the movement of the circle/spherical Whole.

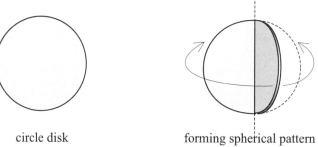

circle disk   forming spherical pattern

Above. The first reflective movement of the circle is a spherical pattern folded in two directions. The unity of the circle/sphere is unbroken.

Below. Multiples of circles generate the same pattern as the reformation of the sphere. Straight lines are connecting the centers and the points of tangency with both the circles and sphere. The major straight lines are the three axes/diameter lines of spherical division. The diameters always line up the inside centers of the circles to tangent points. One sphere reformed is four spheres without separation. This is the pattern for the circle.

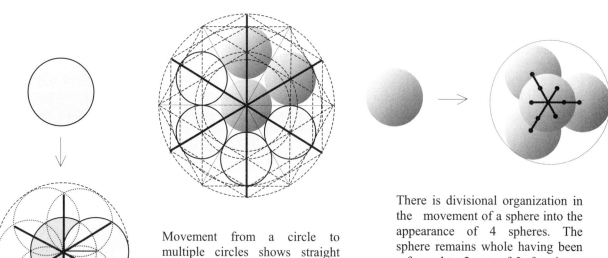

Movement from a circle to multiple circles shows straight lines indicating primary patterned connections between the center points, points of tangency and points of intersections (p.120, 253). This is compressed spatial information.

There is divisional organization in the movement of a sphere into the appearance of 4 spheres. The sphere remains whole having been reformed to 2 sets of 2, forming a tetrahedron relationship. There are now 4 center points of small spherical location, and 6 location points of continuous surface connection. These 10 locations define the reformation of a single sphere.

30

<u>Below left</u>. These drawings show 19 primary points of connection that reveal the hexagon pattern grid of triangles. (1+9=10 the diameter and circle) The primary line connections show 3 diameters divided into 4 equal divisions. As another level of connection is made between the points, the 3 diameters are again divided showing 8 equal divisions. To show the cross connections between points becomes very complex even though the diameter remains at eight-frequency. It is important to go all the way from point to point on the circumference. This acknowledges the whole circle. To draw the line short would leave missing information.

<u>Below right</u>. This shows the equivalent of increasing layers of spheres from *2 layers* of 4 spheres, to *4 layers* of 20 spheres, 5 sets of 4 spheres each (p.124). With the number of tangent points connecting 20 spheres the connective net is extensively complex.

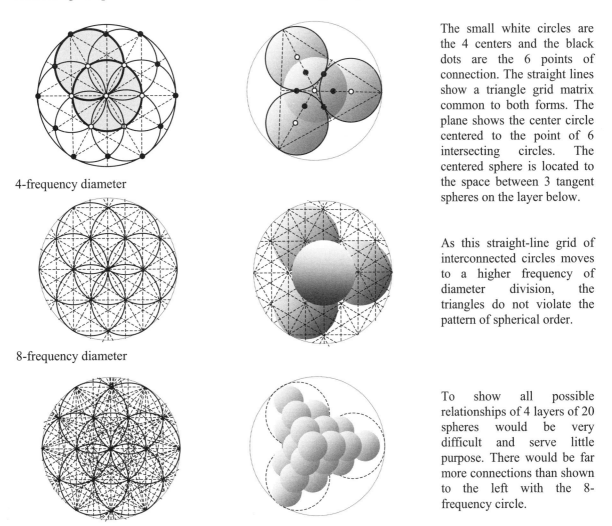

4-frequency diameter

8-frequency diameter

The small white circles are the 4 centers and the black dots are the 6 points of connection. The straight lines show a triangle grid matrix common to both forms. The plane shows the center circle centered to the point of 6 intersecting circles. The centered sphere is located to the space between 3 tangent spheres on the layer below.

As this straight-line grid of interconnected circles moves to a higher frequency of diameter division, the triangles do not violate the pattern of spherical order.

To show all possible relationships of 4 layers of 20 spheres would be very difficult and serve little purpose. There would be far more connections than shown to the left with the 8-frequency circle.

<u>Above</u>. With 19 primary points of intesection in the 4-frequency circle there are then 342 individual relationships with a very large number of combinations. The drawing on the bottom left shows all of the line connections between every circle intersection of the seven circles This makes clear the need for starting from the simplicity of the circle and through division moving into higher frequencies of information step by step. The hexgon pattern of 3 diameter/bisectors of spherical order provides stability and continunity to all possible combinations of development.

31

Below. Here is another look at how the two layers of spheres lay within the same hexagon matrix. It doesn't matter how many layers of spheres there are, from the same point of viewing they will all compress into a hexagon grid that is revealed in the three diameter folded circle. The grid drawing below shows the circle inscribed in the hexagon, which is the reciprocal function of the hexagon inscribed in the circle. This does not suppose equality between the circle and the hexagon because it is the circle/sphere ordering that generates the hexagon in the first place. The bold lines show the primary triangle-hexagon grid forms from points that appear as both centers of spheres and the intersections in the spaces between the spheres. This is a function of the compression of layers of spheres to a single plane. These spaces between circles represent the depressions that hold the next layer of tangent spheres. The triangulated pattern is seen in the spaces between the spheres, as it is with the spheres themselves. Together the centered and non-centered systems form intra-divisional, straight-line relationships that are determined by spherical origin.

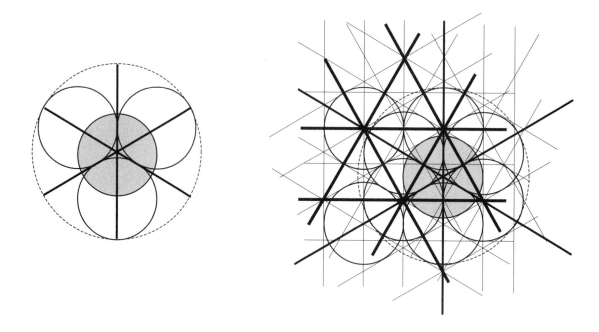

To best understand the formation of the grid and how it is derived from the circle you need to fold the circle and to make a drawing for yourself. To just look at pictures and read the words won't give you the understanding that comes from folding and drawing. It is equally important to explore spheres, to arrange and stack them, to observe layer by layer what they do, to see what the closest packing of spheres is and then to draw what you see. Draw the straight-line connections between the points of intersections. The folded spheres allow you to see the inside straight-line relationships and the linear connective grid that is normally hidden inside the outer surfaces of the arranging of spheres. Every different way of forming the same pattern reveals something different about the nature of pattern and a little more about the never-ending process of formation (p. 90-94).

32

These words, *pattern*, *form*, and *design* are often used interchangeably with only a general sense of difference. I will give reference to how they are used in this book to give an idea of the order and sequence of development between them. As with all words, it is the context that gives meaning to the definition.

In a simple arithmetic analogy; the pattern is *division*, the form is *multiplication*, and design is the *adding and subtracting*. Without first division there is no multiplication of parts to add and subtract. There is no first division without unity and purpose of the Whole.

Pattern has only one function and that is to replicate itself exactly in as many different forms as possible. Form is the finite expression of pattern and source for intelligent arrangements of design. Often design is not obvious and the appearance is a random commingling of parts, and assumed disassociation. Sometimes there appears random transitioning during the reformation movement from one coherent state to another. There are many design possibilities in these transitioning moments of formation. To see the order of changing relationships within the frame of the folded grid is to expand design potential. Pattern is parent to form, even when not apparent. Considered form arrangements of parts is intention to design. There is evidence of design in all life-forming expressions; in all reconfigurations through moving the circle. While pattern is formed and designed to local influence, form and design are not pattern.

The tetrahedron pattern is formed within the first fold of the circle. That first fold of division gives directions to folding that will form the tetrahedron eight creases later. The first opening of the tetrahedron generates a new pattern of the octahedron; adding another open tetrahedron gives form to the octahedron (p.129). Open an octahedron and subtract one half, close it up and there is one tetrahedron. This spatial adding and subtracting is similar to genetics, where a stem cell holds the full pattern of information for forming specific designed parts for an entire system by adding and subtracting through various parts becoming active or inactive.

Order is inherent to multiply packing of spheres (p.194). Order suggests the organization of a plan formed through pattern of design arrangements. The sphere is a primary form pattern. Through compression, spherical pattern is reformed to the circle and triangulated through principled movement, where the circle is reformed through the active/inactive participation of parts. The circle holds divisional memory of spherical movement in its circularity.

Let's back up for a moment. There are in fact 3 differently proportioned foldings generated from that first fold in the circle. This is another level of triangulation. The triangle pattern is in the same ratio of 1:2, each proportionally different. Three diameters is first, then 4, and then 5 diameters. The total patterned ratio of division into the circle is 3-6, 4-8, and 5-10; three basic symmetries patterned to the circle. The 4-8 and 5-10 divisions are covered in my book *The Geometry of Wholemovement: folding the circle for information*, as well as a more extended treatment of the 3-6 folding. Here we are only exploring 9 folds of the tetrahedron. Three comes before 4 and 5. Knowing what 3 diameters do, we will know what 4 and 5 do. They form to the same principles, patterns, and the same process of development, generating different symmetry and formations. Exploring the 9 is to understand what 3 does. Three to itself is 3^2 or 3x3, is 9.

There are no physical properties to pattern; therefore cannot be directly experienced. Only as replicated in form, can pattern be designed to function in all the ways that it does. The creation of life forming is the truth of patterned origin, providing consistency where each forming becomes part of the neural networking that connects each to all. Pattern endlessly reveals greater truth through the forming and interactions of unlimited design diversification.

Basic geometry functions and vocabulary comes up when we first begin to talk about our observations of the circle. The words are used to identify functions and relationship between parts. When we work in classrooms, particularly in primary grades, we do not have the math vocabulary to talk clearly about what we are doing, so we use the words we do know. Talking about our observations gives us words to help us think about our experience. Vocabulary helps sharpen our ability to observe and make conceptual connections. We use common words and precise geometry and mathematical terms are introduced only when they help clarify. Words accurate to our observations allow a greater depth to our thinking. In the following text the underlined words refer to what I expect students to observe and talk about. Even if students do not know the words they should be able to talk about the concepts that arise from their observations by using their own words. New words can always be added when appropriate. Because we are using words, and it takes more words to explain things; you have noticed we are into a smaller font size.

Observation starts before we even know what to do with the circle. We need to understand what we mean when we say the word circle. The drawing is without dimensions or movement. It is a symbol that represents a concept about the circle. We talk about what it means as a symbol, as something in space, the differences and similarities. In discussing the properties much can be discerned about students vocabulary, level of understanding and awareness of observation. Talking about the circle requires appropriate use of common words to describe and reassess what we take for granted. It helps us to look critically, adding greater meaning to what we already know, and to fill in connectives for ideas that have been separated by abstracting definitions into word fragments.

As we describe what we see in the circle, other ideas about what a circle is will come up. We talk about the line or edge as a boundary or circumference; it has no beginning or end, the same everywhere with the curve being consistent all the way round. We can talk about *three* planes (as surface or face) having *two* edges inclosing a volume of space. With these descriptive parts we can talk about coins, compact disk, or paper circles. We compare the description of the image with the thing. It is important for students to talk about what they observe and not what they have learned to think about. Usually size comes up and we talk about scale and the different between a very small circle as a point and a very large point as a circle. The word whole comes up in describing the circle. In that context the circle can be described as self-referencing and inclusive. By reflecting on the information in our discussion we can find the direction to know what to do next.

Before folding the circle in half we have already identified a number of geometry concepts and establish a basic vocabulary simply by talking about what we observe of the circle. With the first few folds of the circle we always talk some about *what has been generated* that was not there before we folded the circle. There is no prescribed order, in that students will see whatever they see and that is where we begin. It is all interconnected in the circle, which makes it easy to address different levels of information at any time. Usually this is not done all at once, but over time throughout the folding process. Older students should know most of the words; the young students learn them through usage as we talk about what we do. I have included examples of questions I ask students in directing their observations towards greater clarity.

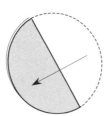

The circle is moved in a self-referencing way so the circumference touches all the way around. The circle is now divided by a line. (What is the line called?) The diameter. (What does diameter mean? It is sometimes important to have them look up the word in the dictionary.) The diameter divides in two equal parts. (How do we know they are equal?) Because they are the same. (How do we know they are the same?) Because they are symmetrical. (How do we know they are symmetrical?) Because we folded the circle in half. (How do we know it is in half?) Because they are the same. (Why?) The two parts fit together one onto the other when we folded the circle. (Yes, the two shapes are congruent.) They are called semi-circles. To divide the circle in two parts means the diameter is also a bisector. The circle now reveals two points, a straight line, two curved lines, individual areas and angles. This kind of questioning allows students to observe more closely, always leading them back to the folding they have done. The experience is what gives meaning to the words. With this level of discussion it is then easy to introduce algebra as another kind of vocabulary to talk about the relationships between parts (p.80).

Let the students discover the parts and functions from their own observations and the words facilitate understanding. They discover the diameter is also an axis when the circle is folded in both directions forming a pattern of a sphere in space. (How do we know it is a sphere? What is pattern? How is it formed? What do you see?) Axial movement in both directions forms a reciprocal function. (What does reciprocal mean?) There is an inside and outside surface. (How many inside and outside surfaces? Are they the same or different?) Through movement they change places. What each side will do for itself it does for the other. This is a dual function. Is it ethical? What does that mean? (Tell me more about the diameter). The diameter is a line of symmetry. (How do we know that?) The parts are congruent having a right-hand and left-handed orientation where the semi-circles go in opposite directions. This is problem solving in discussion format.

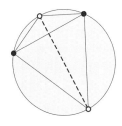

When two points have been visibly marked anywhere on the circumference and folded together there are four points. We can connect the dots by drawing lines connecting all four points. A polygon called a kite shape is formed. The diameter intersects perpendicular with the line between points. There are now points, lines, shapes and angles. (What kind of angles?) (How many right angles? How many left-hand and how many right-hand right angles?) The two perpendicular straight lines are the diagonals to a quadrilateral shape inscribed in the circle. All lines going from one point on the circumference to a second point on the circumference are called chords. Why? (Can they be called line segments. Why are they called segments?) (What is an infinite line?) All lines are diameters to unseen circles. (Are there any complimentary angles?) (What do you think complimentary means?) These are the kinds of questions to guide students to figure out meanings through their own observations and connections. The working level of a student's vocabulary is the place to start building discussion about the folding they are doing.

You can always go deeper. The perimeter of the quadrilateral has a very specific proportional relationship to the circumference. (Is it the same for different size circles?) The circumference can be divided into curved lines called major arcs and minor arcs, with the chords forming different kinds of areas. There are three different kinds of triangular areas; the isosceles, scalene, and right triangle. The lines shown in the triangles are angle bisector, median and altitude. One edge can be established as a baseline for the triangle. As the circle is folded in half the movement forms a polyhedron pattern. Four points in space is a tetrahedron pattern. (Why is it a pattern? What is the difference between pattern and form?) The polyhedron form of the tetrahedron has four triangular surfaces; this has two surfaces and two defined planes (What is the difference between surface and plane?) The biggest generalization we can talk about is folding a ratio; one whole to two parts. All of the rest is about parts. We are still only discussing having folded the circle in half. The more you look the more functions you will find.

A questioning dialogue of this sort can reveal the level of understanding your students have and it allows them to generate higher levels of interest for themselves. Usually it does not happen all at once as related above, but in relationship to the on-going folding process, led by students' observations and how they talk with each other about it. In this way all kinds of generalizations, formulas, theorems, and abstract concepts can be discovered while folding the circle. Everything is connected to everything else showing total interrelatedness. This makes it easy to come back at any time and pick up observations, and introduce vocabulary by talking about different parts. Words are equally as multifunctional as the parts of the circle. Talking is not a substitute for folding and direct observation, it is a part of the process that helps to clarify. Discussion serves to support and expand the experience, increasing observations about what we do and what is generated through interaction. Words are another way to model and share our experience of having done the same folding process and being able to communicate how differently we observed and think about it. Everybody gains something from hearing each others perspective.

FOLDING CIRCLE IN HALF, TOUCHING POINTS

It has been my observation that when people fold a circle in half they touch together two imaginary points on the circumference, look to get the edges even, then crease the circle. No two people ever choose the same two points on the circumference, meaning everyone folds a different diameter. From this we can see that _any two points on the circumference, when touched exactly together will fold a circle in two equal parts._ We can also know that folding the circle is about touching points together and getting a straight line.

In mathematics, imaginary numbers function as "real" numbers, and so it is with points. Use two imaginary points anywhere on the circumference. Mark them so they can be seen. Touch them together and crease the fold. Marking and touching the points is an easy way for young children to fold the circle accurately in half. Making the two invisible points visible reveals more information. Opening the folded circle reveals four points; two marked and two new points where the creased diameter touches the circumference (p.35).

Folding the circle is a process of showing the relationship between two points by touching them together. _In touching two points together a folded line is generated half way between the two points at right angle to the direction of movement between the points._ (To draw a line between two points only shows distance that is already there, nothing is generated that was not there before.)

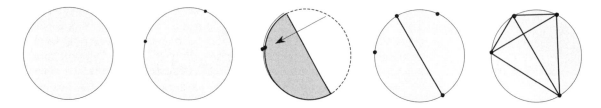

<u>Above</u>. Draw in the lines connecting all 4 points. This maps the paths between the 4 points. A kite shape with 2 <u>diagonals</u> is formed. Each circle has a differently proportioned kite shape. This <u>quadrilateral</u> expression of folding the circle in half is unique for each person. Even though everyone does the same thing they have differently <u>proportioned</u> kites.

Count the number of <u>triangles</u>. There are 8. These triangles demonstrate the <u>multifunctional</u> nature of parts. Everything is a part of each other. Count the number of <u>right-angle triangles</u>, there are 6 unless the kite shape is perfectly a square, then there are 8.

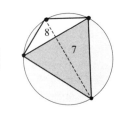

The relationship of any point on the circumference with the two end points of the diameter will always form a right angle. The line connecting the first 2 points intersects the diameter only when the circle is flat. When the 2 marked points are moved towards each other, off of the flat plane, the straight line relationship between them no longer intersects the diameter; it decreases in length until the points touch and the circle is completely folded in half. This decreasing line is the sixth edge of a tetrahedron defined by the 4 points in space. Four points and 6 edges are 10 parts. _Ten is the number of the tetrahedron._

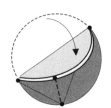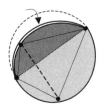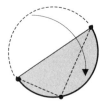

Folding the circle in half generates both a spherical pattern of origin and the structural pattern of tetrahedron formation. This folding goes in two directions forming a reciprocal inside and outside duality. We can then say by folding the circle in half there is a movement pattern revealing 2 individual tetrahedra. One is the inside out of the other (p.82). We can also refer to them as a positive and a negative tetrahedron. The dual nature of the tetrahedra (p.163) is generated in the first full 360° fold of the circle.

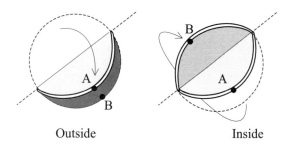

Outside   Inside

The distance separating two points (A and B) is not the relationship between them. The relationship is the right angle touching expressed as a diameter around which they both revolve. If this axis were not centered to the circle/sphere then the points would never touch. This right angle function is what forms the tetrahedron and gives gravity to all kinds of reformations. In traditional plane geometry this line of relationship is the constructed bisector between the end points of a line segment. Two points in a drawing do not touch, so all that is left is to draw a line from one to the other, showing the distance already there. The only movement a drawing can generate is the changing of focus in our minds as we construct more parts. Mathematics has developed through drawing pictures and adding one construction onto another in a linear sequence over thousands of years, expanding the capacity of the human mind towards more abstract and complex construction. The movement of the circle allows us to find an experiential foundation; a meaningful context to this conceptual information. It is a way to redefine the context giving greater meaning and adding clarity to what has been developed.

The circumference is a local boundary. It gives reference to all the points, lines, and the spaces between. When the circle surface is folded the point locations touch and different areas of the same surface come into contact. Lines fall into alignment establishing a balance to the relationships of movement. In space relationships take many paths with always a mid-point of right angle balance between them. When there is a force difference between the two locations, the balance is thrown off center, expressing a complimentary difference between larger and smaller parts. When folding the circle the relationship between any two points is always balanced to a relationship of four points. The attractor of information is always the self-referencing circle moving at right angles.

The diameter is a function of the right angle movement of the circle. It is the longest cord in the circle. Pi represents a ratio difference from the length of the diameter to the length of the circumference. Among other things Pi tells us that the straight-line measure of the circle will never equal the whole circle. Parts can *never* equal the Whole; neither one part individually, or an infinite number of parts combined.

Within the 360° of the circle the diameter serves as the straight edge of each half. This first folding demonstrates how the straight line and the triangle are both 180°. Any two points on the circumference touching each other divides the circle into two equal and opposite right-angle triangles. It matters little that the pattern is formed by straight lines or arcs. Any point on the circumference with the two end points of any diameter form a right-angle triangle. This is minimum definition of half the circle no matter where that one point on the circumference appears. The three angles of a triangle are 180°, one half of a circle.

The triangle will diminish in size the closer the point gets to the diameter from the mid point half way between the two ends of the diameter. The third point will never reach the diameter end point since circles are without scale. The opening of a set angle is always greater the further out towards the boundary of the circle it gets. Pi demonstrates division of the circle without conclusion of ending.

<u>Below</u>. From our point in viewing the picture plane, when folding the circle, half of the circumference and the diameter will line up as a single line. The two semi-circles will be perpendicular to each other. Half of the circle appears to have collapsed into the diameter, leaving half of the circumference, 180°. When the fold is complete it appears as a single semi-circle of 180°. The circle remains a full 360°. The folding of the circle is never less or more than the entirety of the circumference. The interior angles of the quadrilateral will always equal 360°. One half of the kite, a triangle, will always be half of 360°. The angles are a function of marking straight-line paths between one point on the circumference and the two diameter end points. This generalization applies to all triangles and quadrilaterals because it happens with the first movement of folding the circle in half. It is parent to all that follows.

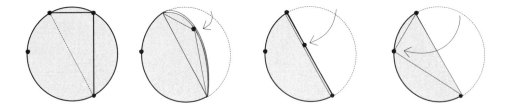

The area of the triangle, half of the circle, is reflected in the other half, making right and left hand right-angle triangles. Compare two circles folded the same to explore combinations of congruency and area reformation using both handed right triangles. <u>Reflection</u> is the flipping and folding we are usually asked to conceptualize. The 180° and 360° and all the other numbers represent angles from a given point of viewing static positions of the circle in motion. We have to look beyond definitions and the logic developed from static images to the dynamics of folding the circle. Straight lines and polygons give expression to the circle; they do not explain or define it. When we look at static images of shapes and angles we lose the greater meaning of movement, the spatial context, and the value of direct experience.

There is a hierarchy of form where the circle is the point, the source of all lines, angles, areas, and countless shapes and forms of spatial configurations. The straight line, or an infinity of straight line segments in a circle path, can never be a circle. Individual parts can be reproduced separate from the circle, but they are always a part of that which is greater, and that is their greater value. An individualized circle of specific measure confined to relative limits of scale always reflects the nature of the circle/sphere pattern.

The difference between points on the circumference, and connecting them with a straight line, is the difference between parts and Whole. Adding more points and drawing lines between them make the lines closer together and shorter in relationship to the diameter. Dividing into shorter straight lines only makes a polygon with more sides. A polygon of infinite straight lines is never be a circle of one continuous line. Any and all polygons are always only part of a circle. There is no summation, only subsets of adding and subtracting within the Whole. The absoluteness of the Whole is infinitely more than the endless numbering of parts.

The most fundamental straight-line expression of the circle is the diameter, 3 times around the circumference, in six equal parts. Pi as a number (3.1415...) represents 3 diameters plus the difference between the measure of the circle and inscribed hexagon. Pi is an irrational number because it is irrational to think that any number of straight-line parts can ever equal the circumference. 3 diameters are 6 radii, the only number of straight lines that can be inscribed exactly in a circle reflecting the ratio 1:2 (one diameter to 2 parts). The circumference equals the diameter times Pi, or C=Dx3.1415..., or C=2Rπ, which means they are not equal. Now that is truly irrational. The difference between the circumference and the hexagon is non-reconcilable (p. 344).

| 3 diameters | circle minus hexagon | hexagon minus circumference (circle folded in) | difference in area between hexagon and circle, 3.14159... |

When the circle is folded in half we find the diameter to be the longest straight line possible in the circle. Three diameters go around the circumference inscribed as six radii. Three diameters lengths fall short of the length of the circumference when compared, because of the fullness of the circle and the diminished nature of the polygon. The same goes for all polyhedra being less than the sphere. Parts are infinitely less than the Whole. Similarly what is infinite is less than absolute. It is the absolute that gives gravity to the effort.

Halfway between the hexagon pattern of multiple circles out and circles in is the hexagon polygon inscribed on the diameter points. (All points are center points.) The nature of the circle in multiples is primary fractal development that shows up not only in the numbers of Pi, but is also diagramed into the symbol as the full radius to the difference.

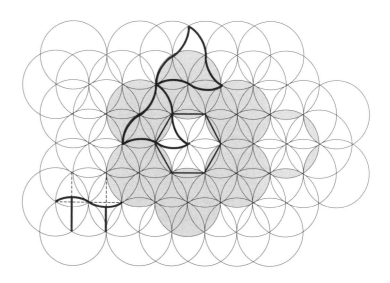

POINTS, LINES, PLANES, NUMBERS

We use numbers to keep track of similar parts and show relationships. The undivided circle functions conceptually as Whole (O) and discretely as one whole (1) part. The first folded division generates three parts showing 1 diameter and 2 semicircles, a relationship of 3 parts from 4 points. Added together 3+4=7. Seven is the greatest number of combinations that can be derived from triangulation. Numbers have no inherent meaning in that they function as symbols, a language created to communicate about experientially physical and mental realities. We count on using numbers to identify pattern and to reveal meaningful relationships. In a similar way letters serve the same function in forming words or as symbols in an algebraic equation. The relationship of numbers and letters reflects specific arrangements about connections and the organization and associations of widely diverse parts.

Below. The triangle pattern is 3 points. The 3 lines mark the distance between 3 points and show 1 area of interaction. (3+3+1=7). The qualities represented by 7 parts reveal the triangle pattern as it relates to 3 diameters folded into the circle (7 points), and the hexagon reformed to the 2-frequency triangle. Three is pattern, 7 is the individual forming of combinations and thus becomes an outgrowth of 3.

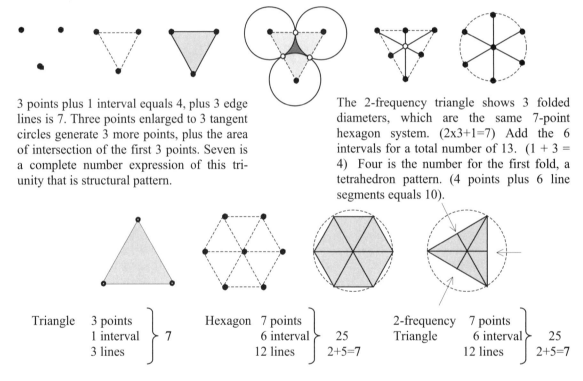

3 points plus 1 interval equals 4, plus 3 edge lines is 7. Three points enlarged to 3 tangent circles generate 3 more points, plus the area of intersection of the first 3 points. Seven is a complete number expression of this tri-unity that is structural pattern.

The 2-frequency triangle shows 3 folded diameters, which are the same 7-point hexagon system. (2x3+1=7) Add the 6 intervals for a total number of 13. (1 + 3 = 4) Four is the number for the first fold, a tetrahedron pattern. (4 points plus 6 line segments equals 10).

Triangle	3 points		
	1 interval	}	7
	3 lines		

Hexagon	7 points		
	6 interval	}	25
	12 lines		2+5=7

2-frequency	7 points		
Triangle	6 interval	}	25
	12 lines		2+5=7

By adding the number of points, line segments, and areas of the triangle, the hexagon, and the 2-frequency triangle we find they all show the same number pattern 7. Numbers reveal connections that are not obvious.

Numbers are a wonderful tool for discovering similar qualities and sameness of pattern between diverse shapes and forms. In counting similar parts, and combining different parts of a system, and then comparing the numbers of different systems, connective patterns emerge that are often hidden within visual forms. Counting parts requires close observation and can provide a physical intimacy with objects not possible in more abstracted ways of thinking about and using numbers. The practical use of numbers is obvious, although we have become quit unbalanced in how we have allowed usage of numbers to control and regulate our lives by the abstract value we give them. As symbols, numbers can give a picture of the dynamics of movement and change, revealing levels of complexity that go far beyond our visual experience. As interesting and useful as numbers are, they are symbols. When we allow symbols to direct our lives we have given over to measuring and regulating and images that offer not experiential understanding.

Three is the triangulation of *points*, *lines*, and *planes*, fundamental to describing any object in space. Four spheres in spatial order are always a patterned relationship of three, again reflected in the 3 and 4.. The 4 points on the folded circle is the tetrahedron pattern of ten. Four tetrahedra forming a two-frequency tetrahedron is a pattern of 10 points, 24 edge lines, and 16 triangle surfaces. 10+24+16=50 (5+0=5). It takes 5 sets of four spheres each to make the spherical equivalent of the 2-frequency tetrahedron (p.173). The tetrahedron pattern is four spheres in the closest packed order.

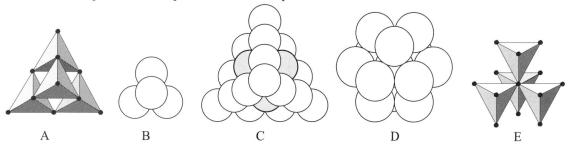

A   B   C   D   E

<u>Above</u>. (A) is polyhedral and (B) is spherical, two different forms of the same 10-point pattern. The third image (C) shows the 5 sets of 4 spheres; 4 sets in one direction with the fifth set of 4 filling the center space in the opposite direction. In (D) the pattern arrangement shows 12 spheres around a center sphere making 13, (1+3=4), the tetrahedron in pattern. Both (C) and (D) are the closest packing spherical order. Drawing (E) shows four tetrahedra sharing the same center point, which is the same pattern of 13 spheres in (D), (4x4–3=13). The 13 spheres are four sets of four where four points, one of each set, are collapsed into one spherical center location. The number 4 as compression of 13 (1+3) shows a relationship between centered and non-centered spherical systems. The difference between centered and non-centered is 9 spherical points, 3 to the power of itself, or 3^2. The relationships between numbers and the correlated nature of fundamental spatial patterns are obvious. Centered (D) and non-centered (C) are two individualized systems separated out from the same spherical order of the closest packing of spheres (p.90).

<u>Below</u>. In dividing a diameter into 4 divisions, 3 points are used. There are 5 points including the end points. Nine is the number of 5 points and the 4 line segments. There are three different size circles all in a ratio of 1:2. This is a circle divided by itself three times (3x3=9). When the diameter is divided into 8 equal intervals (the 8-frequency diameter circle, similar to an octave in music) there are 9 points (a correspondence to the 9 folds of the tetrahedron). The straight-line division is the result of the circle division scaling down across its own measure. This fundamental 8-frequency division shows 4 different size circles. These circle images produce fixed sine waves of endless frequencies of division. Wave functions are observable as energy moving rhythmical through space. When the circle is divided to a sufficiently high frequency where the circles are too small to see, they are perceived as a straight line rather than the self-referencing divisional movement of the circle. This limited perception is relative to scale.

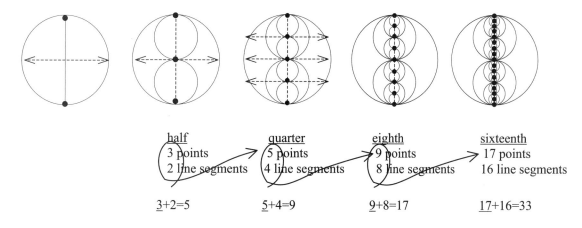

half	quarter	eighth	sixteenth
3 points	5 points	9 points	17 points
2 line segments	4 line segments	8 line segments	16 line segments
3+2=5	5+4=9	9+8=17	17+16=33

41

<u>Below</u>. The next logical step is to compound this scaling progression where each new point of division (2:3) on the diameter becomes the center for another circle of the same division (3:5). There is a ratio of 2 circles developing 3 points, where 3 becomes 5 points, and 5 increases to 9 points, then to 17, 33 and so on.

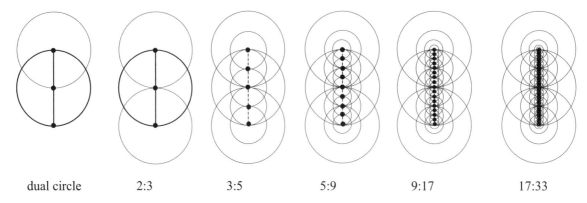

| dual circle | 2:3 | 3:5 | 5:9 | 9:17 | 17:33 |

When each point of division is a circle of the same scale the rate of division doubles minus one. The point is the circle; the circle is the sphere. Lines and planes are partial functions of the point/circle/sphere.

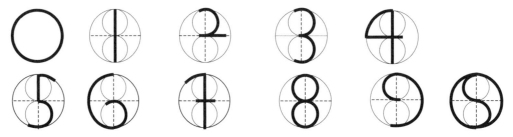

<u>Above</u>. Another way to explore numbers is through the visual function of circle division. Generalized accepted symbols for number formation can be located in the primary division of the circle. There is a certain visual logic to this division. First there is the circle, then one diameter revealing two half size circles of 3 points showing 4 quarters and so on. All symbols are made from combinations of curved lines (circles) and straight lines (parts of circles). Notice the 6 and 9 when drawn together in the same circle make the yin yang symbol. There is no straight line division dividing the circle in a ratio of 1:2.

Starting with the circle there is consistency to the number 9. With numbers one through 10 there is an inconsistency of two ones and one of all other digits. By removing the one from the ten and putting the zero up front where it belongs the ten-digit sequence now consistently reflects the pattern of circle division; a complete octave of spherical harmonics in the 8-frequency diameter division from 0-9.

There are basically two systems used for counting; 1-10 and 0-9. Both have ten places and are part of each other. We tend to slip back and forth without being much aware of the connections between them.

<u>Below</u>. Comparing both systems of number sequencing they are two parts of one binary system.

1, 2, 3, 4, 5, 6, 7, 8, 9, (10)   adding all digits, 55   5+5=10   →   1+0=1

O, 1, 2, 3, 4, 5, 6, 7, 8, 9   adding all digits, 45   4+5=9   →   +9

$$10 → \mathbb{O}$$

1 through 10 reduces to 1.
0 through 9 reduces to 9.
1+9=10, the first fold of the circle, the tetrahedron pattern of right angle movement. Without circle origin there are no parts to count, no numbers to count with.

The mid point between 1-10 is 5. With 0-9 the mid point is half way between 4 and 5. There is a consistency reflected in the 4+5 to 9, that is again reflected in the pattern of 4 spheres in 5 sets forming the tetrahedron within the spherical closest packed order of spheres (p.173).

Ramanujan, a mathematician from Southern India around the turn of the twenty-century wrote; " *Zero, it seems, presented absolute reality. Infinity, or ∞, was the myriad manifestation of that reality. Their mathematical product, ∞ x o, was not a number, but all numbers, each of which corresponds to individual acts of creation.* "[*] This statement suggests that the absoluteness of zero times infinity is absolute everything. It makes more sense, and is a closer fit to our experience, to say everything comes from something rather than everything coming from nothing. The difference is crucial for mathematical considerations as well as for all levels of education. It sets a contextual direction that determines how we experience all the rest. This is a good statement to consider as a class discussion about numbers.

Point/circle/sphere origin contains every number, every interaction, everything. The alternative is that nothing exists, and that is absurd given our awareness of self and other. The unmanifest and perfect Whole is origin and container for infinite numbers of imperfect and transitory parts. The absolute, undivided, and infinite Whole is symbolized in the circle and is thus limited to the relative measure of a given diameter, the first divisional movement of the circle.

The relationship of 0 and 1 contains all subsequent numbers. The folded circle is a binary system. This two-part system had much to do with the development of computers. The bar code is simply multiples of one in parallel where the combinations of varying widths of black lines and white lines are reciprocal dual functions for information storage. The object becomes also the space where the context is replaced by interval lines which function as real lines using all possible space. A number line is usually added to compound layers of information to the black/white ones. This is bottom line reductionism. It generates nothing but is efficient storage of abstract information. As with all codes it is only useful when there is agreement on the arrangement of information. In understanding the circle as Whole, we understand the circle has an unlimited number of parallel lines in the right-angle division of each of 3 diameters. This creates black and white stripes in three directions of a highly organized interference screen.

Placing the two number systems back into the circle as primary divisions along the diameter we can then look at the difference of wave pattern one to the other. On the next page we see some interesting number correlations between the same wave function of each system and the forming of structural pattern that is reflected in the numbers of forms we have already observed. The concentric nature of the circle is a natural way to understand the scale of numbers. The fact that all of these interesting correlations can be traced back to movement functions within the circle should tell us something very fundamental has been missed in how we have been interpreting and separating information.

Numbers are a way to track divisional movement of relationships between parts within a given context. Numbers describe a forming process where consistency is in pattern and potential is expressed in the endless evolving arrangements and changing designs. Without division, there is nothing individual, no relationships, nothing to count.

This circle process is not about constructing abstract information as much as exploring comprehensively the dynamic interactions between parts in the largest context allowing as many connections possible at any given time. Observations can then be generalized and represented appropriately with symbols that represent the clearest levels of information. We get into trouble with formulas where distortions in number compression have not been accounted for. Mathematics is a social language about patterns and relationships and we all understand patterns differently. The more ways we have to demonstrate "math" concepts in an experiential way the greater chance people will have to understand the patterns that are foundational to mathematics. It is often easier to assimilate abstract concepts about numbers when we understand what numbers represent. The examples given as they relate to the circle, is another entrance into this extraordinary number language.

---

[*] Ramanujan, the Man Who Knew Infinity, the Genius Ramanujan. Robert Kanigel. Washington Square Press, 1991

<u>Below</u>. In the numbered circles from 1-8 the sum is 36, which reduces to 9. Nine is the number of points of division, 2 end points and 7 points in between.

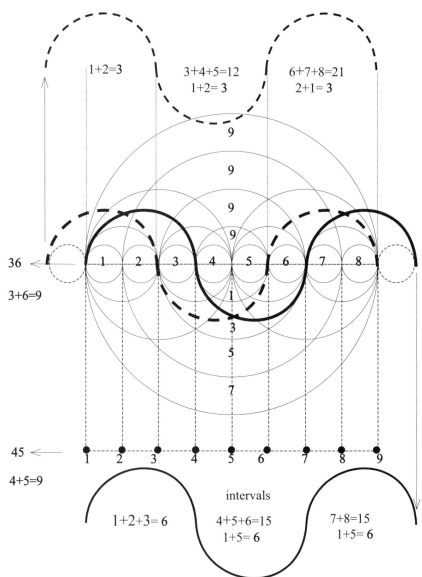

(I sometimes use the O. it represents the zero that is full rather than the 0, a deflated symbol.) When we count points that divide a line (1-9), O is always the point of origin and the container, but usually intervals are counted. In learning to count 1-10 we think of numbers representing the individual subject or groups being counted. One is always preceded by 0, but not often shown or spoken because then there would be 2 ones and 2 zeros, being more inconsistent than it is. Numbers have come mostly to mean things starting with one. Yet often we count the intervals and spaces in-between and give them a number, as with inches and minutes. They all start from a place before one. Numbers tend to separate things, removing them from context. That is the convenience of numbers. They can be used to give context or meaning to anything we want. The problem is when we begin to think everything is relative to numbers, regardless of context.

The number of points (1-9) and the number of intervals (1-8) both show a different quality represented by the number 9; 36 and 45. They are 2 parts of the same function O-9. The 8 divisions of the diameter (4 different concentric circles) reflects the 4 points of the first fold of the circle as they do the 4 spheres of spherical order.

The two numbers in each concentric ring of the divided circle adds up to number 9, shown in the top half.
(1+8, 2+7, 3+6, 4+5, 5+4, 6+3, 7+2, 8+1)
Subtract the two numbers in each circle ring, shown in the lower half of the circle, and they will reveal the

first 4 of the prime numbers, 1, 3, 5, and 7.[*]   (1−8, 2−7, 3−6, 4−5, 5−4, 6−3, 7−2, 8−1)
There is a nice symmetry between these two aspects.

---

[*] Prime numbers are those numbers that can be divided evenly only by themselves and no other number. They come first.

Here is another way to see number sets in graphic wave function factored to the sequence 1 through 10. These number sequences can also be laid out in 9 divisions around the circle and then connecting corresponding number points with straight lines.

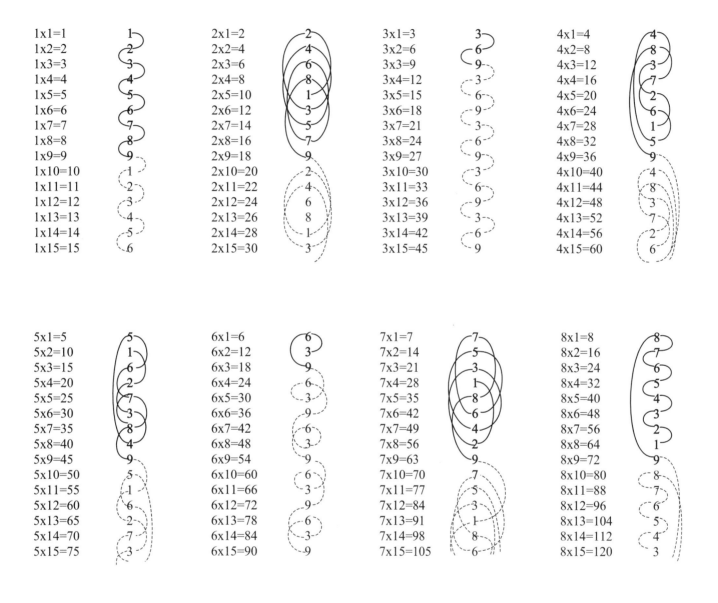

1x1=1	2x1=2	3x1=3	4x1=4
1x2=2	2x2=4	3x2=6	4x2=8
1x3=3	2x3=6	3x3=9	4x3=12
1x4=4	2x4=8	3x4=12	4x4=16
1x5=5	2x5=10	3x5=15	4x5=20
1x6=6	2x6=12	3x6=18	4x6=24
1x7=7	2x7=14	3x7=21	4x7=28
1x8=8	2x8=16	3x8=24	4x8=32
1x9=9	2x9=18	3x9=27	4x9=36
1x10=10	2x10=20	3x10=30	4x10=40
1x11=11	2x11=22	3x11=33	4x11=44
1x12=12	2x12=24	3x12=36	4x12=48
1x13=13	2x13=26	3x13=39	4x13=52
1x14=14	2x14=28	3x14=42	4x14=56
1x15=15	2x15=30	3x15=45	4x15=60

5x1=5	6x1=6	7x1=7	8x1=8
5x2=10	6x2=12	7x2=14	8x2=16
5x3=15	6x3=18	7x3=21	8x3=24
5x4=20	6x4=24	7x4=28	8x4=32
5x5=25	6x5=30	7x5=35	8x5=40
5x6=30	6x6=36	7x6=42	8x6=48
5x7=35	6x7=42	7x7=49	8x7=56
5x8=40	6x8=48	7x8=56	8x8=64
5x9=45	6x9=54	7x9=63	8x9=72
5x10=50	6x10=60	7x10=70	8x10=80
5x11=55	6x11=66	7x11=77	8x11=88
5x12=60	6x12=72	7x12=84	8x12=96
5x13=65	6x13=78	7x13=91	8x13=104
5x14=70	6x14=84	7x14=98	8x14=112
5x15=75	6x15=90	7x15=105	8x15=120

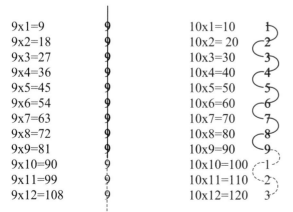

9x1=9	10x1=10
9x2=18	10x2= 20
9x3=27	10x3=30
9x4=36	10x4=40
9x5=45	10x5=50
9x6=54	10x6=60
9x7=63	10x7=70
9x8=72	10x8=80
9x9=81	10x9=90
9x10=90	10x10=100
9x11=99	10x11=110
9x12=108	10x12=120

By compressing all multi-digit products to a single digit and drawing connective sequential paths from one through nine, each multiplier displays a unique visual tracing of repeating number arrangements.

Paths 1, 3, and 10 are the same movement paths of connection. These are numbers that surface in the first fold of the circle. The 8 path is the only function where the 1 and the 9 are next to each other. The 4 and 5 are similar in connecting every second number spanning eight places before continuing. The 2 and 7 have similar movement tracings. The numbers 6 and 9 are uniquely different in wave pattern. Nine is consistent to the diameter.

45

THE FIBONACCI NUMBERS

This series of numbers is about proportional growth sequencing that starts with the Whole (O) and through movement a division of one (1) generates a duality (2) that is formed in triangulation (3). The number sequence of formation is O, 1, 2, 3. This is the only number development that consistently describes the first movement in the Wholemovement process. Origin and an infinity of numbers are progressively accounted for. The triangular sequence of parts 1, 2, and 3 is a linear description of a principle function of division represented in the number 3. We read it in sequence of discrete parts; one part plus one part is two parts, plus one is three parts. This shows a consistent first progression of parts that has origin. This is a number form that represents fractal growth in a self-similar proportional sequence. O, 1, 2, 3, 5, 8, 13, 21, 34, 55, 89, 144, 233, 377, 610… This consistency is reflected in the O, 1, 2, 3 process observed in the first fold of the circle in an endless folding sequence where the process continually builds on the accumulation of all that has come before. *This series is an evolving movement where all previous function is collectively utilized to move forward leaving nothing behind.* In this way the unknown is generated in a specific proportional relationship to all that is has been, continually carried forward, and combined in the present towards evolving the forever-unknown and infinite future. This number sequence has no meaning without the context of origin and a finite starting place. Inherent to origin is destination, endless in process.

There are other sequences that can be generated from the same triangulated pattern that change the primary proportional progression of forming numbers. By starting from the finite we can change the directives in many different ways and still the process remains endlessly consistent to origin.

Other examples starting with 1, 2, 3 are;

1, 2, 3, 4, 5, 6, 7, 8…	adding one each time
1, 2, 3, 4, 6, 9, 13, 19…	adding second back and bring sum forward
1, 2, 3, 6, 12, 24, 48…	adding all previous numbers forward (doubling)
1, 2, 3, 6, 18, 108, 1944…	multiplying by previous number
1, 2, 3, 3, 6, 18, 54, 324…	multiplying by second back number… and so on.

One diameter does not exist without the circle. We can draw a single line but it does not exist spatially without the circle/sphere. Number progressions are abstract proportional relationships without context starting with a straight line, number one. Traditionally the Fibonacci numbers start with one and add another one getting two. Add one to two and get three, add two to three to get five, three to the five gets eight and so on. There is no origin, only putting two individual parts together and building on the inconsistency of counting two ones, whereas all the rest of the number groups generated are without duplication. What is the context of duality, two ones that are used to begin this series? The meaning is in the proportional accumulation between the numbers that is principled to the Whole.

Traditionally the Fibonacci sequence shares the same inconsistency of two ones with the base ten counting system. Number one is not origin, it is finite. The context is the Whole manifesting progressive positions in-finite where the intervals between positions reflect proportional interaction. This is illustrated below.

0   1   2   3   5   8   13   51   34   55   89   144   233… consistent development from origin
     1   1   2   3   5   13   51   34   55   89   144…   finite numbers counting intervals

The bottom row of numbers cannot happen without the top row of numbers first. The double one sequence comes from counting the intervals between the numbers where the starting would be one and not zero. There are many ways to geometrically show this proportional division of the Whole reflected through the growing development of periodic accumulation of sub-wholes, where each grows in proportion to where it has been and where it is going. Most often it is called the *golden ratio* (p.167). In drawing polygons and dividing them by construction, the most common is called the *golden rectangle*, a proportional rectangular growth coming off the division of the square (pp.47, 159). Starting with the circle, the square, as the pentagon, show relationships of triangle function that determines self-referencing growth.

GOLDEN RATIO

The traditional construction formula for drawing the golden rectangle is to draw a square and bisect the base line. With a compass set to that bisected point on the base line open to the diagonal going to an upper corner, swing an arc down to intersect with the extended base line of the square to get the correct proportions for constructing a golden rectangle.

Where does this construction formula come from? Given it is a flat construction; we can trace it back to the dual circle image. One radius is always a duality of circles, similar to the two circle faces of the circle disk. Starting with dual circles, we find the square is a relationship of triangles, where the triangle grid is the self-referencing circle matrix. The circle determines, as origin, this proportional growth ratio formula.

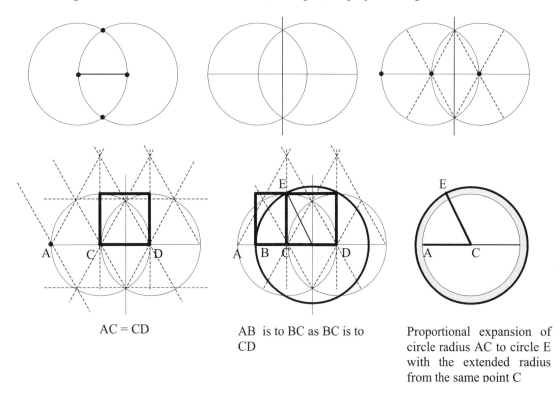

AC = CD

AB is to BC as BC is to CD

Proportional expansion of circle radius AC to circle E with the extended radius from the same point C

<u>Above</u>. Straight-line radial connections between two circles with two radial lines perpendicular to the first will reveal the remaining side of radial length outside of the two circles connecting them to form a square. The division of the square base is a function of the bisection of the rhomboid inherent in the <u>vesica</u> of the two circles. Use this diagonal as a new radius from the center point on the first radius and draw circle E. This locates the intersection on the base line of the square to extend it to the rectangle. This diagonal 1:2 perpendicular measure is fundamental to constructing the golden rectangle by using the diagonal as radius to the point of circle intersection. Using measure EB above a pentagon can be inscribed to the original circle. It is all a circle function, (p.71). This diagonal expands the original circle CA to circle CE. We can now restate the formula; *One half of the radius at right angle to the full radius generates a diagonal radial expansion that is proportional to the circle*. This growth proportion is called the golden ratio.

The original dual circles are expanded from both centers in a concentric proportional waveform. It can move infinitely out from and into any point location. This image of two-directional movement is reflected infinitely throughout the circle/sphere matrix and is reflected in the Fibonacci number series.

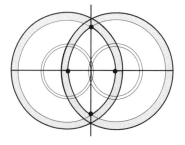

This drawing represents another view of the Fibonacci growth ratio shown by the dark wavy line. Each number is the sum of the preceding divisions of the circle based on the ratio 1:2. The half intervals of the octave division of the diameter; the sixteenth, is the measure. The circle division comes first and the numbers are descriptive of proportional division. As smaller intervals are defined, more mid points are revealed upping the frequency and complexity of this proportional progression that starts with the first fold of the circle in the ratio of, one Whole: two parts.

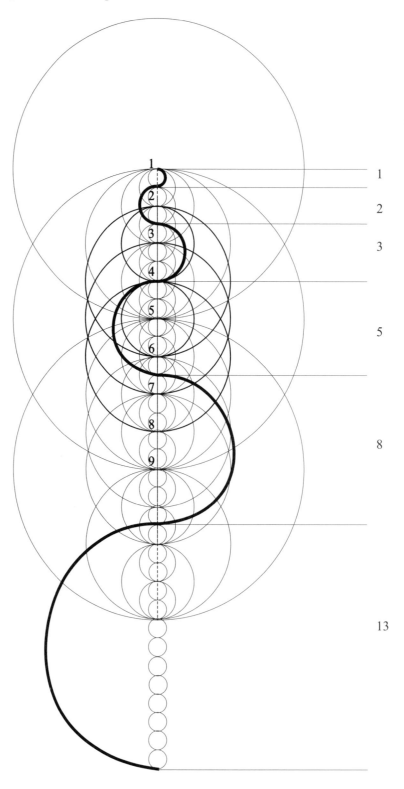

Here is another way to visualize the Fibonacci progression in a spiral form. Begin with 3 diameters folded into the circle showing 6 sectors (p.72). Near the center on a line start drawing a spiral towards the circumference. Make enough space for numbers to be written between the lines. It is not necessary for this to be an accurate spiral drawing.

Write number 1 in the first area starting the spiral. Move to the next area and write number 2. Now we have two areas, number 1 and number 2. Move to the next area and write 3. Continue to fill in all the areas adding one number at a time following the spiral until all areas have been filled. Notice the first three numbers come as a unit of 1, 2, 3. From 3 add 2 and it is 5. Count three more and add that to the five and there are 8 areas covered. To the 8 count five and there is 13. All numbers are accounted for as the progression develops. Continue the same as far as you can on the spiral you have drawn. Color in the areas in which the Fibonacci numbers occur. Look at that design of intervals. They all happen in some place.

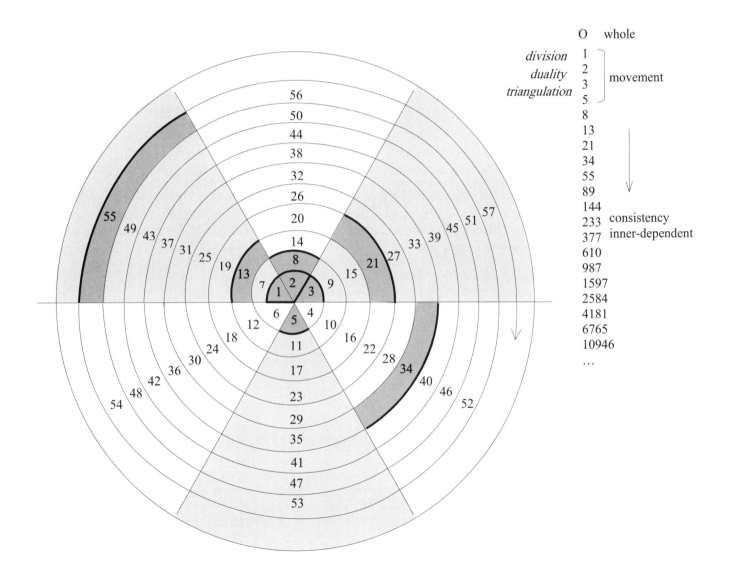

O   whole

division   1
duality   2
triangulation   3
   5   } movement

8
13
21
34
55
89
144
233   consistency
377   inner-dependent
610
987
1597
2584
4181
6765
10946
…

DRAWING CIRCLES

It is useful to include drawing circles since we do that anyway. Numbers, letters, symbols, images; drawing pictures and writing all require the same observational skills of proportional mind/body organization and coordination. The feel for moving a marking tool in space is developed in much the same way as walking or getting a feel for folding the circle. It takes doing a thing a number of times to develop a feel for it. The circle is an absolute; no one ever draws it perfectly, and we all do it differently. It is good practice for beginning students and people that don't think they can draw, to draw circles.

First, with a finger draw a circle in the air. Use your entire arm; to the right and left, with both hands. Consider what you have done. Feel the circles you have drawn. As a class talk about it. Where did you start and where did you stop? How far has the earth traveled by the time you finished drawing and what was the actual path of your finger in space? Now do the same thing on a piece of paper with a pencil, freehand. *Start by making a point on the paper. Then draw a line in as smooth a curved path as you can, ending up at the point where you started.* Just like drawing in space. Nothing in space is in the same place where you started. Did the pencil actually take the path that is marked on the paper? The circle image is closed; the reality of the movement in drawing the image is an open spiral. No two people will have drawn the same size or shape circle. Discuss how simple the shape is, how difficulty to draw the image perfect.

Draw a small circle so it looks like a point. This is easy using the magnification tool in a computer drawing program. What is the relative scale of a circle and a point, at what point are they one or the other? The concentric nature of the circle shows they are the same thing.

Below. 1) Draw a point. 2) Draw a circle starting from that point, making the curve as smooth as possible and ending at the same point. 3) Approximate the center of that circle and mark it with a point. 4) Starting from that second point draw another circle around the first point as center, as close to the same size as you can make it to the first circle, ending at where you started at the second point. This forms two intersecting circles sharing the same radius (p. 47,192). Accuracy is not a concern; that will come later. 5) There are 2 center points and 2 points of intersection and 3 areas, 7 parts, similar to the first fold of the circle (pp.72, 102). By counting the 4 points and 6 arc segments the number 10 becomes apparent.

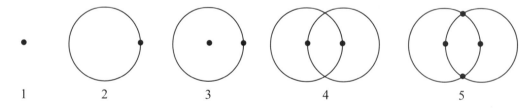

6) Draw a straight line connecting the 4 points all the way through the circle from circumference to circumference. Do this free hand, no straight edges or rulers. 7) There are now 10 points of intersection.
8) Connect all 8 points around the outside with a straight line. How many triangles are there? How many are pointing up and how many pointing down? What other combination of shapes can be found? Five and six-year-olds are good with this exercise. It's the old connect the dots game. How many ways are there to connect the dots? (p.31) How many circles can be added, given each point is a center point.

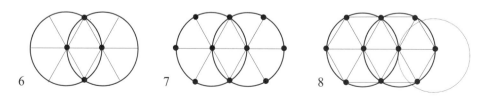

Once folded the circle shows 10 parts that define the relationship of the tetrahedron (bottom p.36). Here is another form that shows the same numbers of counting parts. The circle shows 2 circle planes, and the drawing shows 2 circle planes. This yields much more information than does drawing one abstract isolated circle. Nothing exists by itself. Nothing is generated from nothing. Each point is a center point for more circles. In that way this is a self-referencing, self-expanding matrix generating infinitely large and small circles ordered to the pattern of spherical packing (p.31).

Below. The dual circle in the previous drawing is multiplied using one circle as a centering for 6 levels of concentric expansion. The circle multiples are in alignment with the point locations along an axial line. Concentrically expanding the boundary has generated precessional symmetry dividing the diameter into a 12-frequency circle without using straight lines. There are 12 intervals generated along the horizontal centers with 13 points of divisions and 6 sizes of circles. All this is a function of the 1:2 ratio of the first fold in the circle.

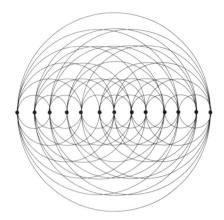

There are many drawings in this book that suggest different directions of 2-D design that can be explored using this triangle grid of 2 intersecting circles. It is the same grid with the same information that is folded into the circle, which is explored, by folding. There are many connections to be made between the 2-D and 3-D since they are the same pattern and the same divisional forming process.

In Euclid's book The Thirteen Elements, Proposition 1 in Book 1 starts with two circles sharing the same radius to prove the construction of an equilateral triangle on a finite line segment. He uses them again in Book 3, Proposition 15 to prove the construction of an equilateral hexagon. Nowhere else does he use the dual circle. In the Euclidian system there is no context larger than the logic system of connections based on abstract postulates and definitions to prove what already exists. There is a need to continue to re-evaluate, to up date, and move through past accumulations, keeping what has lasting value and letting the rest be absorbed back into the developing future. While acknowledging past developments we cannot afford to let them hold us back. Imagine Fibonacci getting stuck on number 13 and replaying variations of 5 and 8 because they got him there. He might never have reached 21, being only vaguely aware of the infinite progression of development that lays ahead.

Loa-tse (p.23) and Ramanujan (p.43) both tell the same creation story, the same one revealed in folding circles. Young children need to hear and to fold this story. It is no longer enough to only draw pictures of circles and learn construction proofs and logical connections. It is time to enlarge the truth of geometry by considering the movement of the Whole as the most advantageous place to start understanding the infinite numbering of interrelated and multifunctional parts.

Drawing circles freehand educates the fingers, the arm, the entire body, and imprints the mind. There is the physical experience in space that informs the body. You feel the movement, which is an important part of understanding the circle. Standing on one foot and dragging the other foot in a circle as you turn your body in a compass movement of balance. Change feet and drag the opposite foot around and you have a dual circle. It is easy to make a circle image on a computer, but you gain nothing beyond the image. Using the compass is more/less accurate but little is learned from the tool about what to do with the circle.

In drawing a circle image you have to know something to get information from it. The visual, mental image without physical experiential understanding remains abstract. The mind must feel what it is to know.

The drawing at the top of the preceding page looks complicated to draw by hand, but is easy to draw using the computer. However, the understanding of the image comes from the experience of drawing it by hand. The computer provides a level of abstract accuracy that would take a skillful hand to draw, but accuracy at first, is not always important. This is a system of proportional scaling of relationships and that is what is important to understand through the process of drawing. Drawing develops hand-eye coordination that leads to visual clarity in providing the skill to understand, construct, and to organize systems. Anyone that can render even the most rudimentary drawing of a circle can do the step-by-step drawing suggested on page 50. Accuracy comes from understanding the proportional relationships between the parts within the circle context. This is particularly important for young students and those who believe they cannot draw. Drawing is a form of recording information. Writing, making numbers, and all other forms of symbol-making require the same skills of organizing and visual translating developed with drawing. Just doing it increases the skill towards greater control. Perfection of accuracy is unattainable, but from any relative level it is a worthy pursuit.

Drawing by hand is essential for observation and for understanding the pattern and proportional development of forms, rhythms and the symmetries that are woven through the movements of divisional relationships. Drawing by hand is about moving through space. Computer drawing programs offer a new kind of drawing where we have all but physically been removed from the process, no longer being a spatial experience. The computer is not a place to learn to draw. The simplicity and Wholeness of the circle is the most instructive place from which to freely draw.

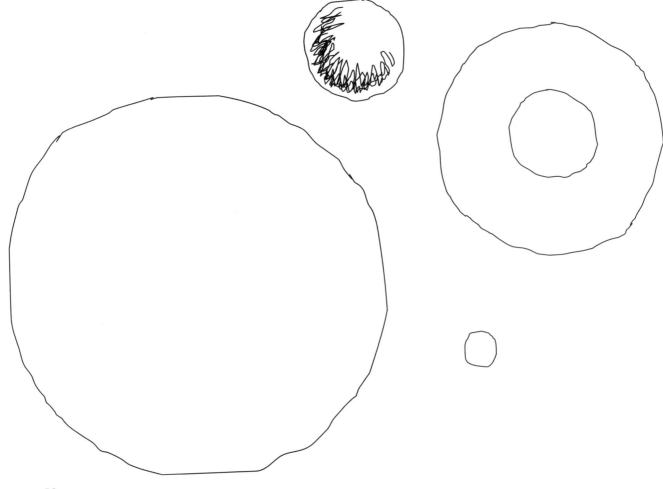

DIAMETER

Looking at how the diameter functions as the axial symmetry in the development of polyhedra gives a hierarchical order to polygons determined by the number of diameters rather than the number of sides. The diameter is the first fold of the circle and primary to the forming of all polygons and polyhedra.

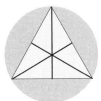  

| No diameters | Two diameters | No diameters | Three diameters |

<u>Above</u>. The 4 primary inscribed polygons are arranged by increasing numbers of sides. Only the square and the hexagon have bisecting diameters. With the other two polygons the diameters have been shortened and serve only as bisectors when confined to the polygon edges. The square with two diameters is the tetrahedron fold into the circle (pp.36, 90). The hexagon with 3 diameters is structurally and proportionally primary before all other polygons. The 7 hexagon points are the full expression of the relationship of 3 diameters. It makes sense that from endless diameters of the circle, the first pattern to emerge is a division of three. The three primary polygons emerge from within the context of the hexagon division (p.74).

The image of the square shows eight individual right angle triangles, four in and four out. The square is a special case of the first fold of the circle in half (p.112). The square is a compressed relationship of the four triangles of the tetrahedron where the two diagonals are the opposite edges perpendicular in crossing (p.122). Another way to understand this is to reconfigure the three diameters in the circle where the hexagon is the polygon parent. All the vertices are defined through movement of the three diameter/bisectors (p.74).

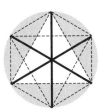  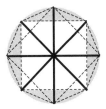 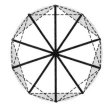

<u>1 diameter</u>	<u>3 diameters</u>	<u>4 diameters</u>	<u>5 diameters</u>
Circle folded in	Circle folded into	Circle folded into	Circle folded into
a ratio of 1:2;	1:2 three times;	1:2 four times;	1:2 five times;
a 1-2 symmetry	a 3-6 symmetry	a 4-8 symmetry	a 5-10 symmetry

<u>Above</u>.
The numbers of diameters show a structural hierarchy starting with the circle with one formed diameter. The two diameters are inherent in the duality nature of four and the three in the six. Two is not spatial, rather it is a principle quality (p.17). The most economical way to fold a square polygon showing the triangle divisions is to fold the circle 4 times in a ratio 1:2. The symmetry of each polygon reflects the 1:2 formation in the 3-6, 4-8, 5-10 descriptions. All folded symmetries come directly from the 1:2 ratio of the first folded diameter.[1]

---

[1] *The Geometry of Wholemovement: folding the circle for information,* B. Hansen-Smith, Wholemovement Pub. 1999

BRANCHING

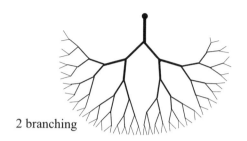

2 branching

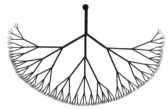

3 branching

Branching is a spatial growth extending through periodic division starting from a given location. This phenomenon of spreading out is represented in a generalized linear concept through stages of successive division or periodic junctions of divided growth. Branching starts with 3 diameters informing both the 2 and 3 divisional systems. When viewed spherically as a scaling function, branching is a fractal development.

The two types of branching are odd and even; one divides evenly in two directions and one divides with odd numbers in three directions. With the three-branching system the center direction creates sub-major roots or spines that are consistent for divisional outgrowth from each successive point of origin. Branching traditionally is an out growth function, but comprehensively is both an in and out development.

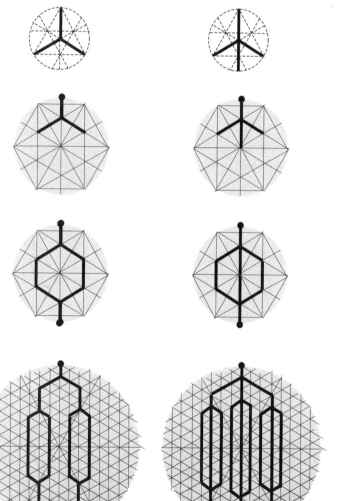

Here are the 2 and 3 branching symbols overlaid as parts of the three diameters in the 9 creases.

Using the 4-frequency division of the circle with the starting points on the circumference the growth path follows the folded creased lines in. The branching points are where the vertex appears.

Acknowledging the circle with the line starting on the circumference the growth path will end on the circumference. The growth is formed along the diameter line of symmetry to a 90° movement reflecting a duality across the circle (p. 180).

A higher frequency grid will allow extended branching and greater diversity of subdivision to the diameter line. The width of the line is relative and frequency of branching is endless.

As the branching expands inward from two opposite end points on the diameter a precessional movement is generated. This forms multiple divisions that become increasingly dense at right angle to the diameter. This reflects the same right angle movement of the first folding of the circle. 2+7=9, 8+1=9, 2+4+3=9, 7+2+9=18 (1+8=9) and so on.

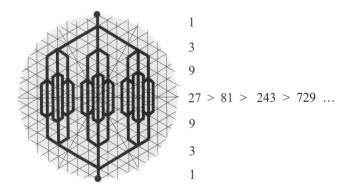

1

3

9

27 > 81 > 243 > 729 …

9

3

1

 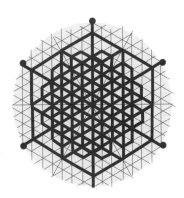

Left. Two-branch system in hexagon pattern

Right. Three-branch system in hexagon pattern

Below. Branching not only grows from the outer boundary in, it also grows from the center out when you consider the starting point generates multidirectional growth within a larger system. The circle is not only the starting point but the containing boundary as well. This is an example of the two-branch system overlaid on the grid net moving out to fill the circle. These can be thought of as pathways or circuits of a non-local energy grid. This is not unlike what is observed in the diversified designs of snow crystals.

Above. The two-branch system moves out in one direction. The next image shows the same path triangulated in three directions, then in six directions. These of course are not the only branching lines on the grid, there are other designs and in many scales where the lines will vary in length, in thickness, and in angulations. With a higher frequency grid there are more path options for joining, moving in parallel, spiraling around, over and under each other, or canceling out. There are endless variations to the diameter branching into and out from the circle location.

The single two-branching system doubles on each stage of growth; 1, 2, 4, 8.... The single three-branch system shows a numerical development of 1, 3, 9, 27.... Counting the junctures in each concentrically circling out, or in, gives us the accumulative divergence of growth. Using the two-branching function in three directions the numbers will be 1, 3, 6, 12.... The three-branching in three directions will be 1, 6, 24, 72... This is another example of creating multiplication through the function of division where accumulation can happen quickly.

This picture shows the same two branching from above moving out in twelve directions on a flat plane where paths overlap. Consider this is a compression of what takes place spherically in space. It is the same pattern-based diameter matrix in which many different polyhedra can be developed. Viewed from a different angle it would appear as a square lattice or spatial grid (pp.253, 336).

What is observed about branching on the hexagon grid of three diameters can be observed spatially in the isotropic vector matrix, or the vector equilibrium infinitely expanded, where every point location has six diameters. When the six diameters are viewed from the triangular perspective and the sphere compressed, the circle will have twelve equal sectors (pp.59, 109,a,b). This is a straight-line description of the closest packing of spheres (p.195).

Right. The three divisional nodes in the hexagon form, moving in from six locations at the end of each diameter. This represents the compression of internal spherical branching that builds mass that eventually will implode, moving back out. There is much about branching that multiplies into many areas of geometry and mathematics, revealing multiple connections. Branching indicates that everything starts from the same place of Wholeness, patterned to spherical form, compressed to a circle that is reduced in scale to a point from which local branching occurs reaching out towards the Whole of where it is. Connections move through 6 degrees of freedom formed by three diameters in a centered system of 7 and 13.

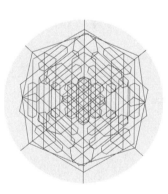

The quality of growth in the two and three-branching is reflected in
the number sequencing in both an odd and even, complimentary mode of expression.
When two-branching is divided out through many generations, the number pattern (when multi digit numbers are reduced to one number) show a sequencing of 1, 2. 4, 8, 7, 5, [1, 2, 4, 8, 7, 5] [1, 2, 4, 8, 7, 5,] [1, 2, 4, 8, 7, 5]... There is a repeating unit of six digits. The sum of each unit is 27, which reduces to 9.

The three-branching, when totaling each level of division and reducing to a single digit, develops a unique sequencing of 1, 3, 9, 9, 9, 9, 9, 9.... . What is complimentary, being different in form, is in the larger context the same in origin.

LETTERS AND NUMBERS

Symbols evolve as our minds give meaning to our experiences and find connections in reflective understanding. We formally learn letters and numbers in the first grade. What follows is a good activity for first/second grade students as they become familiar with folding and joining the tetrahedron. It gets them into drawing on the folded circle using the creases already there. It is an "anything goes" activity as long as you do not violate the folded lines. Students can subdivide them and combine them in various ways; being inventive with what is already there. This is a natural approach to directed exploration and discovery and has value for older students as well as young ones.

Open the tetrahedron and talk about the creases the students have folded. Have them trace the lines with a pencil or marker letting them become familiar with the creases and the combinations of shapes they make (pp.109-114). Have them look for numbers and letters in the various combinations of folded lines and color them. This can be done individually, in groups, and as a class with collective board work. Later you can discuss proportions, size, orientation, symmetry, style, and forming the generalized characteristics of letters. Maybe color the vowels and consonants, odd or even numbers, in warm or cool colors, or any scheme to keep track of the grouping of characteristics of various letters and numbers.

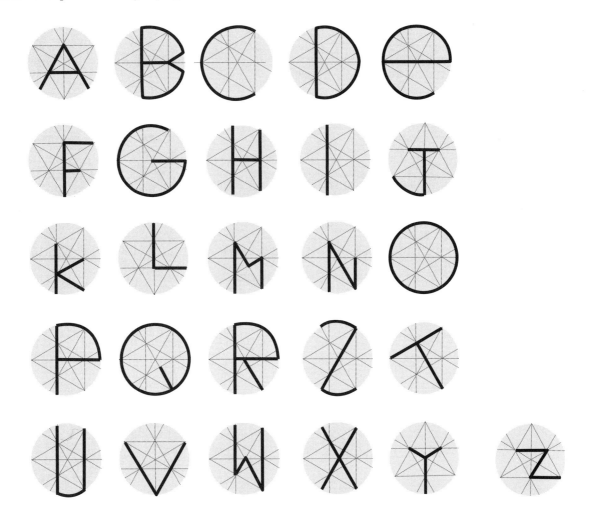

Above. One alphabet variation. There are many ways to form these letters using the nine creased lines and the circumference. Size and placement is not important at this stage, it is more about discernment, observation, and imagination within the limitations of the folded circle.

For variations have students pick out their initials and color them so each letter can be seen individually even if they use the same lines for different letters. Use the letters to color in a design using the creased lines in any way they want, making each letter so they can be read clearly. Some use their name if not too long, some include numbers for their age or grade. Use the entire circle.

Lay all the circles out and look at them. Discuss how they all look very different and at the same time are all the same folded circles. Discuss all the different ways each individual circle was designed. This is a good exercise in using the abstraction of symbols to design the circle. After they have all colored their circles lay them out on the floor so the creases connect to form a large triangular grid and look at them all together much like a quilt of circles (p.70). Talk about what it is. There is no right or wrong, better or worse in this exercise. They are all different. It is about observing differences of individual designs and the pattern they have in common, and then talking about it. How does each design reflect the person that made it?

Later the students can fold their individually colored circles into tetrahedra and observe what happens to the 2-D designs when they get folded into 3-D configurations. Join the colored tetrahedra into larger frequency tetrahedra systems (p.122). This literally takes the grid off the floor and into space. In extending the project fold the VE spheres (p.89), and octahedra (p.129), using the variety of designed colored circles to explore the many different combinations.

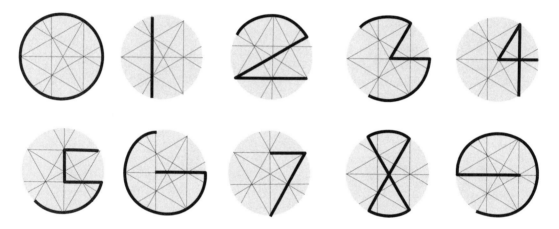

<u>Above</u>. Combinations of the creased lines and the curves of the circle can be used for locating numbers.

<u>Below</u>. When students are familiar with folding and coloring the circle, have them draw the circle and the folded lines. Start by drawing a circle. Then have them draw in the three diameters freehand using only their eyes for the proportional relationships, no measuring or rulers. Have them draw in the large triangle by connecting every other end point on the circumference. Then draw a line between the three midpoints on the large triangle, extending the straight lines all the way to the circumference. This is the pattern they have folded. For young children it is often harder to draw a straight line than a circle, but they can do it and accuracy is relative and will increase with practice. It is important that you draw this with them as you give verbal instructions. With older students it is interesting to only give them verbal instructions and see how many different ways they interpret what is being said in the images that they draw.

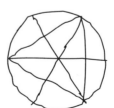
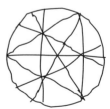

After students have made a few drawings of the folded circle have them draw one on a piece of paper. Extend all the straight lines out to the edge of the paper. Have them observe and talk about what they have done. By playing with our observations we are likely to make any number of connections; as an example (p.119). Have the students color their drawings using the divisions of the lines and shapes they have drawn in any direction that makes sense to them.

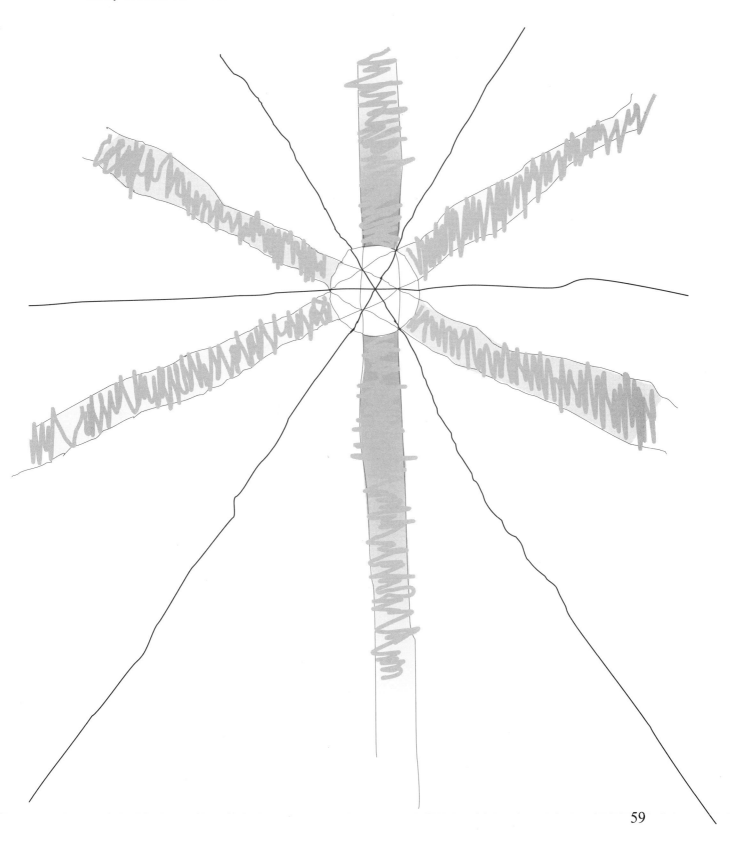

59

PYTHAGOREAN THEOREM

This formula, $A^2 + B^2 = C^2$ represents a triangulated function between parts of the right angle triangle. The first fold of the circle in half is a right angle movement that generates relationships of right angle triangles (p.36). That it happens first makes it an important concept in mathematics. It is the first touching of any two points on the circumference. Two points generate four points, two lines perpendicular to each other. The pattern of movement does not show the right angle triangle but it does give evidence of right angle function.

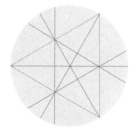

In these nine lines of the folded tetrahedron there are three rhomboid shapes sharing the same center triangle (p.111). Two perpendicular bisectors divide each rhomboid; three diameters and one large triangle. All we need is one rhomboid shape and the two bisectors to give proof to the correctness of this theorem. The proof lies within the proportional division of spherical order.

 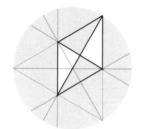

There are two right-hand and two left-hand right triangles. Each is reciprocal to the other in the same way the inside and the outside of the folded circle are to each other. We shall only consider the 2 right hand right triangles since the left hand triangles are a reciprocal function. (Notice how the rhomboid with two bisectors is the image of a compressed tetrahedron.)

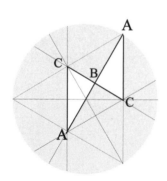

In the context of the flat circle we can label these five points defining the rhomboid by using only the 2 right-hand right triangles. They are the same triangle if you see they are rotated around one point. Only two lines are needed to show a four-triangle relationship. They are the two diagonal bisectors; *AA* and *CC* with *B* the point of intersection.

Line $AB = BA$, $BC = CB$, and $AC = CA$.; each line division is equal to itself because it was folded proportioned 1:2. This can be written as $(AB+BA) + (CB+BC) = (AC+CA)$. The square (4) is a dual function of two to itself, times two. (Two is the only number we have a choice to add or multiply to itself getting the same results.) We can then restate this to read AB times itself plus BC times itself generate AC times itself, or $AB^2 + BC^2 = AC^2$, or re-symbolized this relationship is expressed as $A^2 + B^2 = C^2$

By substituting the letters A, B, and C to designate each edge we then have $A^2 + B^2 = C^2$. This is a generalization about all right triangles and will always work because it is the first movement of folding the circle in half. When one right triangle gets removed from context —missing three triangles— and abstracted from the circle, we lose understand of what the formula means. We don't know why it works. Numbers do not prove its functionality. By understanding the context we can easily see why it works with all right triangles. It also explains why two squared, two to itself, cannot be any number higher than two and still function in the same way.

SUGGESTIONS FOR PRIMARY GRADE TEACHERS

The circle is a gift because we do not have to know anything to get information from it. The circle is the only form that models a comprehensive process where what is formed is within the context of everything else. The circle has little value if you see it only as the mathematical equivalent of zero, as nothing. When the circle is also taught as Whole, endless information is there to be discovered and it becomes fulfilling. What students learn must serve to enrich and deepen their understanding and meaning of their lives as parts of this wondrous and extraordinary universe. This must be experiential and not just formalistic.

Much information in this book looks like it is beyond first grade level; it is not! Young students need to experience activities that take them beyond their level. Quality of understanding is the result of first encounter, expectations and contextual presentation. For young children everything is in a large context. When we are born everything is enormous and there is so much of everything, and that stimulates our brain growth. We learn by observing how things relate in the larger context of everything else. We do not learn with limited amounts of information. We model what we see others doing without first understanding. It is important to present material of the largest possible context that contains not only the information that they are ready to absorb, but what they will need to learn later on. What is first given must hold what we expect of them later. That will allow them to make their own connections in their own time, and it will lead to understanding beyond our teaching. In that way they may teach us, and, more importantly, they will become more conscious of how to educate themselves through looking in as well as looking out.

The fundamental circle-folding activity is not beyond 5-and-6-year-old students; nor is discussion about what they have folded, or putting multiple reformed circles together. If a teacher is familiar with the folding process and can make the connections to the mathematical information they want their students to learn, they will not have trouble guiding their students through this process. By making connections you can take students to levels of abstract thinking beyond what is usually required at the primary level. Throughout this book are indications of some exercises I have found to work well with young children. With your own experience you will find other connections that will work for you. Do not be put off by the look of complexity. It is always easier to hold a tetrahedron you have made and understand it, than to read about it or look at a picture while somebody talks. Primary grades should present principled information.

If you are not familiar with geometry, or mathematics, a standard math textbook can be helpful for making connections to specific information not covered in this book. Exploring and talking about what is generated in the folds of the circle can easily lead to the discovery of the fundamentals of mathematics. It is equally important to freely explore the "art" expression of the two and three-dimensional nature of the folding process. The patterns in art and math are observable in the circle and identifiable as the same that come up in all other subject areas, but not in forms we would necessarily recognize. There are suggestions throughout about introducing 2-D and 3-D forms without making separations in the process.

BEGINNING FOLDING

Familiarize yourself with the folding process. Particularly helpful for first grade students, and sometimes older students, is understanding that two imaginary points anywhere on the circumference of a circle when touched together will fold the circle exactly in half. This makes the first fold easy and accurate as a place to start (pp.36). Have 5-and-6-year-old students actually make the two marks right on the circumference, then touch them right together. This tells us the entire process is about touching points. Older students want to change the line rather than to correct the accuracy of the touching points. If the points are accurate the lines will be where they need to be. Five-year-old students are very capable of folding three diameters and discussing some of the information generated (pp.34-39).

It is important to explore reforming the circle using only the three folded diameters to become familiar with the way the circle moves (pp.73-75). Students work individually, in groups of two, then in groups of four, making the vector equilibrium sphere (p.89). The circle—through folding, reformation, and multiple arrangements—generates far more information than you will use, but it is there when the students are ready for it. Using masking tape they join together the forms they make, exploring the point, edge, and surface joining. It is instructive for students to experience the process of generating fundamental patterns, polyhedral forms and interrelated systems out of nothing but paper circles, some tape and hair pins. This is an easy way for young children to make for themselves and discover in the process without using templates, elaborate instructions, cutting, measuring or gluing parts together. This process is instructive for all grade levels and does not leave them dependent upon materials and black lines made for them by somebody else. The learning is in doing, not in redoing what someone else has done. This is a process where they do everything themselves and then through observation learn the inherent properties of the forms they have made. Some students cut circles out of paper they chose, to the size they want rather than using paper plates. The forms all develop directly from the folding patterns of the circle. The information about the properties of the forms and systems emerges in observing and describing the forms they are folding. The more familiarity young children have with the Wholeness of the circle the more connections they will make with other materials, information and experiences, and the more they will learn.

Some first grade students will not have the strength to make good creases in a paper plate. In that case cutting circles out of thinner paper allows them to fold well with a good crease. Tracing around a bowl with reasonable cutting will be sufficient. The circles do not have to be absolutely accurately cut, although it is important to have accuracy as a goal. In folding the circle students learn it is unnecessary to measure, they learn to see proportionally. Proportional understanding is usually ignored because we teach them to measure first. Make it the other way around. Seeing proportionally promotes understanding towards fundamental principles and relationships. A child is well measured who develops proportional eye/brain coordination before going to the ruler.

Each grade level is different and has its own approach, as does each class; but the process and the information is the same. The presentation is adjusted for what is appropriate for the experience and education of the students being worked with. The most meaningful connections will be the ones that make sense to the teacher and what comes directly from the student's own experiences.

Below are suggestions for consideration as a beginning outline that may be used in what every way makes sense to you and your students. It can easily be integrated with other subject areas with many mix and match folding options and with drawing and discussion activities. You will also find other information that will be appropriate as you go through the book.

- o Discuss circle image, symbol, spatial. Talk about the idea of whole and parts. (pp.7,10,19,21) etc
- o Fold the circle in half; talk about the information that is generated. (pp. 34-6,71)
- o Fold 3 diameters. Explore shapes and forms counting parts in hexagon. (pp.72-5,83)
- o Make the sphere working individually and in groups; count parts, discuss properties. (p.89)
- o Fold tetrahedron, count parts, and discuss properties. Explore joining. (pp.122,127,201,203,323)
- o Observe differences of open and closed tetrahedron. (pp. 126-7,129,329,369)
- o Make octahedron. Explore joining in different ways. (pp.129,133,137,140,147)
- o Explore using tetrahedra, octahedra and sphere together. (p.136)
- o Draw pictures of the folding and the constructed objects and draw circles. (p.50)
- o Stellate the tetrahedron and octahedron, stringing the end points. (pp. 67,163,165)
- o Explore nets of tetrahedron and octahedron, (pp.127,141-3) and expand grid (pp.70,340)
- o Fold the icosahedron from spiral net. (p.149)
- o Stellate the icosahedron and string end points. (p.166)
- o Go back when it is appropriate to the first fold, introduce algebra, as another way to explore the relationships in the circle (pp.80-89). This is good for first grade on up.

This is plenty of material to begin exploring with a sense of direction, and a good experiential base from which to begin to talk about and introduce much more abstract information. This offers a tremendous amount of exploration of both 2 and 3-D without the usual fragmentation and separation. Explore the art and design by using the folded lines. (There are a good number of suggestions throughout.) This is all about developing "hand-eye" coordination, mental skills of observation, communicating, working together, making connections, pattern recognition, and problem solving; all qualities for developing self learning. It is easier to blend concepts and subject areas at the primary level than to teach them separately later when students are less open to making connections. All symbol making is a visual art composed of combinations of straight lines and curved lines; the circle and the folded diameter.

ASSESSMENT

General testing in schools is a linear function. Teachers give students information and students give it back to teachers. If the information is congruent we have a successful teacher and a successful student. There is little room to find out what the student knows in reflect-back information. There is barely enough circulation of information to keep the system going. At lower grade levels this kind of testing is not only unnecessary, it is detrimental to the self-learning process, which I must hope is the aim of education.

Some teachers have shown me how they use the circle for testing for a student's understanding rather than for right answers. As you go through this process you will begin to see the inclusive nature of the circle, which means that literally all the discrete geometry information a student is required to know is in the circle. To bring out that information is the job of the student with guidance by the teacher. A teacher must know that the information is there in the first place to be able to guide the student. The information that students discover is through their own observations and classroom reflection. The meaning is in the connections they make to what they already experientially understand, how comprehensively they can open their minds, and how far their imagination will take them into what they are able to observe.

One middle grade teacher handed out paper plates and asked the students to fold a tetrahedron, open it flat and write down everything they know about it; to label all the parts and give accounting for all the relationships of parts and functions they could find. This would include formulas, theorems, generalizations, words and numbers, 2-D, 3-D, anything that describes whatever they can see about what is going on with the creased lines in the circle.

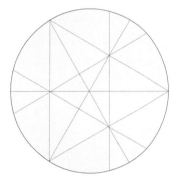

Do this for yourself and see what you come up with in 30 minutes.
Fold a tetrahedron. Open it to the flat circle and label everything you can about the two and three-dimensional information in the circle. Write down anything else you can about this circle, the folded lines and any important relationships between any of the parts. (Have students compare results.)

At the younger level evaluation is observational; what is of interest to the child, the connections they are making, and the attention and meaningfulness of the activities in which they are engaged. This is always in relationship to the value of the material and circumstances that are presented to the students. If the tools and materials are not principled and are insufficiently informative to hold their attention then any observation has little meaning in regards to the child's learning and their development.

It is generally obvious that students understand by observing what they do, and that understanding is revealed in how they talk about what they are doing. Other ways to assess what students are understanding is how much time they are spending doing what they do and what they discover; the kind of focus they exhibit; and what kinds of connections they are making to other things in their lives and to the other things they are learning in school. Because of the principled, multi-layers of information inherently in the circle, folding it becomes a rich learning experience and seeing what a student is getting from it is not difficult. They explore by expanding on what they already know. In the midst of folding activities, one 4th grade girl just sat looking at a tetrahedron. I asked her what she was doing. She said, "Pondering". Had I paid more attention I would not have needed to ask.

The circle provides something like an open book test since all the information is there in the circle to be observed and discovered. This enables the student to make connections not seen before, to discover things not specifically addressed. In this way the test becomes a learning experience as well as an indication of how much a student understands and the kinds of connections they make, the vocabulary they use. This is not the "you tell me what I told you" approach.

There are other approaches for using the circle. Don't tell me what you know, tell me what you don't know. Fold a tetrahedron and open it to the flat circle, write about what you consider to be the four most important things about the circle and explain why. Use only algebra to fully describe as many relationships between as many parts as you can find. Locate a function, give the formula, and describe in written words why it works.

Write a one-or-more page description of how to fold a tetrahedron. Describe how you would use that information to then make a 2-frequency tetrahedron, or one octahedron, so that anyone reading your paper would be able to follow your instructions and make what you have described. Draw illustrations to go with the instructions if desired. Technical writing indicates the level of observation and writing skills along with mathematical understanding.

Fold the circle in half and write a paper on everything you know about what has been generated that was not there before you folded it.

Below. This is an example of some information that might be identified in the circle that has only been folded in half. Front and back can be used as well as additional writing paper. The more one knows about the basic parts and how they are interactively multi-functional, the more one will have to write about. The student's observations, the kinds of connections they make, and their ability to express what they understand is evident when starting with hands-on folding. Whatever abstractions are going on in the head will always be expressed in some kind of concrete demonstration.

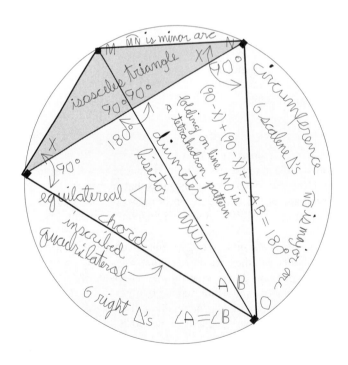

64

Mechanical suggestions

CREASING

When folding the circle it is important to be as accurate as you can. Accuracy is about placing points exactly together; it is not about where the lines will go, or if angles are sharp. _If the points are accurately touching, the lines will always be in the right place. If the lines are in the right place the angles will be correct. If the angles are correct, circles will move, reform, and easily join together_. A strong crease is important to accuracy. It is easier to see a sharp fold than a weak or incomplete fold. Crease all the way across the fold from end to end in both directions. Good creases in-crease folding accuracy. A straight edge of plastic or wood will give a strong crease. I use paint stirring sticks cut in half. A bone creaser works fine.

Fold any 2 imagined or marked points on the circumference exactly together.

Use straight edge to make a sharp crease. Crease the edge in both directions to the ends. Unfold.

TAPING

When taping edges together, _tape the entire edge length with one piece of tape_. Put one half of tape on one side, bring the sides together and fold the other half of tape onto adjoining side of the second edge. Better to take the time to tape well than redo later. One piece is stronger and looks better than a number of small pieces.

Put tape half onto one edge. ⟶ 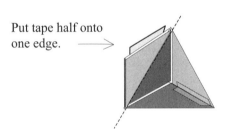 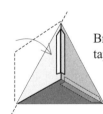 Bring edges together and fold tape over onto adjoining side.

<u>Below</u>. 1 & 2) When making a <u>point-to-point</u> connection use a single length of tape across the joining edges of both units. Each part covers half the tape's width and length. 3) Fold the other half of the tape length to the other side of the edges on both sides of the joining point. This works for triangulated point–to–point joining, as shown using tetrahedra.

point-to-point taping

triangulation is the strength; tape just holds it together

fold tape over and squeeze together

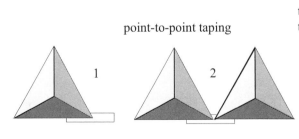 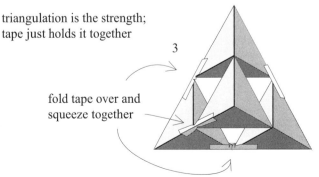

Tape does not stick by itself; pressure has to be applied for it to stick well. Some students do not know this. There are wide varieties and qualities of tapes available. Three quarter inch masking tape is my preference. Clear tape works well. Some tapes will lift off in a couple of days. Good sticky tape can hold for years before it dries out and comes off. The quality of tape has nothing to do with price. Often the cheapest tape works the best. Sometimes you don't want sticky tape, but wish to easily remove it after the glue is dry.

HINGE JOINT

Tape can be used to make a hinge joint that has full rotational movement around the axis of two adjoining edges, allowing a maximum range of movement to the limits of the configurations joined. Taping full length on both sides of the joint makes a strong hinge that moves freely around the axis. Sticky tape is best.

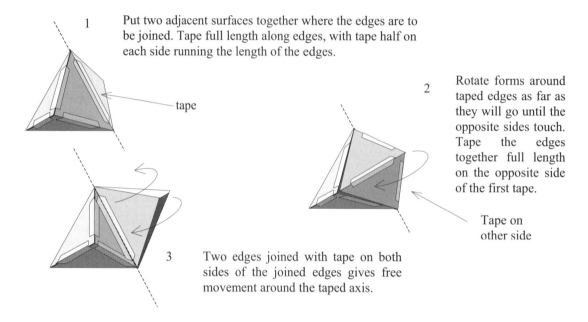

1 Put two adjacent surfaces together where the edges are to be joined. Tape full length along edges, with tape half on each side running the length of the edges.

tape

2 Rotate forms around taped edges as far as they will go until the opposite sides touch. Tape the edges together full length on the opposite side of the first tape.

Tape on other side

3 Two edges joined with tape on both sides of the joined edges gives free movement around the taped axis.

GLUE

When models are complex and take a lot of time to make, apply glue to the joining edges or surfaces before you tape. Removed tape when the glue is dry. White glue works well. Poly vinyl alcohol (PVA) glue is much better. It is archival quality, a bit more expensive, but then it takes less glue and less drying time. It sets in about a minute. That eliminates some taping and is a cleaner way to glue. It dries clear

HAIR PINS

Hairpins are sometimes called bobby pins. I use them when there is tension with edges coming together and tape will not hold. They allow quick and easy assembly and disassembly for exploring different reconfigures and transformations. 1) Hair pins are often the most direct way to join edges. 2) Hair pins can be used to join points together. Interlock two pins through the looped ends and slide each bobby pin to the edges being joined. This also works for three and four edges coming to the same vertex. If a system gets heavy and pins slip out, a little tape will hold them.

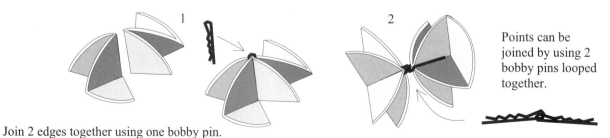

1

2

Points can be joined by using 2 bobby pins looped together.

Join 2 edges together using one bobby pin.

STRINGING POINTS

Use masking tape as a linear element to connect the points on a stellated or star configuration. This will show the edges of a plane relationship formed between the points. Tape is quick and easy. Different colors of tape are available. Yarn and string can also be used to show the edge length between end points.

1) Tape from point-to-point, forming the edges of a single plane; overlap the end points. 2) Fold tape in half by squeezing the sticky side together, sliding finger and thumb along the entire length of tape. This makes the tape a thinner line and gives it a bit more rigidity, leaving no sticky side exposed. Do the same stringing and squeezing, connecting all the points in the same way. Overlap tape on point ends to hold a taut position.

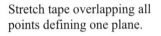
Stretch tape overlapping all points defining one plane.

Squeezed sticky sides of tape together.

Add more tape until all points are strung with tape.

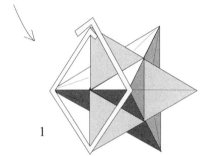

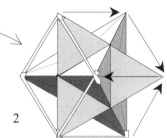

1

2

TYING

Tying is another way to make a point-to-point connection using string or twisty ties. Twisty ties are much easier than tying string and they can be taken apart more easily. The string looks better without visually detracting, and it moves better if it is a transforming model. Use a hole punch to make a hole just in from each point to be joined, then twisty tie the points together. This works well for point rotation, joining the midpoints of edges where the length of the edges is not to be joined. Tying provides a strong point joining where movement of each part is important. Do not tie tight or movement will be restricted. Sewing with a needle and embroidery thread is a nice clean way to join points together. A little dab of glue allows you to cut the ends short. Kite thread also works well for sewing. It is cheap and strong.

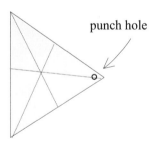
punch hole

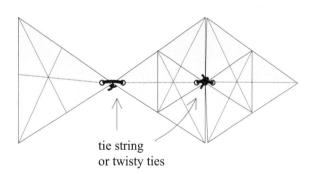
tie string
or twisty ties

COATING AND PAINTING

To make models more durable, apply two coats of orange shellac to stiffen the body of the paper. Glue size can also be used. White glue and a little water cause a slight distortion to the paper. To retain the purity of the original white paper apply two coats of Gesso over the shellac or size. This also gives a well-primed, substantial surface that can be painted with any type of paint. This process makes the paper stronger and much more permanent so that it will keep for many years. This is recommended for larger and more sculptural art constructions. Acrylic matt varnish can be used to keep the natural look of the uncoated paper surface. This also increases surface protection, but is not as durability as shellac.

WAYS OF JOINING

There are three ways to join forms in space; by *end points*, *edge lines*, and *surface planes*, and in combinations. The regular tetrahedron will be used as demonstration. It has 4 points, 6 edges, and 4 planes. There are 4 <u>points</u> to each tetrahedron with 16 possible combinations of joining point-to-point; each set of 4 to the 4 points of the other. Joining <u>edges</u> shows 36 combinations in two directions giving 72 possibilities. <u>Surface</u> joining shows 16 combinations but in three directions giving 48 possibilities. There are more edges than points or faces making more combinations of edge joining between congruent forms.

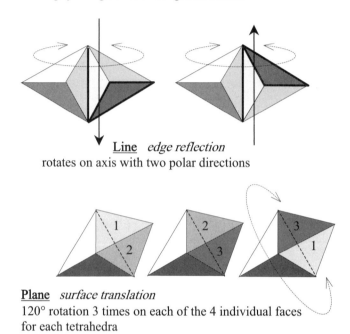

<u>Line</u> *edge reflection*
rotates on axis with two polar directions

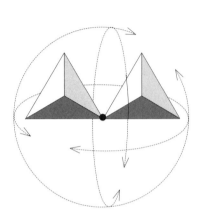

<u>Point</u> *spherical rotation*
fewest possible combinations
of joining with greatest degree
of movement

<u>Plane</u> *surface translation*
120° rotation 3 times on each of the 4 individual faces
for each tetrahedra

<u>4 Points</u>	4 x 4 = **16**	possible combinations of joining on points
<u>6 Edges</u>	2(6 x 6) = **72**	possible combinations of joining on full edges
<u>4 Planes</u>	3(4 x 4) = **48**	possible combinations of joining surface planes

16+72+48 = **136** possible primary combinations of joining 2 tetrahedra together.

With nothing distinguishing between each of the edges, between planes, or the difference between points, we don't see the possibilities available. If each point, line, and plane were individually marked then each of the 136 different combinations would be discernable, although difficult to keep track of. This becomes important when designing surfaces where they are differentiated in some way; different combinations make a difference. When exploring the tetrahedron it is not necessary to know all those combinations, but to know that they exist.

Joining two points is a spherical movement of angles of circle planes. There are specific angles of arrangements regulated by the edges and planes that form specific positions of coherent proportional intervals. The "in-between" relationships must be considered when looking at the full potential of the possible relationships between parts joined together. Nature is continually moving into and out from coherent relationships, exploring the formations and reformations of the space occurring between things as they join together and divide. To look at forms without the context of spatial intervals is similar to exploring the possible combinations of seven musical notes without considering the space between the notes. There is no modulation without articulation of the space between things.

68

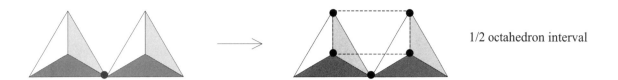

1/2 octahedron interval

Above. Two regular tetrahedra joined point-to-point in the same orientation, in alignment, generates a square relationship between them, forming one half of an octahedron. This one-two relationship is fundamental to the generative nature of pattern, (1:2).

2 tetrahedra intervals   formed tetrahelix

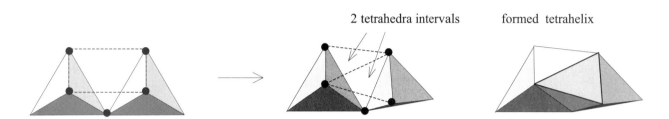

Above. From the square arrangement, one of the tetrahedra is twisted and tilted slightly closer to the other forming a tetrahelix pattern where distances between points are congruent. The square relationship is collapsed to where the diagonal becomes the same length as the edges of the square making two equilateral triangle planes. This is an interval of two regular tetrahedra formed between two regular tetrahedra. This arrangement of four tetrahedra is formed in a linear joining, showing a helix joining of face-to-face tetrahedra. This can be a right or left hand helix depending on which diagonal of the square is twisted.

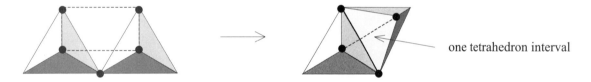

one tetrahedron interval

Above. By bringing two edges together, collapsing one of the open sides of the square, one tetrahedron interval is left. This forms the edge-to-edge axis of rotation. When the width of the interval is the same as the edge lengths of the tetrahedra there are three regular tetrahedra, two formed and one interval. This symmetrical pattern is the basic unit for the tetrahelix, (see above). Adding the next tetrahedron to extend the helix pattern will determine right-hand or left-hand direction of spin depending on where it is joined.

bi-tetrahedra

Above. By using a hinged joining tetrahedra can be rotated and closed together face-to-face in either direction forming a solid bi-tetrahedron.

This progression of movement from a point, to edge, to face joining shows these three individual relationships are not separate. Points, lines and planes are useful abstractions that help define general properties and the symmetries of individual forms. They are important in forming models. The separation is conceptual. Every point is a circle, each line has two end points, and every plane has both points and lines. It is through the interaction of these three elements that systems can be formed and reformed through extraordinarily simple movements.

69

COLORING

Color gives surface planes greater visual interest. Designs on the surface can mask or accentuate the shapes and forms. Color can be used to keep track of design elements of the underlying pattern that would be difficult to see otherwise. Stay with the creased lines when using color. Color the surface in an arbitrary manner to see the difference. Color before and color after folding to see the difference. Cut circles from magazines with colored pictures, and then fold. Use different colored circles, circles with different colored sides. Whatever the design and colors are it is always interesting to see how the flat images change as the circle is reconfigured. Many unexpected rearrangements happen moving between 2-D and 3-D with changing symmetries through reforming and joining in multiples the color-designed circles. While this book is in black and white there are many examples of 2-D designing to the circle grid that can be colored, as well as what each individual has to bring forth (pp.39, 54, 56, 113-15, 326, 340-43).

Coloring different designs is a useful way to explore different kinds of symmetries. It can be used to show the inter-related and multi-functional nature of the many combinations of polygons that occur in the folded lines. Add lines between points subdividing into smaller irregular areas. Add texture and various design elements to differentiate areas and planes. Use the creases as guides to explore your interest, but do not violate them. The design possibilities in coloring the folded grid of the circle are inexhaustible when you get into the higher frequency folding. The drawings throughout the book are a small indication of what can be done by systematically subdividing various symmetries and discovering the design possibilities of the circle.

Below. Join everyones circles together on the floor point-to-point and see how they all form a single grid, like a circle patch work quilt. Observe the larger designs that begin to appear in the larger grid. Overlap layers of circles choosing what combinations are pleasing and tie them together to make a big wall hanging of colored paper plates. Punch holes and twisty tie or string tie tangent points together making strong connections. This fun to do with young students.

Folding the circle

 ONE DIAMETER

FIRST FOLD

We have already addressed folding the circle in half (p.36), and a good part of the information generated in the vocabulary section (p.34). At first you might not want to talk about all of the information in that one fold, but it is important to talk about some of the information. It is instructive to the primary nature of the circle and reveals what to do next. It is important to come back to the first fold and talk about it in depth. Don't lose what is there by ignoring it or forgetting to come back, observe and talk about what is there. That first fold is principle to everything else. What we understand about where we start has everything to do with where we end up. There is greater benefit to giving attention to starting than to the finished product. *The first fold is where accuracy starts*.

<u>Fold</u> any two imaginary or marked points on the circumference together. Crease the fold with a straight edge.

Any two points on a single plane when touched together will generate a fold perpendicular to and half way between the direction of movement between the two points.

Open the folded circle and explore how it has been changed and what has happened with one diameter folded into the circle.

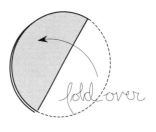

This movement is the largest proportional relationship the circle can have to itself. Everything that can be done with the circle is inherent in this self-referencing first fold. Each circle you fold starts here.

THREE DIAMETERS

The biggest thing that happened in folding the circle in half is the ratio 1:2. That is what happens first, that is what we do again. There are 3 ways to fold the folded semi-circle into a 1:2 ratio. They in turn generate 3, 4, and 5 diameters.[1] We are only folding 3 diameters.

1) <u>Fold</u> the circle in half.

2) On the circumference move the end point on the right halfway between the point on the left and the new point on the right. Slide the right fold back and forth along the circumference until the two parts look equal. The new end point is formed by the folded over edge. (Use your eyes to see proportionally; do not resort to measuring.) Keep the circumference together. **Do not crease yet**.

3) When the parts *look* equal, fold the unfolded left hand point *behind* to the new point on the right circumference. The symmetry of folding *one over and one under* makes it easy to check accuracy of alignment. <u>Look on both sides to see that the end points are even.</u> When they are even the all the straight and curved edges will also be even. If your folding was off, adjust by sliding the end folds back and forth until everything is even. *When <u>end points</u> are even the edges will be even and creases will be where they belong.* <u>**Only after all points and edges are even, do you crease.**</u> Crease with the straight edge. This folds the semi-circle into thirds. If creases are in the wrong place they do not erase, and the proportions are off making further folding more difficult.

Open the circle out flat. By folding the circle 3 times to the ratio of 1:2, three diameters are generated showing 6 interval spaces. There are 6 points on the circumference with one point in the center. 3+6+6+1=16 parts, (1+6=7).
The 7 points that define the hexagon are primary.

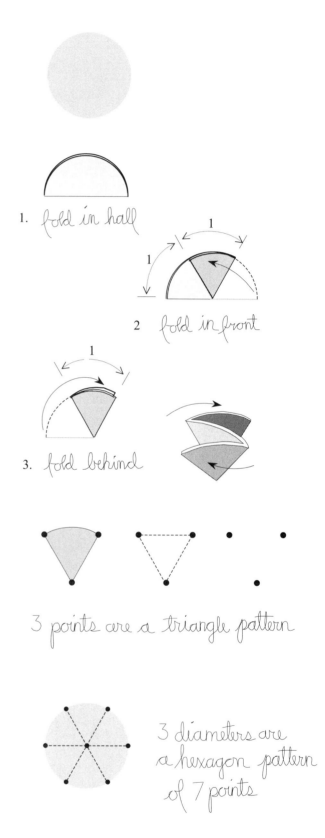

1. fold in half

2 fold in front

3. fold behind

3 points are a triangle pattern

3 diameters are a hexagon pattern of 7 points

---

[1] *The Geometry of Wholemovement: folding the circle for information*. B. Hansen-Smith, 1999 Wholemovement Publ, Chicago, IL

EXPLORING THREE DIAMETERS

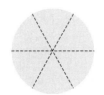

What can 3 diameters do? How many ways can they be arranged to reconfigure the circle? Create a dialogue with the circle as it moves; first in one direction, then in the other. Get a feel for the movements of the circle. Look for the symmetries. Feel the 6 radial lengths as they move around each other. Observe the intervals that are formed as you move the circle. These movements need to be explored and discussed. If reformations are not talked about, come back later and explore them all.

1) Open the triangle cone part way so the circumference looks like a "Z". There are 3 double planes in a zigzag that make a spatial pattern of 2 tetrahedra joined by a common plane. (The tetrahedron is 4 points in space.) By changing the amount of opening on each side the tetrahedra change proportions.

2) Slide the 2 in-folded points out, bringing them together. (One diameter, opposite radii folded together.) This forms 2 tetrahedra joined by a common edge. These 2 formed tetrahedra show 2 more tetrahedra as intervals; a pattern of 4 tetrahedra. The number of points has increased from 5 to 6.

3) From the crossed over "X" position above, slide the 2 points past each other flattening to a semi-circle. Using the center of the circle as a folding point, bring the edges of the bottom diameter together. This forms a tetrahedron with a doubling of surface, one half of the plate folded in and one half to the outside. The center of the circle is opposite the opened triangle plane. It is important to notice the dual tetrahedra, one inside of the other (p.143).

The first few opening movements of 3 diameters reveal the tetrahedron, reflecting the tetrahedron movement in the first fold in the circle.

1.

open triangle to a "Z"
two tetrahedra joined
by common surface

2.

move the "Z" to an "X".
the common face has changed
to a common edge

3.

fold the diameter to itself
using creased radii

inside

this forms the tetrahedron
as an open 3-sided
curved-edge pyramid

73

4) Flatten the single tetrahedron by opening one side and closing the other over the center section. Open between the larger of the folded spaces forming a 4-sided pyramid, half an octahedron. This parallelogram will collapse perpendicular in 2 directions. When all angles are the same the interval opening is a square.

5) Open the inside flap halfway joining the opposite crease with the flap edge. This forms 2 tetrahedra joined by the inside flap. The parallelogram has become stable by changing into 2 tetrahedra sharing a common plane.

6) Open the 2 tetrahedra hinging on the middle flap that holds them together. When the 2 tetrahedra are a radius distance apart they form a pattern of 3 tetrahedra of the same size; 2 formed and one formed interval between them. Three tetrahedra face to face form the seed unit for the tetrahelix. Opened further and we see the symmetry of the circumference crossing on one side and the creases crossing over on the other. Opposite radii come together as a shared edge.

7) Open the circle into a flat hexagon pattern of 3 diameters. From the center out, fold one radius to form a radial width interval. A pentagon pattern is formed (6 minus 1 is 5). This is a pattern of 5 tetrahedra in a pattern of 7 points with a central axis. Continuing to decrease the radial interval until the edges of 5 come together into 4, form the square-sided pyramid (6 minus 2 is 4). Subtract one more and we have 3, (6 minus 3 is 3), the single tetrahedron. Simple arithmetic.

8) The radii remain constant while the perimeter diminishes with the altitude increasing perpendicular to the open hexagon plane/circle. This is the movement of the stellating process (p.163). The 6 moves into space forming 5, 4, then 3, flattening into 2, then folded into a stacked, compressed, triangle pattern. One, 2, and 6 are flat shapes and the 3, 4, and 5 are spatial relationships.

4.

this changes to half an octahedron

5.

2 tetrahedra

6.

radial length 3 tetrahedra   *4 tetrahedra*

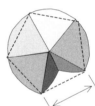

7.

7 points

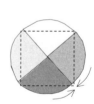

8.

6   *5*   *4*

3   *2*   *1*

9) By alternate folding the radii of the hexagon in and out a 3-armed star is formed where 3 radii become a vertical center axis perpendicular to the triangle plane. This shows 3 irregular tetrahedra intervals joined around an axial center. Open the radii of the axial center to form a center tetrahedron space with 3 congruent tetrahedral relationships around it. This makes a pattern of 7 tetrahedra as a spatial relationship of the 7 hexagon points (p.102).

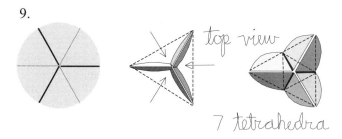

9.

top view

7 tetrahedra

10) As 2 planer arms of the star are flattened down to the same plane the third arm remains perpendicular to the other 2. It can then move to one side or the other, going flat making a rhomboid pattern of 4 points.

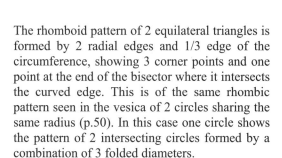

10.

The rhomboid pattern of 2 equilateral triangles is formed by 2 radial edges and 1/3 edge of the circumference, showing 3 corner points and one point at the end of the bisector where it intersects the curved edge. This is of the same rhombic pattern seen in the vesica of 2 circles sharing the same radius (p.50). In this case one circle shows the pattern of 2 intersecting circles formed by a combination of 3 folded diameters.

The circle shows 4 individual polygons. The hexagon reduces to an isosceles trapezoid, a rhomboid, and an equilateral triangle.

The triangle is principle. Everything else is relative to the informed pattern of the tetrahedron. Triangulation gives continuity and stability to all movement, change, and transformations.

It is important to talk about the different shapes that are formed since they will be appearing continually throughout the folding process in the same way all the reformations of the three diameters are fundamental to all folding.

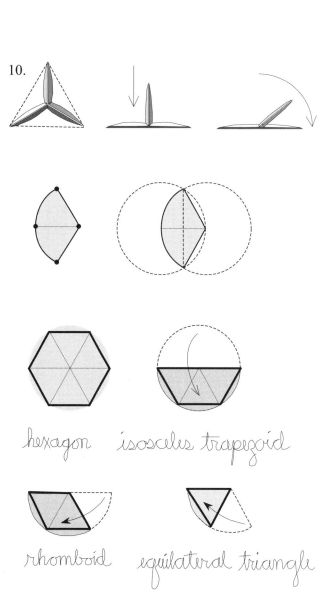

hexagon   isosceles trapezoid

rhomboid   equilateral triangle

When the circle is folded to itself the movement will manifest only one of an infinite number of diameters that give measure to the circle. That one diameter sets the proportional division of the circle to that specific orientation. This one movement will always divide the circle in half and will generate all the information that is principle to everything the circle can possibly do. This half-folding reveals the diameter to be the measure as a straight line between the two furthest points on the circle. Each time any subsequent folds are made it is always in relationship to that first fold, thus the importance of understanding as much as we can about what has been generated with one diameter. While each division in half forms a diameter, they are not all the same. Each is individually unique in relationship to the others but proportionally the same in relationship to the circumference. Diameters are the direct expression of division of the circle because they happen first.

In folding the semi-circle into thirds, folding one side over and the other side under, a zigzag pattern of movement is formed. If the sides were folded over each other the folding would become closed, no longer an open movement. Inaccuracy results from what is hidden from view when folded closed, making it difficult to adjust alignment with all parts. Once creases are in the wrong place, you cannot get them out.

Below. The zigzag pattern is a very old and commonly used design motif throughout the world. This ancient symbol represents life force, moving up and down, much like water as it moves in a forward path. It is open and progressive, not closed into itself. There are many styles and forms this pattern takes with many references to a variety of forms in nature, but always an open symbol of life movement and generation.

Three folds in a 1:2 ratio generate 3 diameters, all bisecting each other making 6 radii, forming 6 congruent spaces all in different orientations to each other. The principle of duality is now expressed in an extended division of parts and in a more complex relationship of 3 sets of 2.

Below. There are 7 combinations of relationships to the number 3; six individual groups of 2 each and one group of all 3 combined. Six and one is 7. (The number of points of the hexagon pattern reflects perfectly all the possibilities.) Three individual diameters plus 7 totals 10, the number pattern of the tetrahedron, observed in the first fold of the circle.

two relationships of one,   1 to 2, 1 to 3 ⎤
two relationships of two,   2 to 1, 2 to 3 ⎥   (3 sets of 2)
two relationships of three, 3 to 1, 3 to 2 ⎦   3 x 2 = 6
unity of 1, 2, 3,   (1 set)   + 1 = 7
individual units 1, 2, and 3   (3 sets)   + 3 = 10

Three diameters embody the number pattern 10 reflecting the first fold of the circle; 4 points and 6 edges, (p.36). The number ten, a one and a zero in separation is also a symbol representation of the diameter and the circle. The one is generated from the circle; the sequence is O then 1. The number 10 is the first in the second generation of numbers following the first zero through nine. Between zero-ten, 010, was all that was necessary to develop a binary system using zeros and ones to represent all combinations of numbers. The zeros were dropped and the bar code was developed (p.43).

The number 7 of the hexagon is also thirteen. Seven points of the hexagon pattern are primary and, by adding the 6 intervals formed by the three diameters, 7 + 6 = 13. By adding the 3 diameters to 13 it is 16. Adding one to six is seven (1+6=7). Seven is inherent in the three. Numbers are important pattern marks.

<u>Below</u>. Counting the ends of the 4 fingers, the thumb, and wrist where it attaches to the arm (the wrist must be considered a part of the hand, the hand does not exist without it). All 6 points radiate out from the center making a seventh point. There are 6 spaces between all these parts. Identify the many things that can be described by the numbers 7 and 13. The most common examples that come up are: the 7 days of the week, 7 notes in the musical scale, 7 primary colors, the 7 wonders of the world, sometimes the 7 major holes in the head, the pattern of snow flakes, chakra locations of the body. Seven has varied symbolic importance. Seven is universal as a structural pattern of spatial organization.

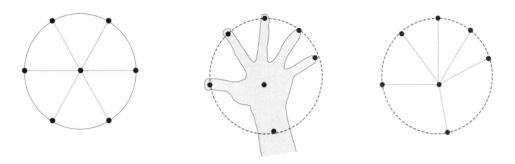

<u>Below</u>. The space allows the hand to move. The 6 spaces separate the 7 points and must be counted. With a finger trace the outline of the hand with fingers spread. Count the intervals as you trace the spaces between the fingers and the wrist. Do in rhythm and vary the interval time. This is a very old form of a child's game of rhythmic movement counting sounds. This is the same pattern found in the clock face.

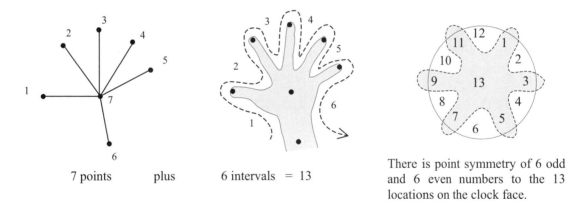

7 points   plus   6 intervals = 13

There is point symmetry of 6 odd and 6 even numbers to the 13 locations on the clock face.

The intervals between notes are essential for making music. The hands on the clock mark the interval movement between the center and outside points, 12 around the center point 13. As with the hand, it is the patterned movement between the center and outside that makes it functional. In folding the circle it is the movement, and touching, between the points on the circumference and the center point that generate the creases that reveal information. The hexagon is a pattern of movement, even though we have been taught to see it only as a static polygon of 6 sides.

Counting the biggest movement joints in the body we find a pattern of 13. With a finger from each hand touch and count; 2 ankles, 2 knees, 2 hips, 2 shoulders, 2 elbows, 2 wrists, and 1 neck joint. You will find this same pattern of major joints in your dog, cat, bird, and all other vertebrate. The proportions will all be different since pattern forming is always coordinated to specific contextual function.

Traditionally we think 13 to be an unlucky number. It is possible that a few centuries ago a small group of people understanding pattern and the "scared" function of what this number represented may have suggested to others that 13 is bad luck. The result would have been a self-regulating condition where people would stay away from anything to do with 13. This would have been effective protection for those who might have wanted to keep the "secrets" of universal relationships for their own advantage, like building cathedrals and such. This is only a conjecture to explain why 13 as the primary, centered, spherical system fundamental to all pattern formation has become "unlucky". Thirteen is 12 spheres closest packed around a center sphere (p.92,98). It is the clock, Jesus and 12 disciples, a baker's dozen, 13 lunar months, the first 13 United States, 13 stripes on the flag, etc. Counting is a necessary part of understanding patterns and a way to see correlation between the diversity of seemingly unrelated and individual formations.

Three sets of 2 numbers, on opposite sides of dice, each add up to 7, the hexagon/cube structurally patterned to 3. By adding the six faces of a single dice you get a total of 21 or **3**.

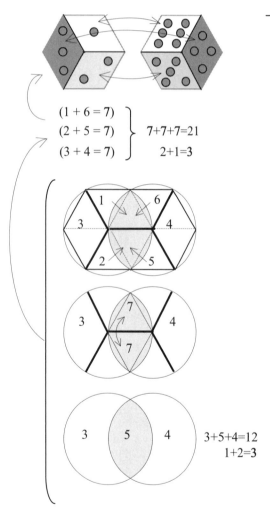

$(1 + 6 = 7)$
$(2 + 5 = 7)$ $7+7+7=21$
$(3 + 4 = 7)$ $2+1=$**3**

$3+5+4=12$
$1+2=$**3**

Adding two 7's is 14 → 5. Five is about growth and generation. It is from the vesica that the patterns of growth emerge, the circle to itself. This is a dual circle (p.47,87).

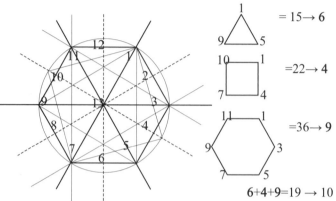

= 15→ **6**

=22→ **4**

=36→ **9**

6+4+9=19 → 10

<u>Above</u>. Starting from 1 and adding the respective numbers that form the individual *triangle*, *square*, and *hexagon*, the total is 19 (1+9=10). The *3*, *4*, and *6* represent the shapes formed evenly in 12 with the center point, 13. (3+4+6=13→4) Four is about formation, reformation, transformation; movement, the first fold of the circle.

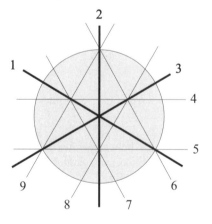

<u>Above</u>. There are 3 sets of 3 triangulated lines. Three to the power of itself is 9. Adding 1 thru 9 is forty-five is 9. It takes these 9 lines to fold the tetrahedron. Add all points of intersection and there are 19, which again is 10, the number for the tetrahedron.

78

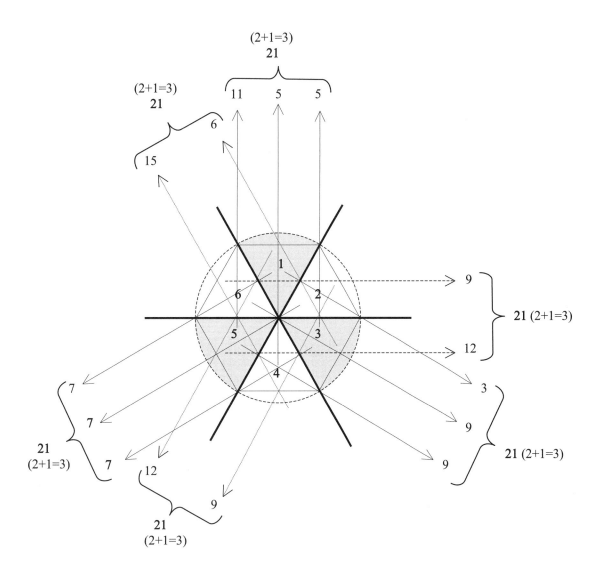

Above. Each of the six sectors of the circle that form the large triangles in the hexagon have been numbered sequentially one through six. It is interesting that the numbers in six individual directions each add up to the same total, 21 (2+1=3). Three is a number that reflects principle in the first fold, and the three folded diameters as structural origin of the hexagon. In the tetrahedron folding there are three diameters and three sets of two parallel lines. The folded ratio of 1:2 is inherently 3 and generator to all subsequent development. The "magic" of numbers is in the pattern of principled relationships they describe.

Adding the numbers one through six total 21, (2+1=3).
Adding the numbers in the gray triangles (pointing down) the total is 9.
Adding the numbers in the white triangles (pointing up) the total is 12, (1+2=3).
The numbers 9 and 3 combined equal 12, the inverse of 21 that is 3.
Numbers always reflect the multi functional nature and inter-association of parts.

When talking with 5-and-6-year-old students about the circle and what they see when folding it, common first grade words are used. Common usage words are generalizations with different coloration depending on the context when learned. This goes for any age. Giving an accurate word expression to our experiences sometimes requires a lot of words. Language is one of our first tools in abstracting expression from the experiential context. Algebra is a language that is more precise than words. It is a code for words that describe observable relationships. Children learn multiple languages easily when they are young. Algebra is another language. Language is a tool for giving expression to our experiences and perceptions, and for sharing our thoughts with others. The basics of Algebra can be learned through folding the circle. It is good fun in abstract thinking for first grade students and they learn a lot about geometry and mathematics.

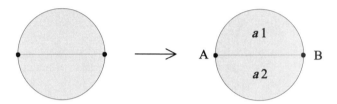

Above. It goes like this; we fold a circle in half (pp.36,71). The diameter is the first line in the circle. It has two end points. We give the individual points names, calling one point Albert, and the other point Bob. To shorten the writing of the names we can just say point A and point B. By naming the points we have named the line. It is between point A and point B, so we call the line AB. We have three names, one for each part from only two letters. We can call the circle O, that's easy. Since line AB divides O in half there are two shapes of areas, they also need names. We decide on area 1 and area 2. To make it short we say A1 and A2. Having already used the big A we use a little a; *a1* and *a2*. (This is about seeing differences.) We talk about this until someone sees we have a line BA. Usually we have talked about movement going in two directions and they see the difference between AB and BA is direction. There is one path, or line, going in two directions. So we see AB equals BA, and BA is the same as AB. By using two arrows in opposite directions between AB and BA we can show they are the same but with different directions, AB ⇄ BA. Or just make two short lines **AB=BA**. We can then talk about the symmetry of the fold in a code language.

Two more paths are observed between point A and point B. Each one of these goes around the circle in two different directions. Now, this presents a problem. We must be able to show the difference between the three different pathways and six different directions. Each half of the circle shows a curved path. So for the top path we can put a curved line over the path AB, and the same for BA. For the bottom path a curved line is put under the AB and BA, just like the circle shows. To make it clear we can put a straight line over AB and BA to show the straight path. We had to all agree on this so there was no confusion about what paths and which directions we mean. We now have $\overline{AB}$, $\overline{BA}$, $\overparen{AB}$, $\overparen{BA}$, $\underparen{AB}$, $\underparen{BA}$

We have now given names to all three paths (two major arcs and a diameter), and six directions (movement between points), two points (vertices), one line (bisector), two areas (major sectors) and one circle (the context). Compare this information to one path and two directions in the straight-line equation AB=BA that is usually given as a commutative function in Algebra books.

Below. The circle is infinitely symmetrical and by folding it one line of symmetry is formed. One half is equal to the other half. All parts come in pairs except the diameter, which has two different directions. We have set up our symbols in pairs to show they are equal. This reflects the dual nature of folding the circle.

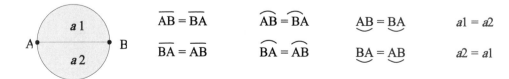

$$\overline{AB} = \overline{BA} \qquad \overparen{AB} = \overparen{BA} \qquad \underparen{AB} = \underparen{BA} \qquad a1 = a2$$

$$\overline{BA} = \overline{AB} \qquad \overparen{BA} = \overparen{AB} \qquad \underparen{BA} = \underparen{AB} \qquad a2 = a1$$

The first fold generates two points. They are the same except they are in different places, so we call them similar. There are two directions because we can start from either point. The two areas of the circle are the same but in different places, so they are also similar. We can say the same about the curved paths, they are the same in different places. Parts and functions that are the same are similar when in different locations. The abstract symbols look the same. When we show parts are the same we put an equal sign (=) between them. If we want to show they are not the same we can cross out the equal sign by putting a diagonal line through it. $\overline{AB} \neq \overset{\frown}{AB}$ shows they are not the same, they are unequal.

Symbols make it easy to add and subtract parts and to rearrange them, which is what happens between multifunctional parts. With symbols we can observe more information and often more accurately than when using words. At this point in development we have no rules other than those on which the class agrees. We play a lot with seeing how many different ways we can arrange the parts to make the two sides equal, or unequal. The more folds we make the more parts there are. It gets confusing so we have to expand our code language. That is when we add more name symbols to identify different parts and find new relationships. Then we can expand the game by discovering different kinds of equations of similarities and differences. It helps students to observe and think in a different way when we use name symbols than when we use common language descriptive words.

How many different combinations of parts can you identify by using the code we have developed so far? This is good to do with the entire class at the board where each student can contribute what they see. This way the entire class can agree if each equation follows the rules. Remember these rules are what the students have agreed upon. When symbols are used to represent the parts in the circle, we can move the parts around which we cannot do in the circle itself. Representations of parts do not have the same restrictions as actual parts, but then symbols are only names we have given to the parts to talk about them. Symbols always represent something else. That stretches our imagination as we make connections to collections of other objects to see if they work in the same way. The use of symbols allows us to see things in space that we would not see otherwise, to see patterns and arrangements that we would normally miss. Symbols help us to make connections, as we have seen with numbers. The usefulness of any language is in the consensual, contextual meaning we share, and the value we gain in collectively sharing our observations and experiences with others.

<u>Below</u>. Some simple equations to be explored, always in reference to the circle O, they have folded.

$AB = BA$	$\overline{AB} + \overset{\frown}{BA} = A1$	$O - a1 = a2$
$2AB = 2AB$	$\overline{AB} + \underset{\smile}{BA} = a2$	$a1 - \overset{\frown}{BA} = \overline{AB}$
$BA = BA$	$\overline{2AB} + \overset{\frown}{2BA} = a1 + a2$	$\overset{\frown}{AB} + \underset{\smile}{BA} = O$
$AB + 1 = BA + 1$	$a1 + a2 = O$	$a1 = O - a2$
$AB + 2 = BA + 2$	$a1 + a2 = \overset{\frown}{AB} + \underset{\smile}{BA}$	$a1 = a1 + a1 - a2$

Remember that the children have decided that zero represents the complete circle, rather than nothing. As we go back and look at the areas of the folded circle, there are in fact four areas to the circle. There are two sides to the circle divided into four individual areas of semicircles. We then have $a1$, $a2$, $a3$, and $a4$. They are all similar and also come in pairs. As $AB = BA$ so does $a1 + a2 = a3 + a4$. Instead of two directions the areas have an inside and outside identified by folding the circle. By observing more about the circle there is more information, and coding can become very complex. We can always choose or not choose to include what makes sense to us. Keep it easy and clear, that is the point of using a code language.

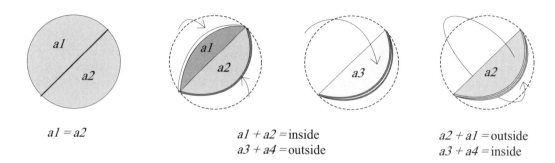

a1 = a2	*a1 + a2 =* inside *a3 + a4 =* outside	*a2 + a1 =* outside *a3 + a4 =* inside

<u>Above</u>. One side is folded to the inside putting the other side to the outside and reciprocally in the other direction. This is a good time to introduce the idea of brackets to keep families together. We can then refer to one group as *(a1 + a2)* the inside, and the other as *(a3 + a4)*, the outside. *(a1 + a2) = (a3 + a4)*. We can now take it further and only use the first letter of <u>inside</u> and <u>outside</u> to say that I = O. As long as the symbols refer to the folded circle the areas move in pairs. It is impossible to get *a1* and *a3* or *a1* and *a4* inside or outside together. It is also the same with *a2* and *a3* or *a2* and *a4*. If we decide all semi circles are the same and not just similar, then we can mix and match any way we want to. The areas have all become abstractly equal and *(a1 + a3)* can equal *(a2 + a4)*, or *(a1 + a4) = (a2 + a3)*. If we keep the context of the folded circle the symbols can only do what they represent.

In using the O for outside we might then want to change the O symbol for the circle to a small o to avoid confusion with the outside O. There must be agreement by everyone for this. In considering the inside and the outside using the symbol names, the reciprocal movement can be described as OI = IO. This is just like AB = BA, it is about movement. There is a wonderful visual symmetry that is reflected in the equations. We can even go further in subdividing parts that are equal because they are the same parts in different locations. That is different than AB = BA where the only movement is in our minds.

$$I = I$$
$$I = O$$
$$O = I$$
$$O = O$$

Since they are all equal to each other there are many combinations. (I + O)= (O + O), 3(I + O) = 3(O + O), 4(O) = (1 + 1 + 2) (I). With this last example if we eliminate the outside and inside the numbers will still be equal. That is total abstraction of numbers without the context. By eliminating the numbers the inside and outside remain equal, which we know by folding the circle. Equations are about relationships between parts of many different kinds and combinations. The balance of equal (=) goes all the way back to the symmetry of spherical Wholeness compressed to the circle. The equal symbol is two parallel lines, the two opposite arrows without the points. To know what is equal there must be measurement in some form, and that is about movement of something more than one.

Fold   two more diameters into the circle (p.72).

We have more points, lines, areas, and now angles to name. The angles where lines come together also need to be named. First we need to count the similar parts to know what we have.

There are 3 diameters. (Write down the number 3 on the board)	3
There are 6 areas. (Write 6 next to 3)	36
There is 1 circle. (Make a O next to the 6)	36O (3 lines, 6 areas, one circle)

This is an easy way to introduce the concept of degrees and to explore divisional relationships from the students' own observations about the circle. We can now talk about angles in relationship to areas that we have already become familiar with.

82

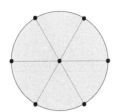
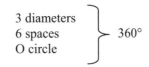

3 diameters
6 spaces   } 360°
O circle

The small circle° on the upper right reminds us that
the numbers 360 are about the circle which we have
changed to a small o.

Above. The numbers 360 describes the division of a circle by 3 diameters into 6 equal sectors. (The radii are not counted, being divisions of diameters.) By putting a small circle symbol to the upper right after the number, 360°, we will remember when seeing this number it is about the circle. We can now find out from this information how much of the circle each angle has in relationship to the entire circle. Figuring out how many parts or degrees there are in a circle corresponds to the actual development of folding the circle and paying attention to what is generated. Gong back to our code names, if the circle is represented by 360° and there are six areas, then 6 areas equal one circle, or $6a = 360°$. Depending on grade level you can develop this further to find the number that represents each area.

Below. The first diameter is already named AB. The second will be CD and the third will be EF. There is a center point where the 3 diameters intersect each other. We will call it O the same as the outside symbol but a difference context. Now everything can be talked about by just using the letter symbols for the 7 points.

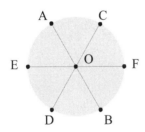

We can now define an area by three points,
a triangular relationship of areas;
a AOCA
a AOCA $= \overline{AO} + \overline{OC} + \overset{\frown}{CA}$
a AOFCA $= a$AOCA $+ a$COFC

To show the angles we can use a symbol that shows two lines coming together at the same point; ∠ is our symbol for angle.

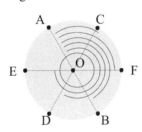

There are six individual angles starting from A going clockwise;
∠AOC, ∠COF, ∠FOB, ∠BOD, ∠DOE, and ∠EOA.

When angles are joined together in combinations they are called compound angles; ∠AOC + ∠COF = ∠AOF. The compound angle is ∠AOF.   ∠AOF + ∠FOB = ∠AOB, a compound angle.

There are an increasing amount of symbols and combinations of relationships. It can be a challenging game to find as many combinations and equations as you can to describe all the relationships you can observe in the three diameter folded circle; all by using a code language the students develop for themselves.

The folding of the three diameters makes all six areas congruent. The equal areas, folding, gives proof that all six angles are equal. From this we can find out how many parts (degrees of the circle) each angle has.

$O = 360°$
$6\angle AOC = 360°$
$\dfrac{6\angle AOC}{6} = \dfrac{360°}{6}$
$\angle AOC = 60°$

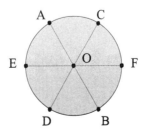

Relationships can be expressed in so many different ways. It can be very exciting to see just how many different possibilities of relationships and combinations of parts there are to be found in a three times folded circle with a few symbols representing points, lines, areas, and angles. This is a good exercise to begin understanding parts in relationship with other parts and the interrelated functions of algebra and geometry. It sharpens our observational skills. Nothing has meaning in isolation and without interaction. In this game there is the only one rule: everyone must agree on what each symbol represents. If new symbols are added or changed, everyone must agree about that. Otherwise there is confusion and we cannot understand each other.

<u>Below</u>. Coloring in the areas of the circle is another way to help see various combinations of relationships. It is a clear way to understand grouping in sets, particularly for young students. This is a good example of how when symbols abstract information from the source, differences are lost. Notice how three number formulas show six different arrangements, and one more set makes seven different combinations. This demonstration is another reflection of the importance of three as pattern in systems development.

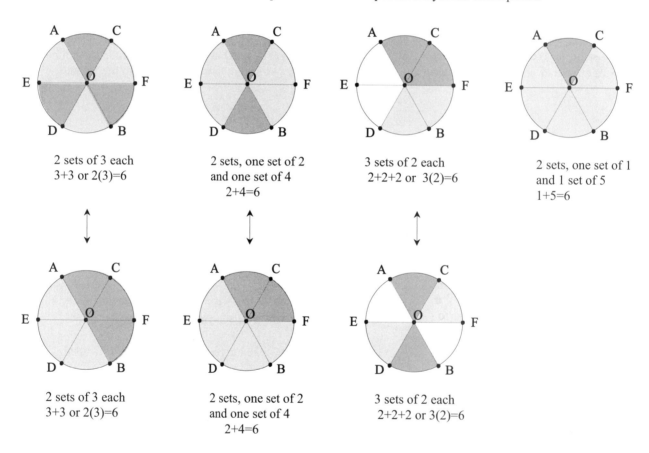

2 sets of 3 each
3+3 or 2(3)=6

2 sets, one set of 2
and one set of 4
2+4=6

3 sets of 2 each
2+2+2 or 3(2)=6

2 sets, one set of 1
and 1 set of 5
1+5=6

2 sets of 3 each
3+3 or 2(3)=6

2 sets, one set of 2
and one set of 4
2+4=6

3 sets of 2 each
2+2+2 or 3(2)=6

There are 7 larger groupings in combinations of sets. Consider that what has been presented in this development of relationships has expanded to three different inside/outside axes of the circle. The increased combinations of relationships have become very complex to describe in abstract symbols and are best understood by the actual folding and reconfiguring of the circle, coloring to track relationships.

The development of an algebraic code language should be fun and controlled by the students themselves, guided by the teacher. Even for teachers who are not sure about their math skills this is not difficult. The more that is folded into the circle the more there is to be discovered.

84

Three equally spaced diameters show 360° for the circle (p.83). There are many ways to approach how much of 360° is in any sector of the circle starting with one diameter. We can simply divide 360° by two. This reflects folding the circle in two congruent parts. The difference is in the circle, nothing is taken away, it is just divided into two parts; $360° \div 2$ or $360° = (a1 + a2) - a2$, or $360° \div 2 = a1$, so we find $180° = a1$, one half of a circle. Another approach is that $(a1 + a2) - a2 = 360° - a2$, or $a1 = 360° - a2$, or

There is more than one way to rearrange and find balance between parts.

$$\frac{a1 + a2}{2} = \frac{360°}{2} \quad \text{or} \quad \frac{a1 + a2}{2} = \frac{180°}{2}$$

We have each half circle divided into three equal areas. With $a1 = 180°$, dividing each side of the equation by 3 we get how much of 360° is in each area: $\frac{a1}{3} = \frac{180°}{3}$ or $\frac{a1}{3} = 60°$.

<u>Below</u>. In touching any two points on the circumference together (A and B), two more points are generated (C and D). Each half shows one point plus the two points of the diameter. From 3 points on the circle we identify 2 triangles. Drawing lines connecting points CAD with the diameter CD forms a triangle. The same goes for connecting points CBD. There are two congruent triangles.

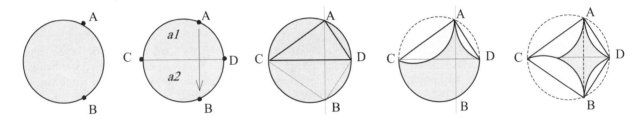

<u>Above</u>. $a1$ and $a2$ are both 180°. One point A anywhere on the circumference with diameter end points C and D will be 180° and form a triangle pattern of three points (p.36). We use the symbol of a triangle, Δ, to show points CDA is Δ CDA. In changing the form of one half of the circle to a triangle, the positions of the points does not change; only the form of two connecting lines. Only 180° remain consistent to the three points. When we fold over the circumference between points CA and points AD we only change the defining edges from curved to straight lines, we do not eliminate any part of 180°. So, without being aware of it, we have added to the vocabulary of our code language by talking about the folding of the circle.

<u>Below</u>. ∠ACD, ∠CAD, and ∠CDA are on one side of the diameter CD. Combined they are 180°, or one half of the circle. We know the triangle formed by one point on the circumference and the diameter is a right triangle of one 90°angle, ∠CAD. That is reflected on the other half of the circle (rotated or flipped 180°). Since 90° is half of 180° then the two angles remaining must total 90°. With this information there are other ways algebraically to show the number of degrees there are in each angle of the triangle. The unlimited symmetry of the circle allows every point to function as a center point (pp.31,105).

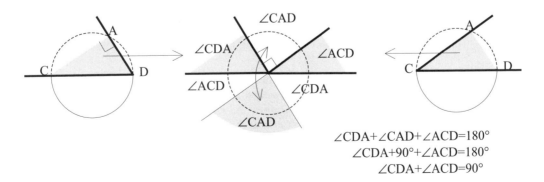

$$∠CDA + ∠CAD + ∠ACD = 180°$$
$$∠CDA + 90° + ∠ACD = 180°$$
$$∠CDA + ∠ACD = 90°$$

The first fold of the circle is a right angle movement between any two points on the circumference. This is principle to understanding the division of the circle in all formulations of angle division.

Given ∠CDA+∠CAD+∠ACD = 180° and ∠CAD = 90°, to find out what degrees the other two angles are we can do the following: (∠CDA+90°+∠ACD)−90° = 180°−90° or ∠CDA+∠ACD = 90° or ∠CDA = 90°−∠ACD and conversely ∠ACD = 90°−∠CDA.

This does not give us a number for the two angles but if we knew the number of degrees of one we could easily find the other. Symbols are a convenient way to express how to think about shifting the relationships between parts within the balance of the entire circle. We will give the symbol of x for the information we do not know. So if we want to know ∠ACD in ∠ACD = 90°−∠CDA it then becomes x = 90°−∠CDA. If we want to know ∠CDA in the equation then it will be ∠ACD = 90°− x or rearranged it might be ∠ACD + x = 90°.

At this point we have advanced beyond most first grade understanding, but then that assumption is based on the idea of teaching isolated abstractions to students who have no context for them. When students are discovering for themselves, in a fundamental way with information that allows them to make various connections and go beyond things they already know, they often move ahead of our expectations. I suggest you go as far as they will go. Each class will let you know when to stop, and often when they are ready to continue. To give *less* than what stimulates the interest and capacity of our children is short sighted.

Below. 1) The longest edge of any triangle is the diameter of a circle. If the third point off of the diameter is on the circumference, it forms an inscribed *right-angle* triangle. This is the proof for every point being a center point. 2) If the third point is not on the circumference but remains inside the circle, the angle will be more than 90°, an *obtuse* triangle. 3) If the third point is outside of the circle boundary the third angle will be less than 90°, an *acute* triangle. 4) If the third point is on the diameter it is a full 180° and called a *straight* angle. 5) The third point on the circumference will always divide the semi circle into three areas, except as it lies inside of the circumference, then the division is four areas. 6a,b) The four divisions can be understood by folding the circle in half and then folding one semicircle twice where the point of folded intersection is inside the circumference. The line segment of a fold is always from circumference to circumference.

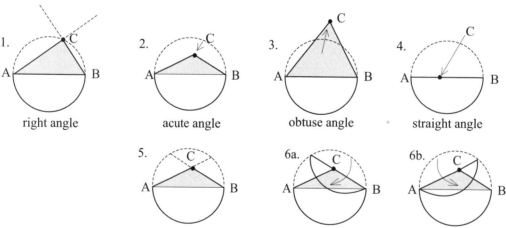

| 1. right angle | 2. acute angle | 3. obtuse angle | 4. straight angle |

| 5. | 6a. | 6b. |

Of the endless proportions and areas of triangles, the three angles will always be equal to the angle of one half of the circle. Two triangles of any kind having one edge in common, of four points, will always be 360°. *All interior angles of a triangle are a reformation of the semi-circle to straight lines and will always add up to 180°.* The 90° angle is inherent in the first fold of the circle and shows up at the third point on the circumference to the diameter. That is the pattern for all triangles because it happens first. It is the triangulation of the tetrahedron.

The triangle is compressed to a single line when point C lies on the diameter. This is the 180° compression of the circle to a semicircle, (preceding drawing number 4). This is somewhat like spherical compression into a circle plane. With the triangle compressed into three points on a single line, the line can be compressed into a single point, which is a small circle. Nothing is lost only the form changes depending on positions and the point of viewing. We begin to understand algebra by examining the patterns inherent to the circle. By folding the circle and discussing the interactions between parts, students can discover fundamentals to algebra with guidance from the teacher; at the same time they are discovering geometry by folding circles. It is all within Wholemovement if we can consider the connections that already exist.

Below. Three approaches to understanding line AB; as an abstract image, a folded line in the circle, and by drawing the radius dual circle.

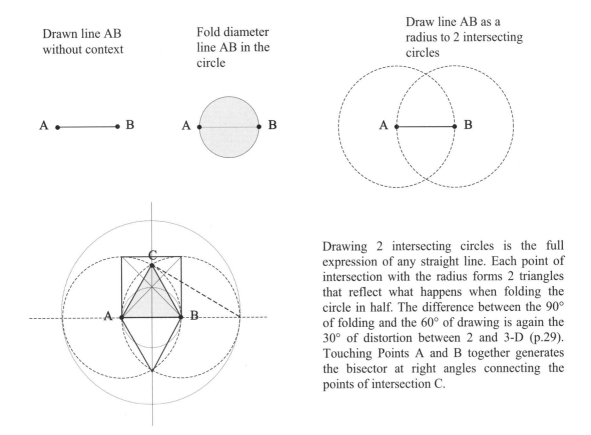

Drawn line AB without context

Fold diameter line AB in the circle

Draw line AB as a radius to 2 intersecting circles

Drawing 2 intersecting circles is the full expression of any straight line. Each point of intersection with the radius forms 2 triangles that reflect what happens when folding the circle in half. The difference between the 90° of folding and the 60° of drawing is again the 30° of distortion between 2 and 3-D (p.29). Touching Points A and B together generates the bisector at right angles connecting the points of intersection C.

Above. Three points, A, B, and C, are always a triangle, 180°; half a circle, no matter what the individual relationships are to each other or how they are formed as long as all points are connected. When a triangle in the system to which it belongs is shown in context it will carry the greatest amount of information and understanding of function. Spherical context always gives the greater meaning even in a compressed and abstracted symbolic form. The triangle, square, and pentagon are different relationships of the same ordering of circle/sphere information (pp.74, 109-12, 194).

FOLDING THE SPHERE

Fold 3 diameters into the circle (p.72). Reconfigure by folding one diameter in half touching 2 opposite ends together (p.73 #2). Use a bobby pin to hold the diameter ends together. The 2 radii of one diameter have become one center radial edge with the circumference crossing on the side opposite from the center.

Fold another circle the same, joining edge-to-edge. Two hairpins will hold the circles together forming 4 tetrahedra repeated in a circular pattern generating a quadrilateral space. The circumferences crosses on one side, the diameters cross on the opposite side.

Make another set of 4 tetrahedra in the same way. Then join the 2 sets of 4 tetrahedra, edge-to-edge, using 4 more hairpins to hold them together. A spherical pattern of open triangles and squares has been formed. This is called the *Vector Equilibrium, or* the *VE*.

Count the parts; 6 diameters, 12 radii, 12 outside points and a center point (13), 6 square and 8 triangle spaces, and 4 great circle planes that appear to intersect each other. This is a spherical pattern of 8 tetrahedra in a 13-point system.

The process of making this sphere demonstrates geometric progression. From the *unity* of the circle we formed 2 tetrahedra, then 2 sets of 2 gave us 4 tetrahedra, 2 sets of 4 gave us 8 tetrahedra. The tetrahedron doubles with each step, consistent to its dual nature. *Consistency inherent to pattern makes it easy to keep track of complexities that come with increasing multiple division.*

Right. There are 13 points of intersection from 6 diameters equally spaced, as they are located in the sphere. Each circle is one fourth of this pattern. The appearance of each of the 4 *great circles* is a combination of all 4 circles that form the sphere.

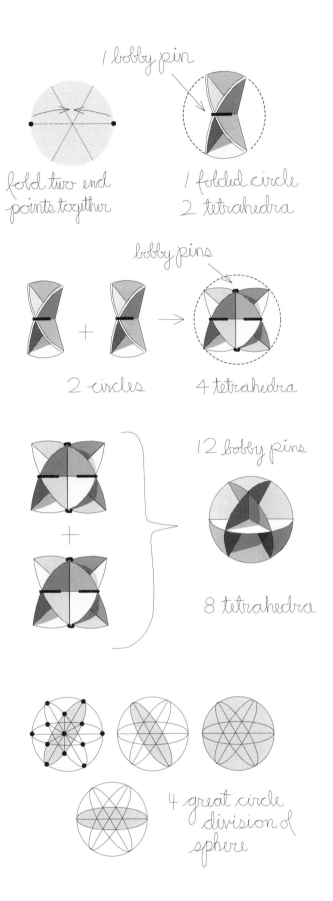

fold two end points together

1 bobby pin

1 folded circle 2 tetrahedra

bobby pins

2 circles

4 tetrahedra

12 bobby pins

8 tetrahedra

4 great circle division of sphere

89

With multiples of this sphere we can explore the *closest packing of spheres*. Each sphere has 12 circumference points of intersections that define squares and triangles. Each open plane will determine a particular orientation that needs to be considered when joining them together. Not unlike the two views of the tetrahedron.

1)   Arrange 3 spheres together in a triangle pattern connecting at the points; all will be *facing in the same direction*. (With smooth surface spheres there is no differentiation of orientation when joining). The top and the bottom side of the 3 spheres show a difference. One side shows points closer together and congruent to the spacing of three points on the fourth sphere. Use bobby pins to hold individual spheres together. Slip one bobby pin through the other, providing easy joining (p.66)

2)   Three points will form a triangle on top for joining the fourth sphere. The fourth sphere forms triangles with the first 3 forming the tetrahedron order of spheres. There are 4 spheres and 6 points of connection, 10 locations. The far right shows the square viewing from the edge of the tetrahedron.

3)   To form an octahedron put 2 sets of 3 spheres together, #1. Put them together with one set on top fitting onto the depressions of the other in opposite direction. These 6 spheres are all in the same orientation. The far right shows the same 6 spheres viewed from one of the end points. The sixth sphere is directly behind the center sphere in the back. As we have seen with individually formed spheres the square appears as a relationship of tetrahedra. Both the triangle and square are in the same way the relationship of spheres (p.194).

The octahedron is the first anti-prism. There are 4 individual anti-prisms corresponding to the 4 different orientations of 8 sides.

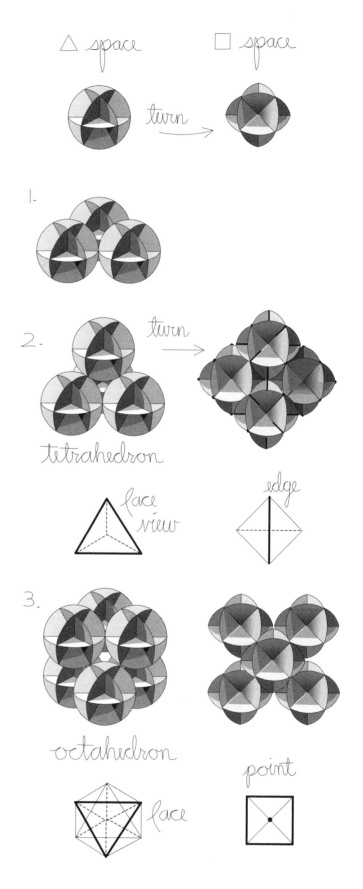

△ space   □ space

turn

1.

2.   turn

tetrahedron

face view   edge

3.

octahedron

face   point

4) By using any one of the 4 spheres of the tetrahedron as the center and adding 9 more spheres around it, attached in the same way all in the same orientation, 13 spheres are formed in the vector equailibrium pattern, reflected individually in each sphere.

The small drawing to the right shows the same pattern with the spheres reduced to points, which is what happens when we make a point and line model. The dark points show the tetrahedron and the light points show the 9 added spheres. There are only 8 possible combinations of arangements with 9 spheres around the tetrahedron, 4 tetrahedron arrangements up and 4 down.

5) Using the octahedron pattern, take any one of the 6 spheres and move it to the opposite triangle face forming a tetrahedron. Adding 7 more spheres around the designated center sphere forms the VE. The spheres are ordered in the same orientation.

Both the tetrahedron and octahedron are non-center systems. When a shift occurs to an off-centered sphere then forming a centered system can happen. Each non-centered system expands, becoming a centered system of the same pattern with all spheres in equilibrium. These 2 individual systems are complimentary within spherical order.

4. 4 plus 9 equals 13

5. 6 plus 7 equals 13

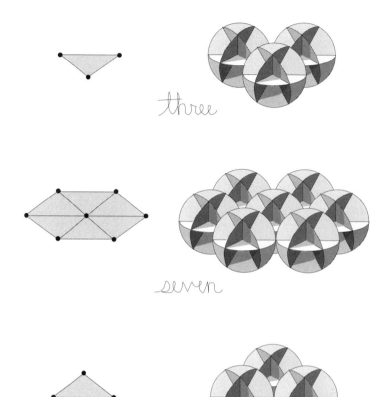

three

seven

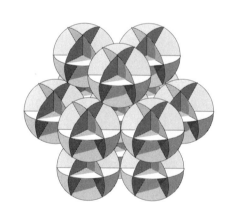

three

The 3 layers of the 13 spheres from any of 8 positions show 7 spheres through the middle (a hexagon arrangement) with 3 spheres sitting on top and 3 spheres on the bottom. The 3 on the top are always in opposite orientation to the 3 on the bottom. Without the middle layer of 7 spheres, the top and bottom together is the octahedron pattern of 6 spheres (p.90 #3). The 2 sets of 3 on top and bottom can be turned to the same orientation which will throw the tetrahedon/octahedron symmetry out of order. There will still be 13 spheres but not in the closest packed arrangement. Pictured bottom right is the VE system and the cube octahedron solid.

thirteen

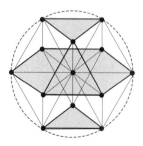

There is one more arrangement found in the closest packing order of spheres, the cube. It is a non-centered system of 14 spheres.

The vector equilibrium of 13 spheres is the only centered spherical system. The tetrahedron, octahedron and the cube are non-centered systems. Individually these 4 systems are all parts of each other within spherical order.

The cube is 3 layers of spheres as is the VE. The cube has a middle layer of 4 spheres with 2 layers with 5 spheres. The centers of the 6 faces of the cube reveal the 6 spheres of the octahedron pattern. Adding 8 spheres into the depressions of the octahedron completes the 14 spheres of the cube. All spheres are consistent to the same orientation of spherical order.

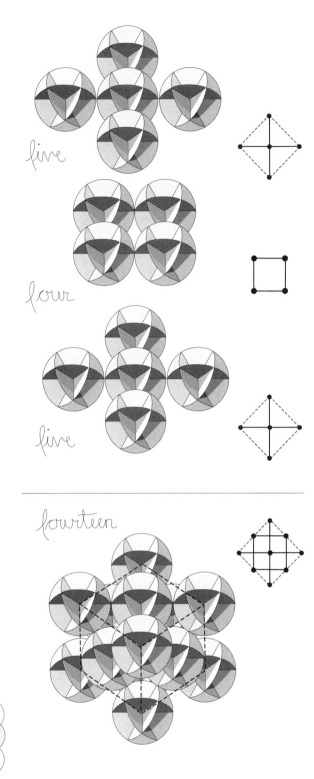

five

four

five

fourteen

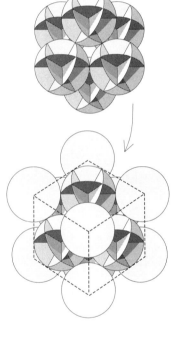

The diagonals of the square sides do not fall outside the circles, they go through the centers and tangent points. The diagonals are the edges of two tetrahedra forming the cube relationship.

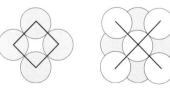

The flat plane of the hexagon shows the 14 spherical points of the cube. The front and back points are compressed into one point showing 13. All edges remain the same length, while the diagonals are distorted into 2 groups of different lengths.

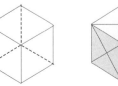

93

The ten-point arrangement of 4 spheres is expanded to 10 spheres in a tetrahedron four-point pattern. The bottom layer is 6 spheres, the middle layer is 3, and the top layer is 1. They are shown individually to the right.

Using only the centers of the 10 spheres a straight-line, two-frequency triangle grid is shown along with the spheres. When considering all points, both center points and tangent points, a four-frequency triangle grid is observed in the 3-layered tetrahedron arrangement of 10 spheres.

The two and four-frequency tetrahedron pattern is the same order of spheres. It is a matter of what we chose to count and what we leave out. The two-frequency tetrahedron is a straight-line form defining 4 spheres. The three-frequency tetrahedron is a function of straight-line division and does not appear in spherical packing. Spherically the two-frequency goes directly to the four-frequency system. Classification of polyhedra is not the same as the relationships observed in spherical order. The sphere is the pattern base for the extrapolation and development of polyhedra. There is not a one-to-one correspondence. The sphere is Whole and all polyhedra are parts in spherical order.

The three-frequency triangle/tetrahedron is a function of planer construction reflecting a flat, linear method of thinking and modeling. Three-frequency does not appear spherically, unless you call a three-sphere line three-frequency, which creates confusion about what is being measured. This all relates to the nature of odd and even numbers (pp.123, 173).

one sphere

three spheres

six spheres

ten spheres

94

COLLAPSIBILITY OF THE SPHERE

To compress the sphere down to a circle, the space intervals must distort in a way that allows for collapse without loss of organization. The arrangement of creases in the vector equilibrium will do that.

The VE sphere has 4 circles with 3 diameters each. There are 8 open tetrahedra, 2 for each circle, there are 4 of 12 bobby pins that hold it stable. The rest of the bobby pins just hold the circles together. There are 6 square relationships of tetrahedra. The squares will collapse if the tetrahedra collapse. By removing 4 bobby pins holding the pairs of tetrahedra with each circle, the sphere will then collapse to a flat pile of 4 circles that are still connected by 8 pins.

How do we find the first 4 out of the 12 bobby pins used? That depends on how the circles are joined to begin with. There are two different ways to join them in making the sphere. When the bobby pins are at right angles to each other, showing only one bobby pin in each square interval, it will not collapse.

Join the first 2 sets of 2 tetrahedra with the 2 bobby pins in the square openings, diagonal to each other, in similar orientation.

Join 2 sets of 4 tetrahedra where the 2 square spaces with the pins are on opposite sides of the sphere going in the same direction. The 4 pins that hold the tetrahedra will be in a great circle alignment connecting the 4 squares.

Remove these 4 bobby pins to collapse the sphere, leaving the 4 circles connected. There will be a slight spiral either to the right or left as the VE collapses flat. You can then spiral back up, reforming the sphere and replacing the 4 pins. This collapsible function makes storage and transportation of many spheres easy. It models the collapsible nature of the sphere and the function of compression.

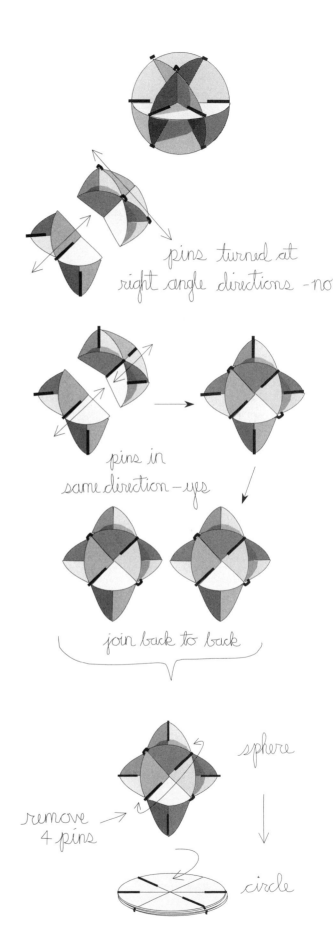

pins turned at right angle directions – no

pins in same direction – yes

join back to back

sphere

remove 4 pins

circle

95

Fig.0a This is a cubic arrangement of 14 spheres from the seen from the square face showing the outer layer of 5 spheres. The diagonals show the edge lengths of the two intersecting tetrahedra. Sections of circles have been colored to show the 4 great circles of the VE and the 2 intersecting tetrahedra. The dark areas are the open octahedron spaces in the tetrahedra. The centers of the circles have been colored black to show point/sphere relationships.

Fig.0b Edge view of the same cube arrangement of 14 spheres with one corner sphere removed to show the intersecting triangles and the internal rhombus. This shows the 4-frequency tetrahedra connections between 3 layers of spheres in order.

Fig.0c Here 8 spheres are arranged in a bi-tetrahedra pattern joined face-to-face. One set of spheres is colored and the other left white. In the colored spheres you can see the 2-frequency tetrahedron that is dark with the octahedron open space.

Fig.1 There are many ways to rearrange the spheres and partial spheres that are not in the order of closest packing. This adding and subtracting can only happen when there is order to a greater context.

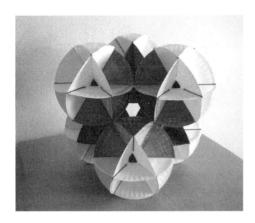

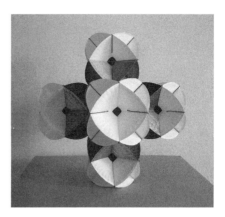

Fig.2a Eight corners spheres of the cube have been removed to show the pattern of 6 spheres of the octahedron. The view is from the removed corner point of the cube.

Fig.2b This view is from the square plane of the cube showing only the 6 spheres of the octahedron. It is sitting on an end point/sphere.

Fig.3 12 paper plates with 3 diameters folded into a pentagon (p.74 #7) have been joined and is held together with 3 hair pins looped together at each intersection. This stellated dodecahedron has 12 triangle openings.

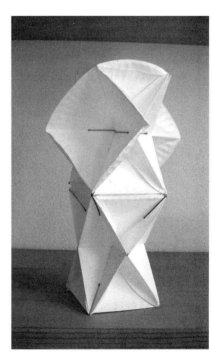

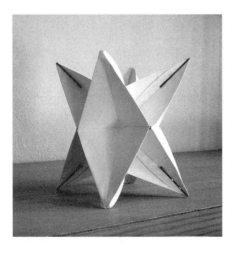

Fig.4 Here are 2 reformed VE spheres joined together with hair pins. The top half shows folding with circumference out and the bottom has the circumference folded into the triangle before assembling (p.347).

Fig.5 This shows the bottom part (*Fig.4*) above where there is more folding in of the circumference forming a star like rectangle prism. 4 circles are joined using 12 hair pins as with the VE sphere.

DISCUSSION: VECTOR EQUALIBRIUM SPHERE

The vector equilibrium, traditionally called the cube octahedron, is a pattern of 13 spheres. It is the only centered system of the 4 different regular arrangements inherent in the closest packed order of spheres. The 3 non-centered systems are the tetrahedron (4 spheres), octahedron (6 spheres) and cube (14 spheres), (p.175). The numbers reflect all four arrangements as parts of the same spherical order, patterned to the tetrahedron, which is a folded circle that is compressed from a sphere.

Below. Twelve spheres are packed around a center sphere where each outside sphere is touching 4 around it. Any one of the 4 spheres, (one touching 3 around it) of the tetrahedron is a center sphere by adding 9 more around it in the same arrangement. Order is an endless spherical matrix of spheres where each sphere is a center to 12 tangent spheres. Each point is a center point. This is the hexagon pattern of 7 spheres, reflected in 4 great circles.

13 spheres in closest packed order.

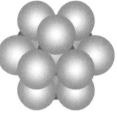

13 center points showing 4 spheres of tetrahedron and 6 of 7 circles are around 1

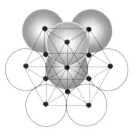

Below. The VE is 4 individual groups of 4 spheres in a tetrahedron pattern, where each group shares the same center sphere. This is represented by 4 regular tetrahedra, equally spaced, all joined to the same center point. It can be represented as $4^2 - 3 = 13$. Four spheres are the two-frequency tetrahedron, a non-centered system that can be represented as $4^2 - 6 = 10$. The difference between centered and non-centered is 3. (The difference between 2 and 3-D is 30°, p.29). Centered and non-centered are different arrangements of 4 tetrahedra from the same order. The centered system is about movement out from, and the non-centered is about movement into. While inseparable within the spherical matrix, they can be individually formed as a one-point-inside system, and a ten-point-outside system. Together they are the endless in/out spatial breathing of spherical order.

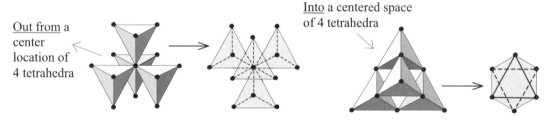

Out from a center location of 4 tetrahedra

Into a centered space of 4 tetrahedra

4 tetrahedral intervals   Octahedron interval

Above. Four tetrahedra sharing the same center point make 4 tetrahedral intervals that are opposite in orientation to the tetrahedra that formed them, otherwise they are the same, sharing the same center point. Of this organization there are 8 alternating tetrahedra, 4 up and 4 down. This is reflected in the hexagon that shows 6 triangles, 3 up and 3 down. The difference between 4 and 3 is the tetrahedron and the triangle.

Below. The VE shows two views; the equilateral triangle/hexagon 3-6, and the square 4-8. There are 8 triangles, 6 squares and 4 hexagons. These shapes mean little without their spherical context.

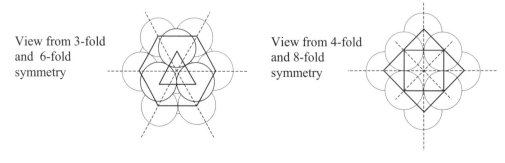

View from 3-fold and 6-fold symmetry

View from 4-fold and 8-fold symmetry

Traditionally the VE is one of the 13 semi-regular polyhedral solids (pp.176-7). By truncating both the cube and the octahedron (systematically cutting corners off through each adjacent edge of polyhedra) the cuboctahedron is revealed; thus its name, cuboctahedron. Truncating both figures arrives at the same triangle and square faces on a static, solid form. This polyhedron form, while traditionally classified as a semi-regular polyhedra, is also, and more importantly, the spherical pattern of order that is the matrix for all polyhedral forms and systems. The old systems of classification are useful but they limit us in how we understand new information; often the information that has been left out, ignored, or have gone unnoticed.

Cuboctahedron cut from both a solid octahedron and a solid form of the cube

8 tetrahedra in a centered system showing squares as relationships of tetrahedra

Pattern of 13 spherical locations in space all in equilibrium to the center and each other

4 individual positioned hexagons

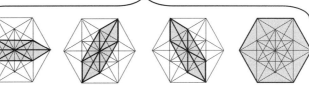

Above. The relationship of 4 circles/spheres is the same pattern of 13 points seen in the straight-line intersections in the hexagon and triangles. This view shows three oblique hexagons intersecting with the fourth regular hexagon, all sharing the same center. Polyhedra are always less than the sphere and polygons are always less than the circle.

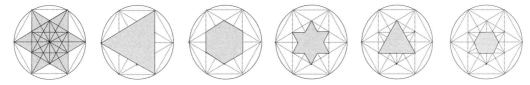

Above. Within the circle is the information to scale up and down in a proportional progression of the hexagon star, the hexagon, and the equilateral triangle. Here we see two different scales knowing that it can be repeated endlessly as a primary pattern of a fractal nature (pp.120-1).

Buckminster Fuller extended our understanding of geometry in many ways; one by simply putting the solid cuboctahedron on its triangular face, seeing it as a spatial pattern of spherical order, and renaming it as the vector equilibrium. This centered unit is the pattern system of what Fuller calls the *Isotropic Vector Matrix*, a straight-line vector description of the closest packing of sphere. His insight has allowed us to approach geometry through exploration from a more practical and comprehensive understanding of movement patterns ordered in space. Spherical order is a much more inclusive reflection of pattern generation we observe in nature as interactive systems rather than the polyhedral "solids" we tend to model. There is intrinsic beauty and truth to the dynamics of interactions ordered throughout the universe that is difficult to get from a static model. As important as polyhedral models are, when they are fixed in concept they offer little to expand our understanding about the dynamics of experiential reality.

By constructing the VE without the center point, defined by only the 24 edges attached with flexible joints, Buckminster Fuller has shown that there is no structural integrity and the VE will collapse revealing the icosahedron, octahedron and tetrahedron. This is not unlike the spherical VE with four circles (p.95). As the VE collapses, the outside configuration of the icosahedron, the octahedron down to the tetrahedron is revealed. It all goes back to the tetrahedron circle, sphere. Collapsing is somewhat like compression; there is spatial distortion.

Notice there are only eight triangles to both the octahedron and the cube octahedron solids. They are the only two primary figures that have four planes coming together at each vertex point. When diagonals are added to the six squares, five edges go to each vertex point showing a five symmetry arrangement.

<u>Below left</u>. By placing a diagonal in each of the six squares, twelve triangles will be formed (p.156). With the eight triangles already there, we count a total of twenty triangles, which is the number of triangle planes of the icosahedron.

irregular icosahedron   regular icosahedron

<u>Above right</u>. When each diagonal is the same length as the edges of the VE then the VE becomes a regular and stable icosahedron (p.171).

When the VE has a center it is stable where everything is triangulated and the squares are just a relationship of triangles. Without the center there are only triangles and squares. The squares are not structural and it collapses. With putting the diagonals in the squares they become triangulated and everything becomes structural again. With the icosahedron triangulation is on the outside rather than to the center as with the vector equilibrium. As we have seen with in the ordering of the closest packing of spheres, the center is not necessary for stability, but triangulation is. That goes back to one of the principle qualities of the first fold of the circle.

2 FOLDING THE TETRAHEDRON

<u>Fold</u> the circle in half.
Fold 3 diameters, (p.72).
Fold 3 alternate points on the circumference <u>to touch the center point</u> exactly and crease. Each fold generates a new point of intersection on the diameters at the mid-point of each side of the equilateral triangle. This forms a 2-frequency triangle.

When the points are accurately placed together each diameter will line up with itself, and the creased lines will be where they need to be.

<u>Fold.</u> Touch the 3 end points to the <u>mid-points on the opposite edge</u> of the triangle and crease. Fold one and crease, open and fold the next, do the same to the third. Do not overlap the individual folds. <u>*Unnecessary over-lapping will cause inaccurate folding*</u>. These folds show the relationship of end points to the new point on the same diameter.

<u>Bring</u> the 3 end points together. Tape the full edge length of each joining edge. You now have formed a regular tetrahedron. Count the number of points, the number of triangle faces, the number of sides. Where is the center of the circle now? How are the sides divided? What else can you discover about the tetrahedron you have just made?

Open the tetrahedron to the circle; look at the pattern of creased lines. They are a consistent development from the first fold of the circle and fundamental to all that follows. There is a tremendous amount of information in the circle to explore and countless formations of systems to be discovered using these 9 lines. This is what this book is about.

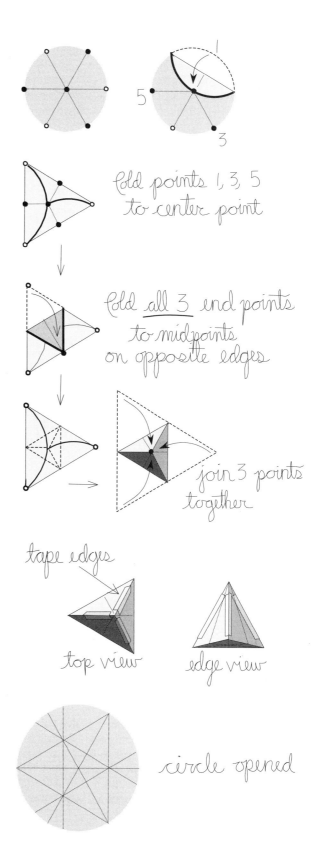

fold points 1, 3, 5 to center point

fold all 3 end points to midpoints on opposite edges

join 3 points together

tape edges

top view   edge view

circle opened

101

After folding 3 diameters and counting 7 points, ask students to fold points #1, 3, and 5 to the center point, or to fold #2, 4, and 6 to the 7th point. Do this without numbering points. This gives students the choice to start with any point and to figure out the sequence. It requires them to look to the activity for the information. How instructions are worded moves the brain towards making different connections. Instructions always have a bias towards particular results, so then play with the multifunctional nature of words to get the most results possible from simple instructions.

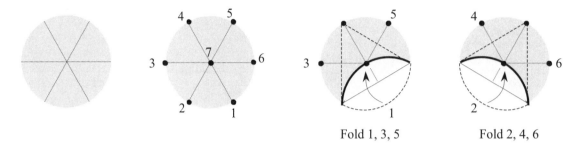

Fold 1, 3, 5   Fold 2, 4, 6

Above. Have everyone number the points on the circle. Half the class folds the odd numbers; the other half folds even numbers. Compare. One is inverse to the other with the numbering giving orientation to the triangle. Everyone has folded congruent triangles with an odd and even numbered symmetry. What is the total of the numbers moved to the center, both with adding the center 7 and without the center 7? Total the two digit numbers to one number.

Continue to fold the odd number points to touch the even numbered points that have moved in to the midpoint of the opposite edge, and crease. The even numbers do the same. Add the pairs of numbers that are touched together. Observe what you get when subtracting the numbers that are touched.

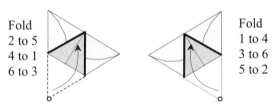

Fold
2 to 5
4 to 1
6 to 3

Fold
1 to 4
3 to 6
5 to 2

This kind of approach is interesting until someone discovers they have folded a pyramid and then you are back into the folding activity. Play with the numbers; they describe the ordering of relationships that are being folded. Sometimes it is good to go back and reflect on what we have done, particularly with younger students who can still get excited about numbers.

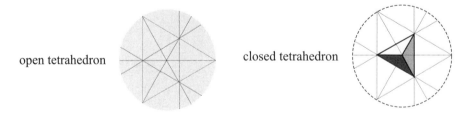

open tetrahedron   closed tetrahedron

The tetrahedron pattern is the first fold of the circle. It takes eight more folds to give an enclosed form to that pattern. There is a logical, systematic, symmetrical, and sequential development from the tetrahedron pattern to the fully formed tetrahedron that is reflected in the numbers. When the tetrahedron is folded, more than half of the circle is hidden within the tetrahedron. *Only half of the circle is ever visible at any one time*, regardless of how it is reformed. Nor is more than one half of the sphere ever directly visible. The two-frequency triangle shows 180° on each side. This does not account for the overlap of the circumference, the difference between the perimeter of the hexagon and circumference (p.39).

<u>Below</u>. Out flat the tetrahedron is a two-frequency, equilateral triangle. This flat grouping of four triangles is called a tetrahedron net. It represents four equilateral triangles, all joined three around one. There are two different flat arrangements of 4 triangles from which a solid tetrahedron can be formed (p.127). This triangle formed net is primary to the circle. Within the circle are 10 points. The open circle net of folds shows 19 points, (1+9=10). Ten is the diameter in the circle of the first fold. Four spheres in the tetrahedron pattern is 10 points in space. Numbers allow us to make pattern connections not obvious when looking at shapes and forms. Information is diminished when shapes are removed from the circle of spherical order.

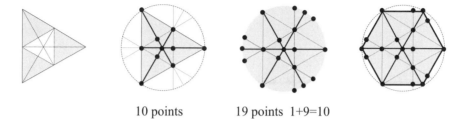

10 points   19 points 1+9=10

The three edges of the inscribed triangle are divided in half, making 6 line segments defining the perimeter of this equilateral triangle. These 6 line segments come together forming 3 edges to the tetrahedron. The center triangle becomes the 3 remaining edges of the tetrahedron. Four equilateral triangles of 6 edges form the tetrahedron (4+6=10) (4 spheres and 6 points of touching is 10). *Every two-frequency triangle shows the number pattern of 10.* The center triangle is in opposite orientation and can be seen as an interval function of 3 triangles; 3 around one. This center triangle reflects the spatial interval of the octahedron pattern as a relationship formed by the 6-point connections between 4 spheres (pp. 122,173).

<u>Below</u>. The 6 points and 6 line segments of the inscribed hexagon (p. 344) shows the same count as the 6 points and 6 lines of an inscribed two-frequency triangle. The hexagon is two congruent equilateral triangles, (odd and even) where the points are all equidistant to each other. Pulling in the points of one triangle to the edge of the other triangle will reduce it to half the original size, as it becomes the fourth and center triangle in the net it then gives definition to the other three triangles. The points remain equidistant. There is an in and out movement between the hexagon and the two-frequency equilateral triangle reflecting the tetrahedron dual (p. 163).

Transformation of the hexagon pattern to a two-frequency triangle

Folding the circumference in shows the relationship of Pi in the petals (p.39)

These kinds of interconnections are lost when looking at isolated polygons and polyhedra. We do not see the process, the movement in-between that changes one form into other forms. The dynamics, the breathing is lost when things are considered individually in separation. Geometry is about the movement and interaction between parts within the Whole.

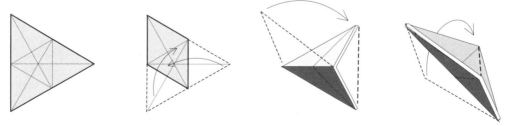

<u>Above</u>: The two-frequency equilateral triangle shows 4 triangles forming the tetrahedron. Fold in 2 triangles to make a rhomboid that will form 2 differently proportioned tetrahedra, one folded from each bisector. Any quadrilateral, when folded individually on each bisector, will form dual tetrahedra.

Below. Reducing the triangle to only three edges is all that is necessary to form a tetrahedron. When the edges of the triangle are independently linked, they can be moved to form the tetrahedron relationship. The rearrangement of 3 edges form the 6 edge positions for the tetrahedron and locate the 4 end points. This is in the form of a 3 directional helix. One half of the six edges show the full tetrahedron pattern (p. 36, 108).

Spatial information could not be opened from a linear defined shape were the pattern not there in the first

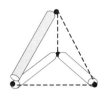

Moving the end of one edge off the triangle plane forms a tetrahedron pattern. This is the same movement as folding the circle in half. One point moves off of the circle plane to touch another point forming a 4-point tetrahedron pattern.

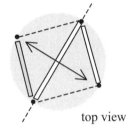 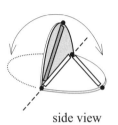

top view   side view

Words, numbers and shapes are multifunctional and take meaning dependent upon the context in which they are presented. The larger the context, the greater amount of functional information available. This allows for more connections, giving us the greatest opportunity to see where principles apply, to eliminate guessing, and to make appropriate decisions. This is not unlike what is done with *fuzzy math*. Local contexts are always shifting and changing in relationship to the stability of the greater system. What is appropriate at any given moment is always for that moment and not necessarily appropriate when the variables change. The objective is to function in the greatest context that is the guide to local and individual action that yields the greatest benefits to all parts. I would rephrase the saying, "think globally and act locally" to say *think comprehensively, observe locally, and act appropriately*. Pattern is universal and sovereign over all changing formations. We must think cosmically to find the global context that gives meaning to local functions that provides individual value. That is where we discover the symmetry of individual to Whole that governs appropriate actions. No matter how complex a form or system becomes it can always be traced back to origin if we look towards comprehensive understanding. This is principle to the first fold of the circle, which is spherical in nature and tetrahedral in formation.

Below. Every point is a circle/sphere, making all line segments both a diameter and radius. When considering the primacy of spherical order it is important to look at the largest, most general and coherent arrangements of pattern first appearing in the associations of circles. Here we can see the proportional relationship between the tetrahedron grid in the circle and the balance of concentric circles as they lay within the tetrahedron folds. The diameter shows divisions from half, to fourths, to sixths. The 3 circle boundaries in the center circle have been proportionally expanded to equal-width circle paths in the third image.

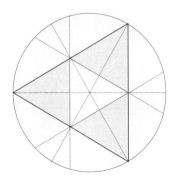 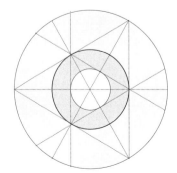 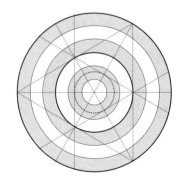

Previous image. There is an in/out relationship that happens with both circles and triangles, reflecting the same 1:2 ratio. The circles are concentric and the triangles alternately reverse orientation. The second circle image shows each circle in a ratio of 1:2. The inner circle is one half the diameter of the second circle, that is one half of the outer circle. From this it is easy to calculate the area relationship between all three circles. This 1:2 relationship moves in and out proportionally expanding and contracting with proportional concentricity of circles that regulates the relationship of triangles into and out from any given point. The last circle is divided into three circle paths each the same width as the intervals between them. This twelve division of the diameter is generated from the 1:2 proportions found in the second image by folding along the diameter in thirds.

Drawing proportional concentric circles forms the relationships between points and lines differently. Straight lines help us locate the circles, and the circles reveal fundamental proportional relationship of lines. Three diameters which establish the seven points are the anchor, the guides, the paths that provides the ways into greater levels of complex systems and relationships without getting lost.

Below. The diameter divided into eight equal parts generates circles of varying frequencies along the axial line. Each circle is a 1:2 division into the next dividing into smaller circles. The straight line is a generalized function expressing primary spherical movement scaling into itself. In folding the circle the straight line is a precessional result of spherical paths of movement of the circumference (p.30).

This drawing shows the eight-frequency diameter division inherent to the folds of the tetrahedron, regulated by the 3 sizes of proportionally concentric circles. This happens to infinite scale with all circles along the three diameters. This one diameter line of circles represents a single layer of spheres showing the same number development (p.42). Copy this drawing and carry the divisional development as far as you can. Draw it freehand for proportional understanding. Then use a compass to draw for accuracy. Each is instructive to the limitations of drawing, and that is reason enough for doing it.

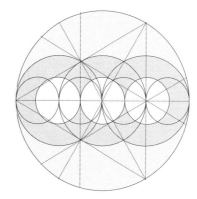

The relationships generate in folding circles reveal angles observed between the creased lines. We have seen this in the first fold of the circle. Folding in both directions is always a full circle path. When all the angles are seen in context of the circle it simplifies the process of figuring out various angles and the relationships of angles to each other. By observing and making connections with everything in context of everything else, the mathematical process takes on greater clarity and meaning. Many possible combinations and associations can be discovered by being consistant in assigning values to each part and observing and discribing the various levels of interaction,. Using Algebra facilatates this process once you see the interconnections that exist between all the parts. What is unknown is always discovered in relationship to what we do know, and is conditioned by the same. It takes time and familarity with the language to make the appropriate connections, but more importantly it takes starting comprehensivly.

It takes time to look for what we don't know about the tetrahedron. We must go beyond what we do know. There is truth to never knowing everything about anything. As we develop and evolve, continually expanding our understanding, our ability to make connections increases thereby raising the level of possiblities. When we approach mathematics using only numbers and symbols we exchange the physical for a conceptual context and lose the grounding, the context that is necessary to move out to more abstract levels of activity. The ability to cross connect is short circuited. The physical and the abstract are not separate and need to be developed with balance. With the tetrahedron nine folds in the circle there is no need to construct, only to move it, see what it does and discribe what is principle that makes the connections already there. The beauty is not in the thing itself but in the movement, the process, the dynamics of interaction principled to that first movement.

In seeing the different angles in the nine-lined folded circle, we notice that all angles moves around a specific circle location of intersection, in relationship to all the other center locations. The more lines of intersection, the more centers and more diferentiation of angles. Angles are complementary by nature; they always come in wholes with a minimum of one diameter/axis with fundamental divisions of 2, 3, 4, 5, and 6. Because angles never occur by themselves, there are many ways to describe their multifunctional nature without having to first know the numbers.

Here is an approach to using the 9 lines creased in the circle to play with and explore the relationships between basic angles knowing only 90°, the first fold (p.36). This is simply an example of adding and subtracting of similar parts. As generalized abstractions angles have no meaning untill applied somewhere which then gives them a context and some meaningful function. Otherwise they are just parts in the transforming process.

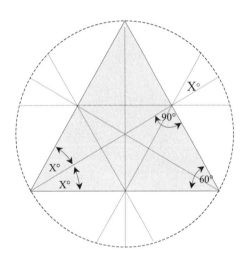

∠X= number of degrees.
Use inverse reflective, translation, and rotation relationships to describe ∠X by what is known.

$$\angle X= 90°- 60°$$

$$\angle X= 30°$$

$$\angle X=60° -\angle X$$

$$\angle X + \angle X = (60° -\angle X) + \angle X$$

$$2\angle X = (60° -\angle X) + \angle X$$

$$2\angle X= 60°$$

$$\angle X= \frac{60°}{2}$$

$$\angle X=\angle X + 90° -(60°+ \angle X)$$

Here is another approach describing the straight line relationships between points using circles.

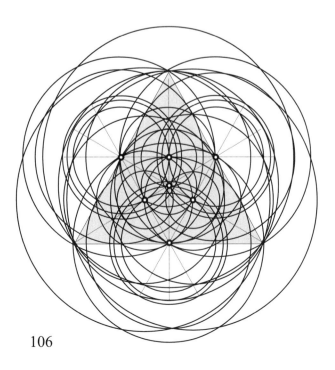

This drawing shows only the circle relationship of the tetrahedron net using the 7 points of intersection. All the possible radial lengths between each point are circled to the various lengths to all other 6 points. Each point has different lengths of radii and is a gravity center for concentric circles. Anywhere lines intersect is another point. A point location attracts relationships to others that are at first closest, then as it moves out to the furthest locations the diameter increases expanding the circular movement. The triangle relationship holds steady throughout.

106

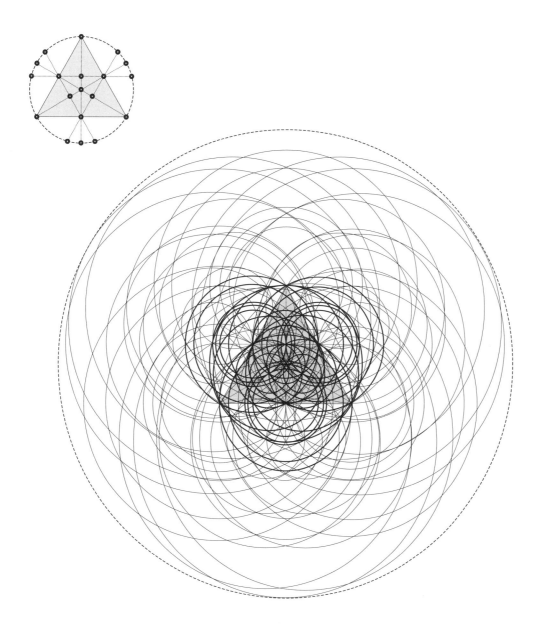

If you were to drop 19 stones into still water in the arrangement of points at the top of the page, all at the same time, this drawing would be an extreme simplification of what you might see. All 19 points as center locations are drawn concentrically out to each radius length between each point and 18 others. Towards the outer limits in the drawing there is more space, and more points for intersections to concentrically expand. Imagine this drawing of intersecting circles representing spherical formation that might be understood as energy moving around gravity points in a triangle pattern that regulates the movement throughout. Everything is in synchronization with total interaction to the original spherical Whole. This is what we are doing when we fold the circle the same way and reconfigure it differently; we are *rearranging the spatial relationships between points that are ordered to triunity.* The points do not change in their position to the circle only in relationship to each other. The potential is in the number of reformations that can be generated between each point location in the circle and the many associations that come from multiple arrangements and reconfigurations between circles. You should be able to locate various arrangements pictured throughout the book in this drawing. This drawing is simply an idea of an expanding circuit of spatial creation that can accommodate endless paths moving between all levels of scaling into and out from all locations. This is demonstrated in folding the tetrahedron folds in the circle.

Let's observe more closely the helix and spiral qualities as they appear in the tetrahedron. The 6 edges of the tetrahedron show two individual sets of 3 adjacent edges going in opposite paths. Any 3 adjacent edges define all 4 points of the tetrahedron in a helix form. There are 12 individual helix paths in a single regular tetrahedron; 6 right and 6 left hand. This is also evident in the 4 points of the first fold in the circle. (The image only appears to be a spiral due to compression of edge lengths in the drawing.)

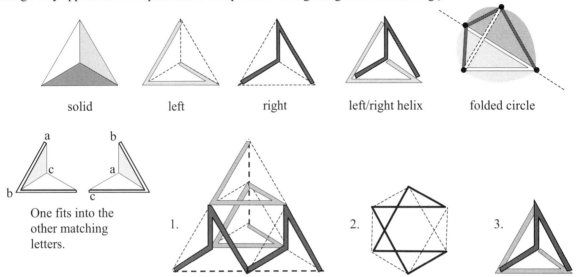

| solid | left | right | left/right helix | folded circle |

One fits into the
other matching
letters.

Above. 1) Here we see two opposite helix forms of 6 segments each that define all 10 points of the tetrahedron. The 2 opposite edge lengths that have not been formed are two helix forms at right angle to each other locating all 10 points. 2) A combination of edges from each helix defines the 6 points of the octahedron. They form a closed loop. The other unformed 6 edges of the octahedron form a complementary loop of 6 edge relationships. 3) One half of each helix shown in the two-frequency tetrahedron when combined together will form the dual helix of the single tetrahedron arrangement. This does not start with one tetrahedron but is observable in one because it happens in spherical order revealed in the two-frequency tetrahedron.

Below. There is an unseen spiral in the tetrahedron when moving from an end point to the center location. 4) Note the 3 edges of the helix path. 5) One of the end points is connected by a line to the center position. 6) From the center point the line is reconnected to the other end closing a loop of 4 outside points and the center point. There are now 5 line segments and 5 points. 7) The opposite spiral loop is formed to complete the dual nature of the tetrahedron. 8) Number 7, compressed to a line image.

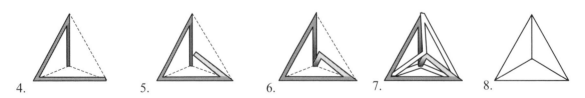

The spiral is part of the loop connecting the inside and outside, the formed and the unformed, the known and the unknown. There is a *golden ratio* relationship between the Fibonacci numbers and the distance between any 2 points of the tetrahedron and that distance proportionally to the distance from any outside point to the center point. This can be expressed spherically as *the relationship between the center of any 2 spheres, to the distance from center of any sphere, to the center space between the 4 tangent spheres*. It is not exact because the golden ratio is infinite in keeping with the infinite centering of the circle/sphere.

108

DISCUSSION: POLYGONS

TRIANGLE

The triangle is formed by three straight edges that define an enclosed plane. This is the minimum number of sides any polygon can have. The paths do not need to be straight to have a triangle relationship (pp. 40, 72). The first fold of the circle generates triangulation, which means it is principle to all the circle can generate. From this principle of a relationship of three there are proportions of triangles that are fundamental because they appear first in the forming process. We can think of the triangle relationship as primary forming of triangular pattern, but it is not pattern (p.33).

Folding the circle in half is a triangulated movement. The circle is divided in 2 areas and the third part is the line of division. Two imaginary, or marked, points on the circle touched together generated 2 more points, making a kite shape that is a relationship of 2 congruent right triangles. Four points form a dual triangulated relationship with 8 individual combinations of triangles (p.103).

The next two folds in this process, making three folds, generate three diameters that are equal distance apart. It is in the relationship of these three lines that we find differently proportioned triangles.

8 *Equilateral* triangles   12 *Isosceles* triangles   12 *Scalene* and *right-angle* triangles

<u>Above</u>. Three folded diameters show 7 reference points for 4 different classifications of triangles. The number of each triangle indicates the number of times it appears in different locations in the circle. While there is no difference with the others, there is a right and left hand right-angle triangle. Notice the interrelated combinations of divisions automatically occurring within each of the triangles. While they can be formed individually, they do not function separately. These divisional constructions are already there.

<u>Right</u>. **a)** Together these triangle relationships form this hexagon pattern of lines. While the pattern is consistent it is not complete, for each of the 6 small equilateral divisions of the hexagon are divided by only 2 of 3 bisectors. **b)** Adding the third bisector to each triangle adds 3 more diameters. These diameters divide the circle into 12 equal sectors, folding a 4-frequency diameter circle. This is not covered in this book, but is used as a drawn reference to some flat designs. The added diameters complete a second

a.   b.

level, 4-frequency diameter division of the first 3 diameters. The full 8-frequency diameter division of the circle is the third level of completion.[1] That third level is where we find the arrangement of the nine folded lines for the tetrahedron (p.118).

---

[1] 4 and 8-frequency level of development is covered in *The Geometry of Wholemovement: folding the circle for information.*
B. Hansen-Smith, Wholemovement Publ. 1999

109

Below. The 10 points in the two-frequency triangle show 3 corner points of 3 diameter/bisectors. By folding on one of the bisectors a tetrahedron pattern is formed. Four points in space is the minimum description of a tetrahedron. There are 5 edges and the relationship between 2 points is the sixth unformed edge. This is the minimum polyhedron formation and the first formed triangles (p.36). The two-frequency triangle can be folded in half on each diameter to individually show 3 pairs of large right-angle triangles.

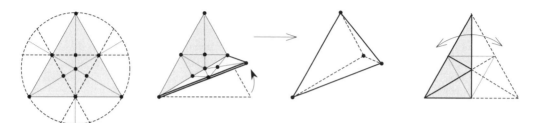

This tetrahedron has two congruent right angle surfaces and two different proportioned isosceles triangle planes. The right-angle triangles are irregular, therefore, scalene. Halfway between the open and closed position there will be 3 right angle triangle planes and one isosceles plane.

Below. 1) The tetrahedron when folded to a flat position forms a right angle triangle. Locate and count the number of right angle triangle divisions in this right triangle. The creased divisions show 6 individual right triangles and 4 in combination, 10 in all. How many other kinds of triangles and quadrilaterals can be found?

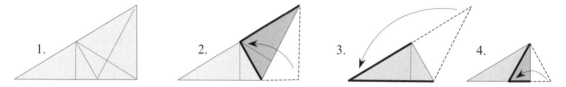

Fold on the creased lines to see the changes in the divisions of the right triangles as the shapes diminish in area and increase in volume. 2-4) Each fold creates another proportioned tetrahedron to a different size, and, when closed flat, changes the triangle configuration. For advanced students figure the perimeter and area ratios between each refold. What are the volumetric ratios between each right-angle tetrahedron folding? What other correlations can be found between them?

When we see the right angle triangle in context of the circle, much more infomation is available than can be derived from a picture of the triangle floating isolated on a page. Generalized mathematical functions and formulas that we connect to triangles work because the circle/sphere context first works by triangulation.

Below. In the 9 folds of the tetrahedron, there are 19 points of intersection, 12 more than the 7 of the hexagon. This adds many more differently proportioned triangle relationships. Look for differently proportioned shapes of triangles and in combinations using the 19 points on the circle. *Every point has a specific triangulated relationship to all other combinations of any two points.*

110

The *quadrilateral* pattern first appears as 4 points on the circumference in folding the circle in half. Connecting these points forms a flat *kite* shape (p.36). Any 4 points with straight-line connections on the same plane form positive angles around the perimeter and is called a quadrilateral.

Below. The primary quadrilateral in the circle has 2 sets of parallel sides where all 4 sides are congruent. This shape is called a *rhomboid*, and is formed by two adjoining equilateral triangles. 1-3) There are three individually positioned congruent rhomboids in the 2-frequency equilateral triangle, each in a different orientation that coincides with the 3 diameters. The rhomboid is the same pattern as the vesica formed by two intersecting circles. Four is found within three in the first movement of the circle. Three is primary to spherical compression. The quadrilateral is a division of the triangle/hexagon pattern. These two equilateral triangles give proof to the Pythagorean theorem (p.60).

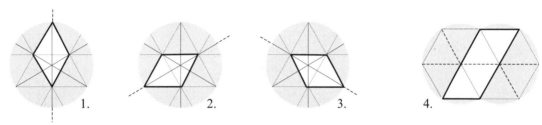

The rhomboid is part of another class of quadrilateral called a *parallelogram*, with two sets of parallel but unequal length sides. 4) A parallelogram is shown in 2 intersecting circles, (p.50). These four equilateral triangles make the second net for folding the tetrahedron (p.127).

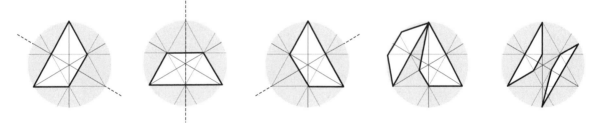

Above. Another classification of the quadrilateral is called a *trapezoid*. This is most clearly seen in the combination of 3 of 4 congruent equilateral triangles in the 2-frequency triangle. Only 2 of the sides are parallel. How many more There are more differently proportioned trapezoids can you find in the folded circle?

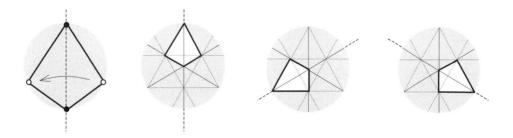

Above. The *kite* is another differently proportioned quadrilateral. It appears in the first fold of the circle in the relationship of the 4 points (p.102). It has 3 different orientations within the two-frequency equilateral triangle that reflect diameter direction. It is the only quadrilateral that has no sides parallel. The kite is bisected into 4 right triangles. In this context there are 2 scalene triangles on one bisector and one equilateral and one isosceles from the other.

The *square* is a special case quadrilateral where all sides and all angles are congruent. The square is a relationship of triangulation. When the 2 touching points of the first fold are a diameter apart, a square relationship is folded where the formed diameter is perpendicular to the unformed diameter. It is also found in the dual circles as perpendicular radii. The square relationship first appears 3 times in the 6 points of connection in the closest packing of 4 spheres in spherical order (p.172). It is the octahedron relationship (p.129).

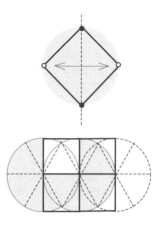

Below. Another classification of the parallelogram is a *rectangle*. The rectangle has two sets of parallel lines; each set a different length, with all angles at 90°. Two parallel diagonals in the circle locate the rectangle. There are three orientations of this rectangle, each perpendicular to each of the diameters. One half of this rectangle is another rectangle of a different proportion.

Any quadrilateral, of any proportion, when one corner point is moved off from the surface plane, will form a tetrahedron pattern. That is because the tetrahedron pattern comes first and every quadrilateral is the image compression of a tetrahedron (p.161).

PENTAGON

The *pentagon* is a five-sided polygon. The first pentagon appears in the reformations of the 3 folded diameters (p.74 #7). In the tetrahedron folds the pentagons are irregular and not obvious.

Below. There are 3 different proportioned pentagon relationships. All 3 have 6 different positions that correspond to hexagon symmetry. There are other more irregular pentagons to be found in this matrix of creases. In these pentagons we can find divisions of trapezoids, rhomboids, equilateral triangles, isosceles, scalene, and right-angle triangles. Three of the 5 diagonals for the pentagon stars in the 3 pentagons are formed, 2 diagonals are unformed. The drawing on the far right shows the stars completed.

The areas of 2 adjacent pentagons occupy 5/6 of the hexagon. The remaining 1/6 is divided into opposite right triangles.

Seeing polygons in the context of the tetrahedron folds makes it easy to calculate perimeters, areas, angle measurements and ratios between similar parts. The spatial context gives meaning to the flat information that includes volume and correspondence between transformed figures. *Everything flat is in relationship to the first 30° distortion of the 60° division into the 90° movement of the 180° fold into the 360° circle, infinitely formed within the sphere* (p.29.) Students do not have to learn construction first; that will naturally grow from the need to reproduce and to explore what they experience in folding the circle. Protracting numbers comes with understanding the movement relationships between parts. Drawing these relationships of shapes freehand in the circle will facilitate students' understanding as they fold.

The primary relationship of the regular pentagon to hexagon is first revealed in the reconfiguration of the three diameters, 6–1=5 (p.74 #7). The first next regular pentagon is found in the forming of the icosahedron and dodecahedron (p.166). This is all reflected in the folds of the tetrahedron but in irregular flat shapes. The regular pentagon of a 5-10 symmetry can be located in the circle folding using five diameters.[2]

Below. 1) Fold tetrahedron and open the circle. Locate the 5 points for the pentagon. Draw in the 2 missing diagonals to complete the pentagon star. A smaller pentagon is formed where the diagonals intersect that is proportional to the first (p.167). Connecting the 5 points of the newly formed pentagon, draw another star inside the first. Continue that process. Use the length equal to the length of the middle stars' arms and go down both sides of the arm connecting them to form another pentagon. Draw in the stars to find the length for the next until they are too small to draw for the scale you are working in. The diagonals of the pentagon star are a local self-centering function of the pentagon. Now fold it into a tetrahedron, or something else.

2) This pentagon star scales down the same way in the arms of the hexagon star as it does in the pentagon star; proportionally different.. Find the length of the inner hexagon star and use that as the length to move down the arm to form the pentagon. Then draw the pentagon star to get the next length down. Continue until it exhausts your scale. When the hexagon moves into the arms of the hexagon star it generates a pentagon, moving from 3-6 symmetry to 5-10 symmetry. The hexagon star is a dual system of two triangles; the pentagon star is a single system of one line. As the pentagons diminish in scale in the hexagon arms they are congruent to an increasingly higher frequency triangle grid where this 6 to 5 growth is in parallel to the 3 diameters.

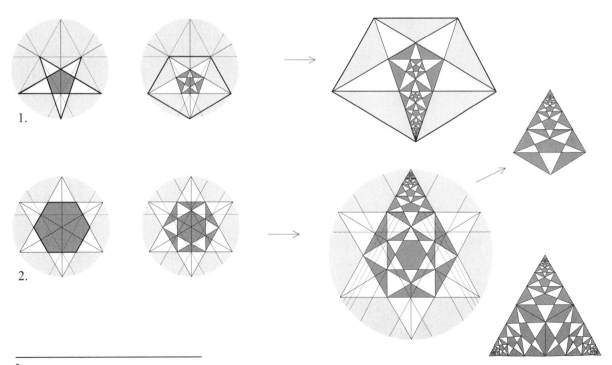

---

[2] Ibid, *The Geometry of Wholemovement*.

Continue drawing the pentagon in the folds of the tetrahedron. Symmetry is a balanced arrangement patterned to order from which many combinations of shapes and forms can be generated. All parts are individually constrained by conditions to location within a specific system. Things that may appear the same have many unseen differences. There are countless differences of arrangements that are buried in the design potential of an ice crystal and in the creased folds of the circle. The beauty of diversity is that each choice yields a different form with different paths of connections combining to the uniqueness of that particular developing system. Each part is always interacting with the fundamentals and generosity of pattern through the process of endlessly reforming.

<u>Below</u>. Using one of the pentagons in the hexagon, the star image is systematically developed as a surface design that will coherently cover the cube. Through consistency of subdivision, more intricate images can be developed. While the process can become very complex it follows simple principles that are systematically developed step-by-step creating a string of choices about what to add and subtract that is appropriate towards intention to direction and location.

<u>Below</u>. Endless design possibilities can develop from the pentagon and star, using the 3-fold symmetry of the hexagon with rotation, reflection and scale. Each step of development increases possibilities giving weight to consistency.

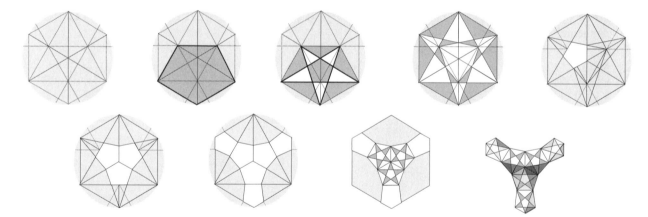

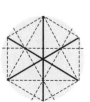

HEXAGON

The hexagon, as all polygons, has been reduced to and defined by outside shape. Triangulation folded into the circle gives context and structural design to the hexagon shape formed to the three axial diameters, (pp.29, 30, 344). It is obvious which yields greater information.

Folding circles is about touching points, but it is the circumference that moves, not the straight lines. The circle as its own compass extends ways to explore 2-D design that is inheret to the tetrahedron folding.

<u>Below</u>. Start by folding 3 diameters. Fold the triangle, one side at a time, tracing the circumference edge with each fold, 1, 2, ,3. Keeping the circle open, fold the inside triangle tracing the circumference edges 4, 5, and 6, on the circle.

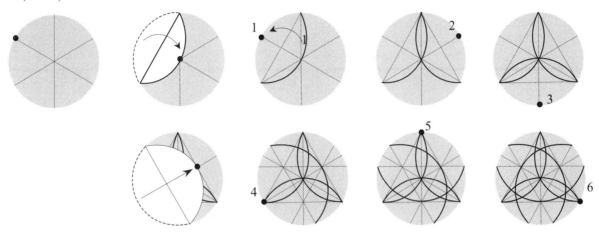

<u>Below</u>. Four shapes are formed by the overlapping of folds, including the full circle in the 3 diameter folds. Each of the 3 off center shapes (*2, 4, 5*) appears 3 times. The two centered shapes (*3, 6*)only once. Each is in a progression of number of points that define the shapes.

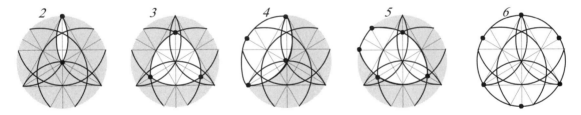

<u>Above</u>. The *2, 3, 4, 5, and 6 point systems* of curved shapes correspond to the more familiar polygons. The small vesica is an exception. *Two* points are usually drawn as a straight line and not a shape. *Three* points are formed by the 3 folds. *Four* is the full width radial vesica, the rhomboid of two intersecting circles. *Five* points is an incomplete vesica defining a pentagon with only 3 curved lines. *Six* curved lines of the hexagon is the complete circle. Here the circle reveals proportionally the 4 primary shapes imprinted by tracing the folded circle.

<u>Above</u>. The 3 arcs forming the small triangle show 10 points forming 3 pentagons (p.167). From that information we can draw the pentagon star and locate the points to draw a picture of the dodecahedron. This is simply another way to decode compressed spatial information. Explore many other relationships of forming that are inherent in the circle tracing of the folded circle.

SPHERE   CIRCLE   CYLINDER   CONE   TETRAHEDRON

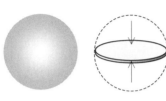 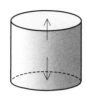

There is a logical, dynamic, and abstract reformation starting with the sphere and ending with the tetrahedron. This is linear in concept, much like starting from a point extending to a line that moves laterally to form a plane that extends perpendicular from the plane forming a cube. It is very tidy but only happens that way as a concept. The sphere compressed to a circle disc can be expanded forming a cylinder, where the two ends can be differentiated by scaling one to a very small circle becoming a cone, where the base circle is reformed to a polygon generating a tetrahedron. This is a concept and not reality, but it has some information that might be useful to consider. The compression of spherical movement into a circle disk is observed in the dynamic reformation of relatively small spherical clusters of matter into huge flat galaxies of immense proportions. The first movement from sphere to circle is inherent in the circle and with the first fold of the circle it forms both a pattern of the sphere (origin) and the tetrahedron (development). The cylinder and the cone are simply two other ways of folding the circle (p.261).

The tetrahedron familiar to most people is usually called a pyramid. The pyramid is often associated with the square base and 4 triangular sides because the Egyptian 4-sided pyramid is most often used to illustrate the concept. Tetra is a Greek word that means four. Not the base but the number of triangle sides, thus tetrahedron. Pyramid is a generic term for any polygon that has its center point raised perpendicular from the base plane forming triangular sides. Any stellated plane will always take a pyramid form. The three-base pyramid comes first, it is structural to all spherical formation (p.101).

The sphere is one point location. The circle is a two-point compression; two sides of equal scale. The cylinder is a two-point linear expansion of the circle where the cone is two points with a large scale difference between end points. The tetrahedron is a ten-point system, with a minimum of 4 points (p.41,90). The numbers tell us there is a big difference between cone of two points and the tetrahedron of ten points. It is similar to the difference between the circle and the triangle regardless of scale.

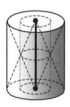 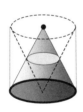 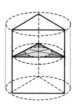 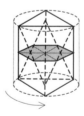 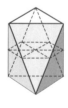

<u>Above</u>. There are a few obvious relationships between the cylinder, prism, and octahedron (antiprism) and how they correspond through specific form changes. The dual circles in the cylinder are inherently two inverse cones (the center points can be located anywhere in the circles). There is a circle plane where the cones intersect. When the two circles are reformed to triangles in the same direction a triangle prism is formed. When the circle/triangle ends are rotated to opposite positions the octahedron anti-prism is formed revealing two inverted tetrahedra where the planar intersection has changed from a triangle to a hexagon. This reflects the two inverted cones in the cylinder. Inherent in these changes of proportional division is the process of *slicing* solid polyhedral forms into parallel planes along an axis. None of the above are isolated functions, they only illustrate a few of the many connections that can be made by reforming the circle, of which only few are shown.

116

Below. This duality of cones can have rotational movement from the center point of intersection as well as vertical elongation and compression. This provides enumerable combinations of angulations that will produce many conic intersections at different angles where the cross sectional plane bisects the shifting angle between different size circles. This same dynamic happens in the polyhedral form where the projected point(s) of intersection becomes another center point of gravity extending the system.

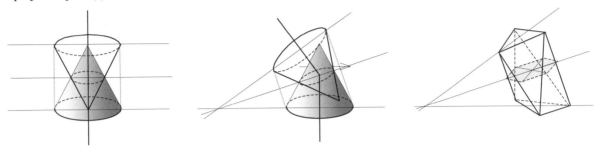

Polygons and polyhedra can best be understood as the reforming of the circle/sphere. Reforming of the circle reveals more potential for generalized pattern expression because it inherently contains all local conditions, and therefore is not limited by any of them. The fluid movement in forming and transforming retains stability in change because of the structurally pervasive pattern base. The sphere/circle/point is the leverage point that does not change. It provides balance between the extremities of formation, reformation, transformation, always revealing information about both local and contextual conditions.

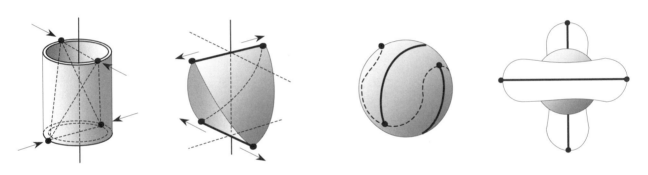

<u>Above</u>. The cylinder open at both ends, much like a tin can, is a tube, a torus form of the sphere (p.203). The two most extreme points on each end of the tube that are not in parallel will form a tetrahedron pattern of four points with six paths of connection. Having the two diameters perpendicular to each other and. squeezing the two end points of each diameter together will compresses the diameters forming an elongate edge closing the tube at both ends in a 90° or precessional movement. The movement reflects what happens when two points on the circle are touched together where a perpendicular movement generates two more points. These four points on the cylinder, as they are moved together in pairs, form the tetrahedron 90° to the four points on the tube before being squeezed. The edges are not defined as lines but rather as a curved surface in a tetrahedron pattern of two opposite edge lines. A square tube will do the same thing (p.329).

The sphere when compressed forms two circle planes. A sphere can be covered by two circles which is observable with the covering on a tennis ball. Notice the relative positions of the 4 points on the closed tube and the ball. The diameter of each circle covering the ball is extended almost twice it's own length, with compression at 90° to the stretching forming a slight bar bell shaped form. When laid out spherically, the identical pieces fit together covering the surface in a tetrahedron, right angle pattern (p.314). This shows a reformation of two circles forming four circles that reflect the unity of one sphere.

The tetrahedron requires only 9 of 24 folded lines formed in the circle by folding each of 3 diameters into 8 equal divisions.[3] This folded 8-frequency triangle grid circle is like a musical octave. With this grid one can compose endless arrangements of spatial complexities with huge amounts of relationship information that can be abstracted and rearranged. This octave grid will be occasionally referenced as it extends both into and out from the tetrahedron revealing a single process. It is important to know the contextual matrix in which these 9 folds occur, for that is the outward expanding expression of the tetrahedron nature of the circle. The 24 folded lines of the 8-frequency grid is just another level of what is in the process of circle formation.

The folds of the octave grid are developed sequentially in a geometric progression dividing the three diameters from 2 to 4 to 8 equal divisions with infinite divisional potential limited only by relative considerations to size of circle and material used. The grid matrix is simply the extended divisional movement of the tetrahedral pattern. It is important to know the 9 lines of the tetrahedron are part of a more comprehensive matrix, but are not necessary to understand the structural unity of the patterned matrix.

Below. The triangular grid is a right angle function of the three diameters. There are 3 sets of 7 parallel lines (21) at right angle to the 3 diameters (21+3=24 creases in all). There is a logical, sequential, divisional process of 1:2 that follows the right angle function of the first fold in the circle.

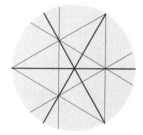

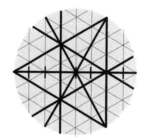

9 lines of the tetrahedron folding show unequal but proportional division of the diameters.

This shows the 9 lines as they are located within the 24 lines of the 8-frequency diameter circle grid. Each of the 3 primary diameters is divided equally in eight sections in 3 directions.

Right. Each diameter is divided by two parallel lines that intersect at given intervals determined by the preexistent folds back to the very first fold of the circle in half. You can see primary diameter divisions in combinations of; 3,3,2 and 2,3,3, or 3,5 and 2,6 of the 8.

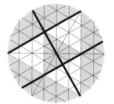
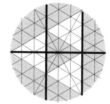
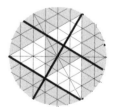

---

[3]Ibid, *Geometry of Wholemovement*

In observing the individual areas formed by the 9 intersecting lines reveals some interesting relationships that reflect something about the nature of the larger grid not apparent when looking at the creased lines

Below. By coloring the shapes alternately dark and light we count 12 of each, 24 individual triangle patterned shapes; the same number as the lines in the 8-frequency diameter folded grid. Similar to the 6 directions of three diameters, there are 3 sets of 2 parallel lines perpendicular to the diameters. There is a balanced positive and negative distribution extending out from the dark/light grid pattern of the areas in the circle. The center crossing develops an off-centered directional band of grid movement out from the triangle-center. There are 3 individual circuits running through this pattern. Think of this as a point of intersection of incoming and outgoing energy (p.59).

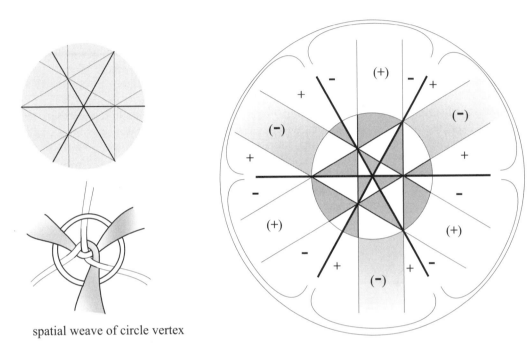

spatial weave of circle vertex

Right. When the folded tetrahedron circle grid is extended out from the circle we see the circle as generator to an endless pattern of a triangular grid. Here the whole circle functions reflectively as a part of an expanded Whole. The circle is inscribed in the triangle extension in the same way that the triangle is inscribed in the circle. Regardless of the reciprocal function of whole circle and whole triangle, the circle remains Whole. The triangle, and grid, is forever a function of the self-referencing movement of the circle.

Imagine this grid matrix as a single pattern event rather than simply as individual triangles, circles, lines, shapes, angles and number images.

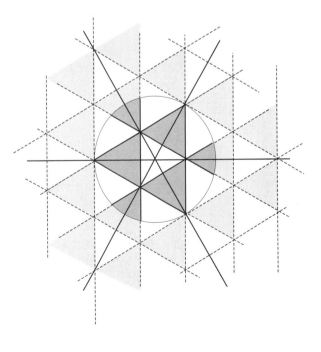

119

The circle/sphere reflecting itself fully within infinitely scaled boundaries, into and outfrom, creating multiple parts of similarities, seems to be a fundamental fractal process. It is not any one form as much as a process about pattern replication throughout all scales. Each individual formation is subject to what has come before and is directive to what is yet to come. Time and space scaling creates similarities of divisional differentiations consistent to principles of self organization, much like scaling observed in the golden ratio growth spirals (pp.108,190). Each part is formed by the last, a combined accumulation of relationships proportionally inherent to the next scale of growth. Everything is cyclical through time and circular in space. Local linear perceptions combined with spherical pattern form spiral growth formations.

Fractals reflect pattern of self-organization in specific local forming of systems. Endlessly different forms self-organize only to the extent they reflect the larger context that is already in place. Every child system is subject to local conditions within the context of directionally parent predisposition.

When we fold 3 diameters in a circle, a pattern of 7 circles is inherent. All 7 are in relationships determined by the three diameters; 6 degrees of freedom for 7 circles/points. Each circle is a center to six others as the pattern of the circle multiplies outward. This matrix of circles is without scale and in movement through time generates spiral forms of a fractal nature. Universe movement is replicated in every part regardless of form and scale giving to the expression, "As above so below. Within the differences of individually formed locations, the closer you look the greater the similarities of underlying pattern. This is ever becoming evident with greater understanding about genetic functioning and tracing back to origin.

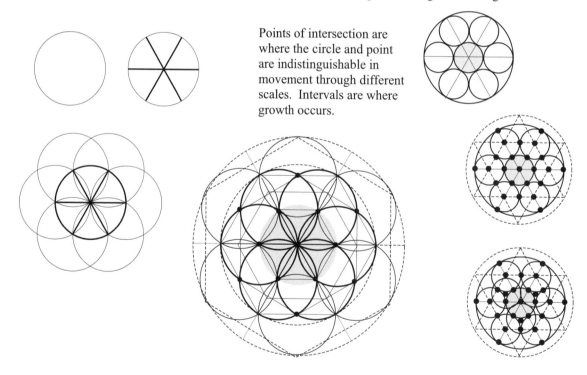

Points of intersection are where the circle and point are indistinguishable in movement through different scales. Intervals are where growth occurs.

Above. The hexagon relationship is inherent to the circle. Usually the circumference and diameters are discarded in favor of the six outside edges that are easier to measure. Separating parts and observing only the outside shape, eliminates all connective informational tissue, causing eventual loss of information and context. The circle defines a hexagon that is without scale of ever-expanding boundaries. Similarly the hexagon divides into itself generating endless multitudes of smaller hexagons, which are all triangles that are ordered to the self-organization of the circle/sphere (pp.106-7).

Seeing the circle as both Whole and parts is seeing the fractal nature of life formation. As with all things we try to explain, most of the context information has been eliminated for the simplicity of abstract demonstration, and we are left with a simplistic and partial understanding reduced to little meaning.

Below. Using the triangle within the circle matrix shows the divisional development of the hexagon pattern. It is all hexagons, starting with the three-frequency triangle that develops into a nine-frequency to a twenty-seven-frequency triangle. The triangle is 3 to the power of itself. No matter what level of development three can never be more than to itself nine (pp.45, 186-7). Far more is taking place inside than what is observed at the ever-changing boundary. There is no separation of one thing from another, except as we conceptualized and categorize differences of discrete properties of individualized parts. The mind gives balance to the apparent physical time/space separation and unity of the Whole.

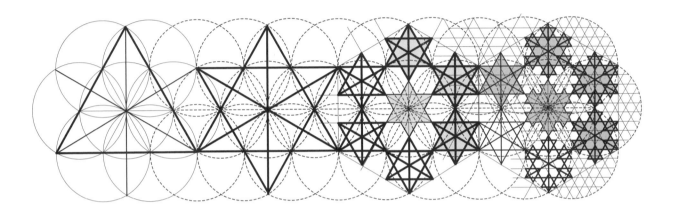

Below. The fractal process can be seen in the down-scaling growth of the tetrahedron. It is very clear and simple. The sphere-circle is the origin of the fractal process. (Nine lines of the tetrahedron plus three of a triangle/tetrahedron are shown in the circle). A single tetrahedron is a duality, in this case the tetrastar or stellated octahedron showing the cube pattern revealing the VE. This is an easy model to make (pp.165, 170, *Fig.14*). Each new level of tetrahedra are made proportionally with smaller-scale circles. The square intervals show one half division of the octahedron creating space for further tetrahedral division creating more one half octahedron division in endless fractal forming in all directions.

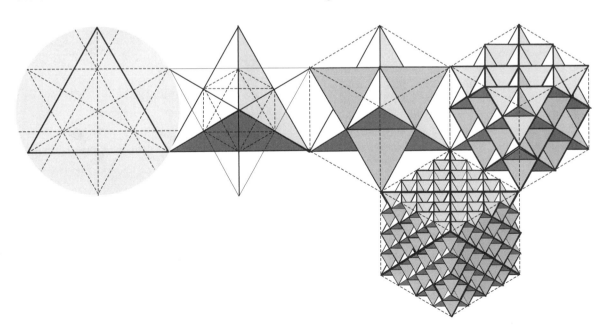

121

FREQUENCY

The triangle and square are 2 views of the tetrahedron. The 1:1 ratio between all 6 sides is compressed, causing different lengths in the square image that are called diagonals.

<u>Fold</u> 4 tetrahedra and tape each closed along the edges. Arrange them with 3 on the bottom layer and 1 on the top. They will be facing in the same orientation forming triangles with each other. As with 4 spheres they touch at points. Tape across the joining points on both sides of each tetrahedron (p.65). This is the 2-frequency tetrahedron, 4 spheres in order (p.31,41).

The 6-point interval formed by 4 tetrahedra touching is the <u>octahedron</u> pattern. This space is defined by one triangle from each tetrahedron, and 4 triangle intervals, 8 triangle planes.

Make 4 two-frequency tetrahedra and put those together in the same way making a 4-frequency tetrahedron. This forms 5 octahedra intervals, 4 congruent and one 2-frequency octahedron. There is an increase in sizes of octahedron spaces in relationship to the frequency of the tetrahedron.

Make 3 more sets of 4-frequency tetrahedra and put the 4 sets in the same tetrahedron pattern forming a 4-frequency tetrahedron. There are now 3 different sizes of octahedra, 64 individual circle-formed tetrahedra. The fractal nature is obvious.

When the tetrahedron gets to be 8-frequency it will accumulate weight, beginning to weaken the taped joints. At 16 to 32-frequency it begins to need support. Some students tape rulers, pencils, or dowels across the edge joints to give it strength, others just use more tape. This is not a sound solution. One group of students made 3 more 4-frequency tetrahedra and taped them onto the congruent openings, giving it strength necessary to support itself without having to add anything foreign. That was a thoughtful solution using the inherent strength of the dual tetrahedron and good taping. It was star-like with a wonderful open spatial quality now being an octahedron with 7 stellated sides. This necessitated expanding what they thought about what they were doing causing a change in direction.

122

end point

edge line

octahedron interval

2-frequency tetrahedron

4-frequency tetrahedron

3 sizes of octahedra

8-frequency tetrahedron

One triangular side of a tetrahedron gives us basic information about triangle division, inside layers, branching, and some interesting number functions as we count parts. The difference between spherical layers and polyhedral division is analogous to the circumference and the diameter, they're different (p.39).

Below. Start with one triangle; add a second layer of 2 triangles in the same orientation. An inverse triangle is generated making 4 triangles. The inverted triangle is the octahedron space. Each edge of the triangle is divided into two equal parts making a 2-frequency tetrahedron (p.124). Adding 3 more triangles on the third layer makes 6 triangles and 3 intervals, a 3-frequency triangle. At any number of equal divisions on one side of a triangle when times itself (number squared) is the total number of triangles. *The frequency number of a triangle squared gives the total number of triangle divisions* ($F^2=n\Delta$). The squaring function is the divisional nature of the triangle. The square grid function is simply the edge view function of the tetrahedron face view (p.122).

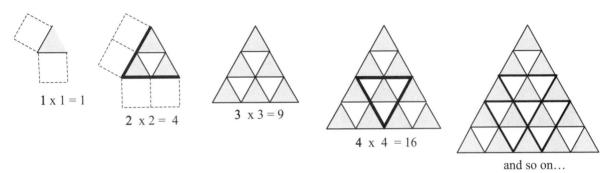

1 x 1 = 1   2 x 2 = 4   3 x 3 = 9   4 x 4 = 16   and so on…

Each new layer of dark triangles is the number of the row above plus one. This is the same for counting white triangles.

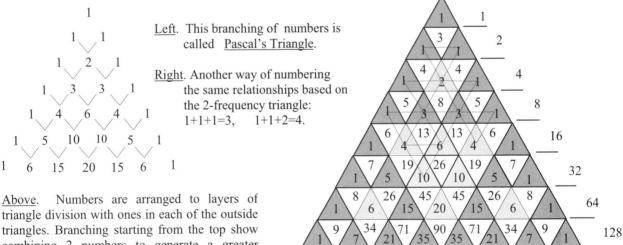

```
              1
           1 \/ 1
         1   2   1
       1 \ 3   3 / 1
      1   4   6   4   1
    1 \ 5  10  10  5 / 1
   1   6  15  20  15  6   1
```

Left. This branching of numbers is called Pascal's Triangle.

Right. Another way of numbering the same relationships based on the 2-frequency triangle:
1+1+1=3,   1+1+2=4.

Above. Numbers are arranged to layers of triangle division with ones in each of the outside triangles. Branching starting from the top show combining 2 numbers to generate a greater number. Add across to get a sum below; 1+2=3, 2+1=3, 1+3=4, 3+3=6…
The inverse triangles from bottom up show the subtraction of the same numbers; 6-1=5, 6-1=5, 15-5=10, 15-10=5, 5-4.=1, 5-1=4 and so on.

Right. This triangle function is the hexagon where numbers in the white cells work the same way in an inverted two-frequency triangle, 7+19=26, 8+26=34. In the 2-frequency dark cells, the numbers add up to the center white triangle, 1+1+6=8, 6+7+21=34.

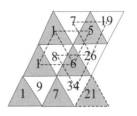

There are many different relationships and various number combinations in this grid that reflect the triangulated functions of the hexagon pattern.

123

Numbers work because straight lines work, because circles work, because of the whole of spherical order works because it is Whole. There is no direct equivalent between straight lines and circles/spheres even though both formed to pattern. Circles can be straight lines, but straight lines are never circles.

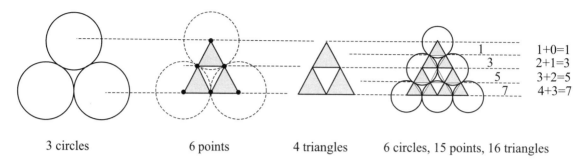

| 3 circles | 6 points | 4 triangles | 6 circles, 15 points, 16 triangles |

$$1+0=1$$
$$2+1=3$$
$$3+2=5$$
$$4+3=7$$

Above: A triangle of 3 circles generates 4 triangles (2-frequency). There are 3 circle points and 3 tangent points. Three layers of 6 circles generate 16 triangles in a grid where with each successive layer there are 2 more triangles than the one above it and one more circle than the one above. The numbers show adding each layer of triangles, those pointing up and those pointing down. These drawings represent one side of the tetrahedron pattern both spherically and in polyhedral form. The 2-frequency tetrahedron is the equivalent of the singularity of the 4-sphere tetrahedron. There is no 1-frequency tetrahedron except as concept, even though it can easily be modeled as an individual polyhedron.

With an inscribed triangle in each circle the spaces in-between the circles are the inverse triangles. This dark and light way of showing triangular orientation is called Sierpinsky's triangle, the same triangle as Pascal's number triangle. Individual functions have been separated and individually named. The more familiar we are with spherical packing the more connections we will be available to make between the many functions that have been separated from each other.

Below. Without removing any balls of a five-layered tetrahedron arrangement of spheres, determine the total number of spheres? Calculating layers of any of the 4 outside faces will reflect what is happening on the inside. Counting layers from bottom up reveals how many total spheres. Layering of spheres comes first, and then is reflected in the frequency development of polyhedra. This is counting in sets.

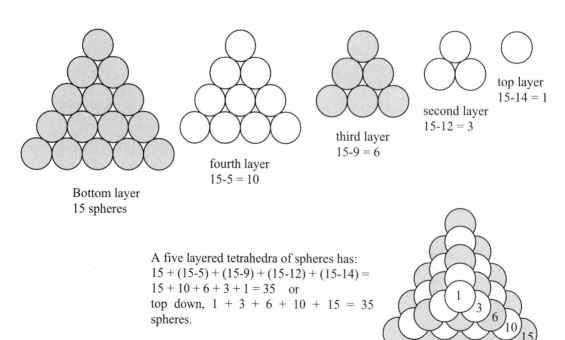

Bottom layer
15 spheres

fourth layer
15-5 = 10

third layer
15-9 = 6

second layer
15-12 = 3

top layer
15-14 = 1

A five layered tetrahedra of spheres has:
15 + (15-5) + (15-9) + (15-12) + (15-14) =
15 + 10 + 6 + 3 + 1 = 35   or
top down, 1 + 3 + 6 + 10 + 15 = 35
spheres.

124

OPENING TETRAHEDRON

<u>Fold</u>   a tetrahedron.
Open one side.
The open side can be pushed in, leaving an open plane. The triangle side can rotate 360° on the folded edge from the inside surface all the way around to the outside surface of the same plane.

When the point is opened to the same distance as the edge length, two regular tetrahedra are formed. One has three surfaces and an open plane, the other with one surface and three open planes. This shows a bi-tetrahedron relationship of 2 regular tetrahedra sharing a common open triangular plane.

Move the triangle flap so the point is opposite the position from where it started. This forms another regular tetrahedral space. The base triangle now serves as the plane common for the bi-tetrahedron.

Completing the reflective movement a single open tetrahedron is formed. The action of the opening plane generates a number of possibilities for joining tetrahedra at different angles forming many combinations of open and closed forms.

Fold one of the triangular sides of a two-frequency triangle flat against the center forming a trapezoid. Hold it down with a piece of tape. Fold the 2 end triangles together and tape, forming an open-sided tetrahedron. The taped open tetrahedron does not limit angle joining, it now does it differently.

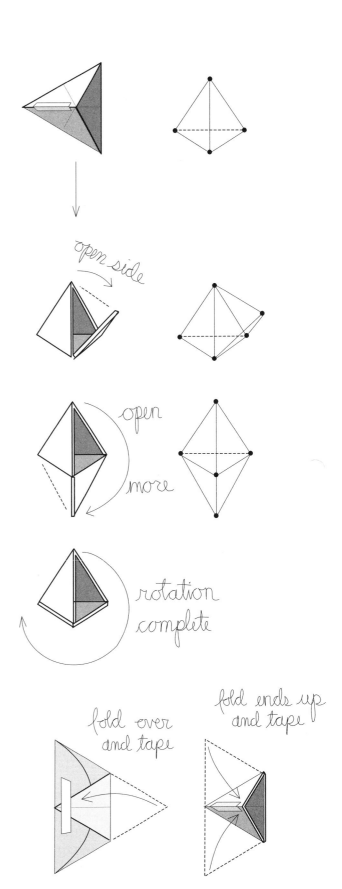

JOINING OPEN TETRAHEDRA

Fold 2 open tetrahedra with flaps out.

Put 2 open tetrahedra together joining one triangle flap edge to the edge of the other tetrahedron. Tape edges together.

Bring the point of the triangle flap onto the point of the other tetrahedron. This forms a configuration of 4 tetrahedra in a line, with 3 common triangular planes joining them. Two tetrahedra have generated two more tetrahedra. This is called a *tetrahelix*. It will either be a right-hand or left-hand helix depending on which edges were first joined (p.108).

Close one tetrahedron leaving the other with an open flap. Bring the loose triangle flap to the edge of the closed tetrahedron. This brings the 4 tetrahedra down to 3 tetrahedra joined by 2 common triangle planes.

Continue to move the connecting triangle down to close the side of the open tetrahedron. There are now 2 closed tetrahedra joined by a common edge length that allows rotation around the joined edge as the axis in a reflective movement.

Move in either direction around the axial edge to form a single unit of 2 tetrahedra joined by a common surface. This bi-tetrahedral unit has 2 different positions relative to each other, similar to folding the circle in half.

Any one of these variations can be taped and used as multiples to explore different arrangements, adding and subtracting, forming systems one to another. There are many combinations of 2 tetrahedra to be explored.

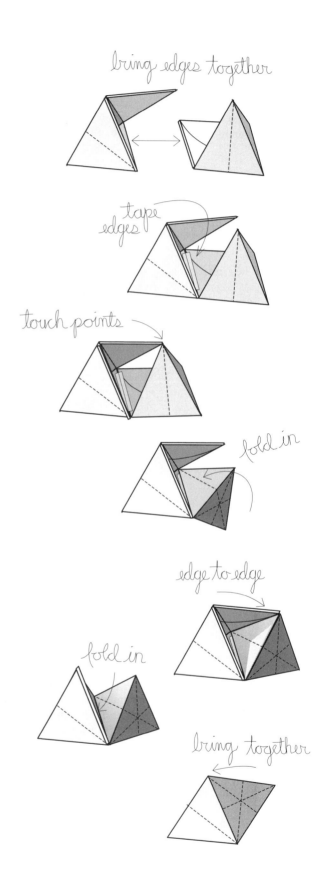

126

TETRAHEDRON NET

The net of a polyhedron is the flat unwrapping showing all sides as a connected system of polygons. The net lays out the entire surface area on a single plane for easy construction. There are various net arrangements for all polyhedra. There are only 2 nets for the tetrahedron, in keeping with the dual nature of the tetrahedron. Circle division is the master net for all polyhedra.

Net #1. To draw a net, trace around the bottom triangle of a solid tetrahedron and roll over each edge individually and trace around each side. This is the same arrangement that is folded into the circle, except the circle and the three diameters are missing.

<u>Fold</u> 4 tetrahedra. Open them to triangles and tape edges together in a centered arrangement. This makes a 4-frequency triangle. Taping on both sides makes it stronger. Because there are more divisional creases in this net there is a lot more reforming to explore.

Net #2. Reflecting each side of the tetrahedron in a line one side to the next, trace each step until all 4 sides have been traced. You have drawn a tetrahedron net in a linear arrangement of triangles. The starting direction will make a right- or left-handed net.

<u>Fold</u> 4 tetrahedra triangles as before and tape them together in a line, either right or left-hand. Fold together taping on the joining edges.

A reflective turning will change the orientation of the net and can sometimes make a difference in opening, transformation, and how it combines in multiples. Both nets, through rotation, are part of the same hexagon matrix, two intersecting circles of the same radius (p.111). Turning the net over will change the orientation, the look of it, and offers interesting variations in using the folded in circumference.

There is much to explore in the multiple creases in each of these nets.

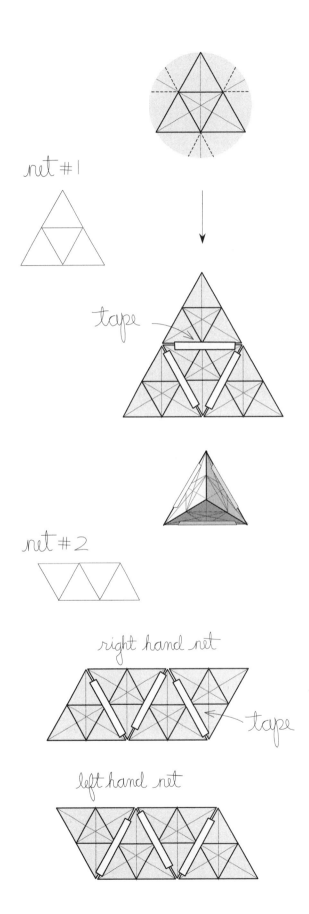

net #1

tape

net #2

right hand net

tape

left hand net

<u>Fold</u> 4 tetrahedra using one circle each.

<u>Fold</u> one open with the triangle side open (p.125).

<u>Fold</u> the other 3 tetrahedra with the open side folded to the inside (p.125).

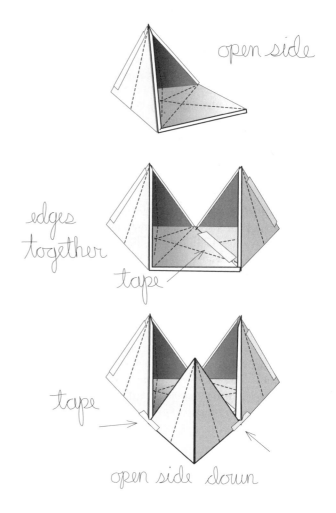

Join one side of the open tetrahedron to the open triangle flap so the open planes face inward. Tape them together.

Add the third tetrahedron with the open side facing to the bottom and tape the edge to the second edge of the triangle flap. This will complete the two-frequency bottom triangle.

Place the fourth tetrahedron in the center of the three with the open side up and the opposite point to the center of the bottom triangle. this completes an arrangement of a tetrahedron with one corner truncated.

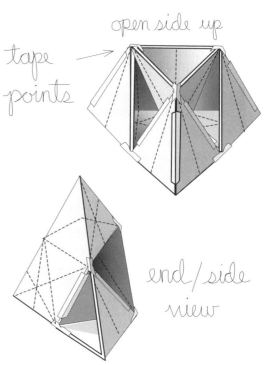

This is another view showing the one open tetrahedron on the bottom triangle. This is only one of many ways to form a partially truncated tetrahedron. Make four the same way and join them into a tetrahedron, or 8 into an octahedron (p.129), or 8 into a torus ring (pp.203-4, Fig.19a-e), or 20 into an icosahedron (p.149), or...

③ OCTAHEDRON

There are a number of ways to form an octahedron. The most direct and simply way, which is consistent with the tetrahedron, is to half open 2 tetrahedra and join them together.

<u>Fold</u> a tetrahedron. Halfway between open and closed 4-triangle intervals are formed between the triangles that form them. This position reflects a 1:2 ratio halfway between the formed tetrahedron and its net. All positions between open and closed are an octahedron pattern.

The 4 points of the tetrahedron become 6 points, the 4 triangle surfaces become 8 triangle planes, and the 6 edges become 9 edges. (The remaining 3 of the 12 edges of the octahedron are there in relationship without form.) Only half of the octahedron is formed but the pattern is complete. *Pattern is always complete*, it is never partial.

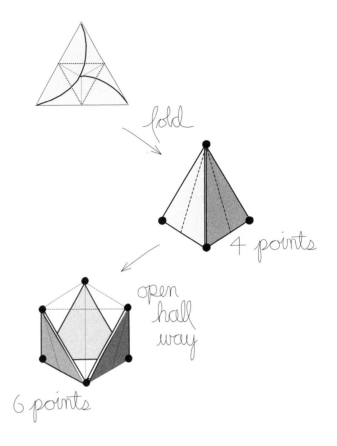

fold

4 points

open
half
way

6 points

<u>Fold</u> another tetrahedron and open it the same as before. Put the 2 half-open tetrahedra together where the triangles of each fit into the triangle intervals of the other. Tape the edges together joining the 2 open tetrahedra. This is the complete octahedron form. Again count the points, edges and planes. Observe the design of the taped edges and the locations of the centers of the 2 circles.

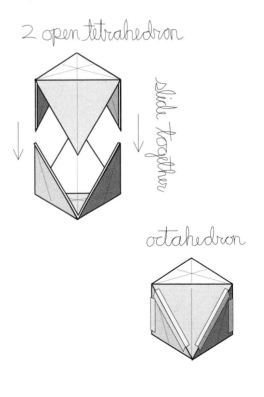

2 open tetrahedron

slide together

octahedron

<u>DISSCUSSION</u>:   OCTAHEDRON

As with the tetrahedron there are 3 symmetries to viewing the octahedron; *end points* (6), *edge lines* (12), and *surface planes* (8). Viewing from the points shows squares, from edges a rhomboid, and from the plane a triangle and hexagon. They all show compressed-image distortions of edge lengths and shapes.

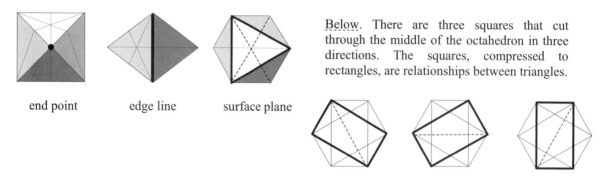

end point   edge line   surface plane

<u>Below</u>. There are three squares that cut through the middle of the octahedron in three directions. The squares, compressed to rectangles, are relationships between triangles.

When looking at drawn images it is difficult to read the compressed spatial information in the flat shapes. The rectangles are squares where 2 sides have been shortened to the picture plane. The square is seen only when drawn from the end point of viewing. All three views are necessary to see the full symmetry.

<u>Below</u>. Each square intersects the other two at the end points locating 3 axes through the center that bisect each other at right angles. This can be seen looking at the end points. This spatial relationship of 12 right angles that meet at the center of the octahedron appear as 6 in the hexagon image of 60° angles (p.29). The rectangles in the hexagon are squares in the octahedron. The 6 end points of the diameters locate 3 axis where the 3 square planes intersect. The 3 diameters of the folded circle serve the same structural function as the 3-axial division of the octahedron; same pattern, different form. These 3 axes model the 3 physical dimensions of our spatial universe. From a position of standing on the earth we define this 6 directional, 3-axis movement as *up* and *down, side-to-side*, and *front-to-back*. This is the hexagon pattern with the seventh point being our own centered perspective; both inside experience and outside other.

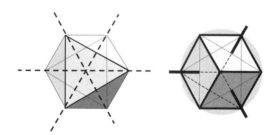

3 axes, 6 directions, show the relationship of the cube to the octahedron compressed to a single image. By emphasizing some parts over other parts, the appearance and our perception change. This happens in space just as it does with flat images.

<u>Below</u>.   The axial diameters in a spherical octahedron pattern show the three great circle divisions of the sphere. The image of the hexagon shows the octahedron and cube as different forms of the same image. The three axial positions to both sphere and octahedron are the same.

3 axes of rotation

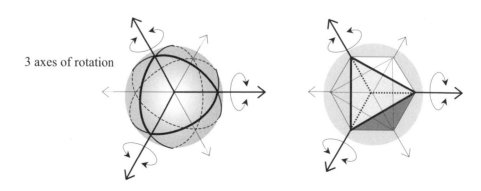

130

Below. The octahedron is a pattern description of the 6 points where the 4 spheres are joined in spherical order. This spatial interval can be formed independent from the tetrahedron as a separate polyhedron with 6 points and 8 triangles. The individualization of the octahedron does not deny the primacy of the tetrahedron and octahedron as two aspects of the same spherical order. They are simply two individualized aspects of what is the same pattern.

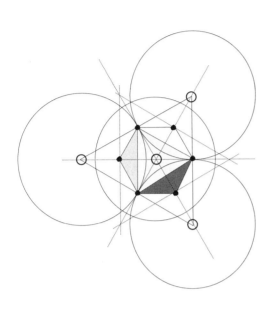

The octahedron is found in the pattern of 4 spheres by the 6 tangent points where the spheres are connected. There is no separation of spheres; only the reformation of one sphere. The sphere can only generate the Whole of what it is through multiple self-reference. All else is relationship.

Below. The octahedron pattern formed with 6 spheres in the closest packed order of spheres (pp.90 #3, 97, fig.2a). The spheres and center points define the octahedron. The 12 connecting points of the 6 spheres when joined by straight lines (6 diameters) form the outside definition of the vector equilibrium without the center point.

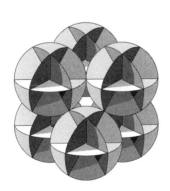 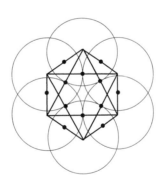 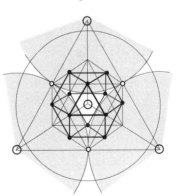

We have been looking at the regular tetrahedron and octahedron. They are expressions of pattern within spherical order. There is a lot to be explored by describing the differences and similarities between the tetrahedron and the octahedron as they are formed from two open tetrahedra. There are many relationships between parts and configuration of each that open to a large variety of proportional arrangements between them that is not regular. We must first understand what is regular before we can know what is irregular. It does not work the other way.

There is a lot of abstracted information derived by observing the difference in areas, angles, and ratios, etc. What is the ratio between the volume of a nine-inch diameter sphere and the volume of the tetrahedron, and octahedron made from two tetrahedra? What generalizations can be made about the observable differences? What is tetrahedral and octahedral in our personal and collective lives? When we form deviations, we develop away from regularity and extend individual differences towards asymmetry and eventual degeneration. All the while the context of consistent regularity is directive, even as we are unaware of it.

When the tetrahedron is opened and 4 points generate 6 points forming 8 equilateral triangle planes, a transformation has taken place. Any 3 of the 6 points of the octahedron form a triangle plane. That leaves 3 individual points to form 3 different tetrahedron relationships with the first 3 triangle points. Eight triangles times 3 individual tetrahedra is how many combinations of tetrahedra there are in an octahedron. These sets of relationships are traditionally demonstrated by cutting into smaller pieces through planar divisions.

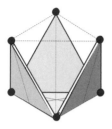

8 triangles x 3 points = 24 tetrahedra

a set of 4 points (3+1) makes a tetrahedra
 for each of 6 points
(3+1) x 6= 24 or
4 x 6 = 24

(p.148)

How many combinations of right triangles are there in the octahedron? Is there more than one right answer?

<u>Below</u>. The 24 interrelated tetrahedra that are generated in the open tetrahedron are not regular. The fourth point is a different distance from the first three, making a longer edge than the equilateral triangle. The gray areas in the hexagons indicated how the irregular tetrahedra are positioned in the octahedron.

1   2   3

Three positions times 8 individual triangles of the octahedron reveal 24 tetrahedra.

The 3 irregular tetrahedra together are the expression of a single open regular tetrahedra. Four triangles (a tetrahedron) times 6 directions (3 axes in 2 directions each) is another way to understand the relationship that forms 24 tetrahedra in the octahedron.

4 triangles x 3 diameters x 2 directions each = the number of tetrahedra in an octahedron.
4 x (3x2) = 24   or   4 x 6 = 24

2 tetrahedra of 4 triangles x 3 diameters = the number of tetrahedra in an octahedra.
(2 x 4) x 3 or   8 x 3   or
2 x (4 x 3) or   2 x 12   or
2 x 4 x 3 = 24

All 24 tetrahedra (2+4=6) are represented in this image of 15 complete lines (1+5=6). There are connections of 6 edges defining the hexagon, the 6 points of the hexagon star, 6 points that define the octahedron, the number of connections between 4 spheres, the directions of 3 diameters, the two-frequency triangle. There are many ways to describe the relationships and qualities of 6 points that describe something about the octahedron relationship.

The <u>tetrahedra</u> is a *triangle* relationship of *four* triangles (3x4=12, 1+2=**3**).
The <u>octahedron</u> is a *square* relationship of *four* triangles (4x4=16, 1+6=**7**).
Together (3+7=10) are the 4 spheres, the first fold of the circle, 10 points.

We need to explore the octahedron in the same way we have the tetrahedron to better understand the difference in form, the sameness of pattern. The first thing we want to do is to arrange 4 octahedra in a two-frequency tetrahedron pattern, (p. 122).

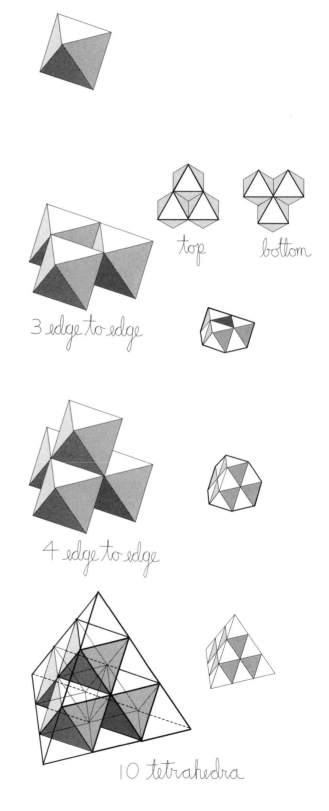

Fold and tape 4 individual octahedra. Put 3 of them into a triangle pattern. You will know when it is correct because everything becomes triangulated and all octahedra are in the same orientation. The top plane will show a two-frequency triangle with open center. The opposite side will show the hexagon with 3 triangles (p.123). All 3 octahedra are joined on edges and form 4 tetrahedra intervals; 3 in the same orientation, one inverse to the center.

top   bottom

3 edge to edge

Place the fourth octahedron on the top with triangle face congruent to the open triangle in the same orientation with edges coming together. These 4 octahedra are in a tetrahedron pattern, three on one plane, and the fourth on another with everything triangulated. There are now 6 tetrahedra intervals and one inverse to the center, 7. The drawing far right shows 6 tetrahedra and four octahedra, called a truncated tetrahedron (p.248).

4 edge to edge

There are now 4 corner triangle faces where 4 tetrahedra can be placed. This completes a 3-frequency tetrahedron by filling in all 10 accessible intervals. A tetrahedral space is left open in the center, which is in opposite orientation to the 10 and reflects the non-centered pattern of the tetrahedron. The tetrahedron is always self-generating.

10 tetrahedra

133

Fold and assemble 4 tetrahedral sets of 4 octahedra each. The 4 units can be tapped together in individual sets. Arrange 3 sets into a triangle, as was done with 3 individual octahedra and with the four-frequency tetrahedron (p.122).

This forming with sets has changed the look of the larger system but the pattern does not change. Observe the inside edges of the top 3 triangles; this is where the fourth set will be joined.

Add the fourth set to the top in the same orientation with edge-to-edge joining. The edge line is not straight as with the tetrahedron, it is in/out and zigzags the spatial boundaries. This is a fractal expression having everything to do with division and reformation of the tetrahedron using octahedron thus changing the form.

a). Viewing this system from the edge definition, we can see the edge change to 6 rectangular planes that form triangle intervals at each of 4 corners. This is not a tetrahedron form but clearly a tetrahedron in pattern in the arrangement of an irregular vector equilibrium (6 quadrilaterals and 8 triangle spaces).

(b). Filling in the 18 intervals with tetrahedra completes the straight edges, changing the form to a truncated tetrahedron. Each rectangle plane has become a line edge.

(c). Adding 4 more tetrahedra to the 4 triangle corners completes the 5-frequency tetrahedron. This tetrahedral system is now in a linear form that is open to the inside and defined by the 6 outer edges and 4 end points.

This is now both a 5-frequency tetrahedron and a 4-frequency octahedron system with an open center from all four sides.

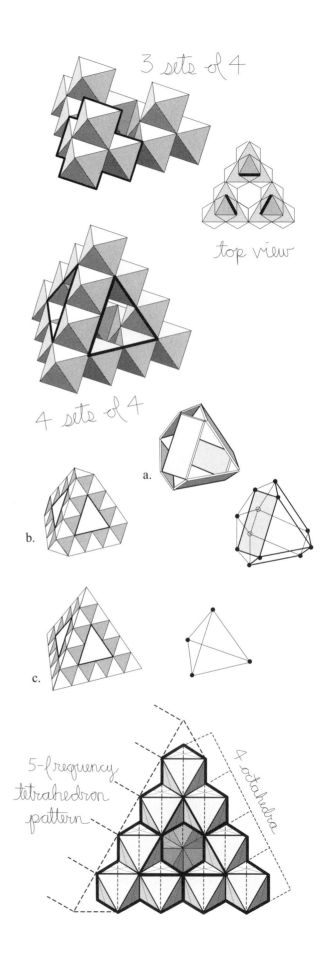

134

Remove all tetrahedra leaving only the 4-frequency octahedra. Remove the top set of 4 octahedra, turn upside down and insert the set into the open center. The internal space of four sets of 4 octahedra is similar to that of 4 sets of 4 spheres (p.173). The tetrahedron pattern is self-generating, intuitively it makes sense that the interval will be tetrahedral made from a set of 4 octahedra.

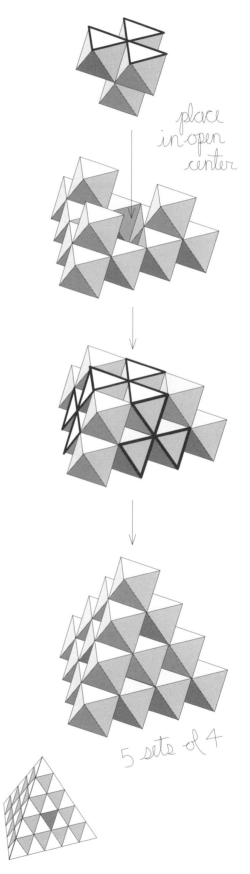

place in open center

The inserted set of 4 octahedra will fill in the first and second layer of octahedra completing the hexagon on 3 sides and the center triangle on the bottom, thereby completing three more interrelated hexagons.

Fold a fifth set of 4 octahedra. Place the fifth set of 4 octahedra on the top, triangles to open triangles, edge-to-edge. There is now a consistency in the alternation of octahedra with tetrahedra intervals, the isotropic vector matrix (p.98). There are 10 octahedra on the first layer, and 6 on the second layer, and with the last set of 4, the third layer is 3 and the top layer is one (p.124).

There are 5 sets of 4 octahedra reflecting the 5 spherical tetrahedra in the 2-frequency closest packed ordering of spheres. (pp.41, 173)

5 sets of 4

How many tetrahedra will you have to make to fill in all tetrahedron intervals including the four corners, making a solid tetrahedron? (p.124) What is the volume of the tetrahedron? What is the volume of just octahedra? What comparisons of volumes can be made to numbers of each?

TWO-FREQUENCY OCTAHEDRON

Start with the VE sphere using 4 circles folded on the 3 diameters (p.89) It has 6 square open planes. An octahedron will fit in each open square.

vector equilibrium sphere

1 octahedron added to sphere

Fill in all square intervals with octahedra changing the sphere into a 2-frequency octahedron. The circumference of 4 great circles can still be seen as they show 8 tetrahedra intervals in the center of each triangle face of the octahedron.

6 octahedra

Here we see 6 octahedra in the two-frequency octahedron without the 4 circles.

2-frequency octahedron

The half line symmetry of the vector equilibrium shows two layers of 3 octahedra alternate to each other. One set is opposite the other, twisted 60° to an edge-to-edge joining, hexagon face to hexagon space. Compare this to the arrangement of spherical order (p.90).

3 octahedra triangle face up

3 octahedra triangle face down

Eight tetrahedra will fill in the intervals making this a solid 2-frequency octahedron using 20 circles. Make more spheres, octahedra and tetrahedra and expand the exploration of this system.

fill in 8 tetrahedra

136

OCTAHEDRON FACE JOINING

<u>Fold</u> 2 octahedra.

1)   There are 168 combinations of how 2 octahedra can be joined face to face. Choose two center faces and join them together. The creased lines will show the difference. Tape around all 3 joined edges. The octahedron with 6 points shows 4 sets of opposite pairs of triangle faces making 4 intersecting triangle anti-prisms. When one prism is joined onto the other, 3 pair of connected edges that go in opposite directions line up in a helix form. Two connecting edges are shown to the right.

Make another pair the same way. Put the 2 sets of 2 octahedra together, forming a line of 4 octahedra. It is now easier to follow the connected edges to see 3 twisting to the right and the 3 twisting to the left in a helix. This helix can be endlessly extended by adding more anti-prisms (pp.212, 222).

2)   This time put the 2 sets of 2 octahedra in a circle around a common center point joining face-to-face. This makes 4 parts of a pentagon leaving an opening for a fifth octahedron.

Add the fifth octahedron, completing a pentagon star system.

This can be expanded into a spherical, dodecahedron-patterned system by adding eleven more pentagon units joined to each other on their edges. To the right is a triangle set of three pentagon stars joined on the edges. You will need 60 octahedra, 120 tetrahedra to make a ball. There are a number of variations to this arrangement (p.300, *Fig. 73*).

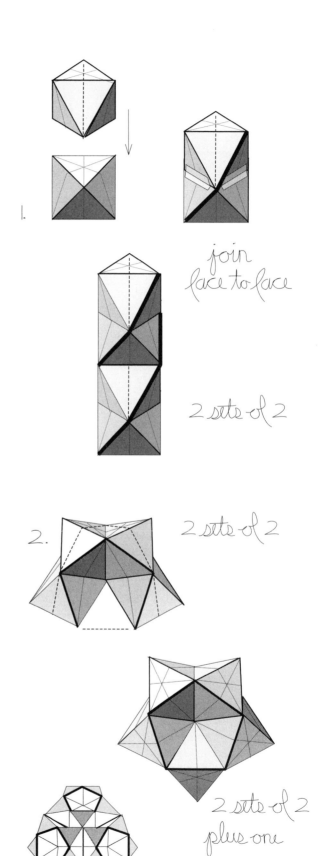

join face to face

2 sets of 2

2 sets of 2

2 sets of 2 plus one

3) Fold 2 octahedra and join them face-to-face as before.

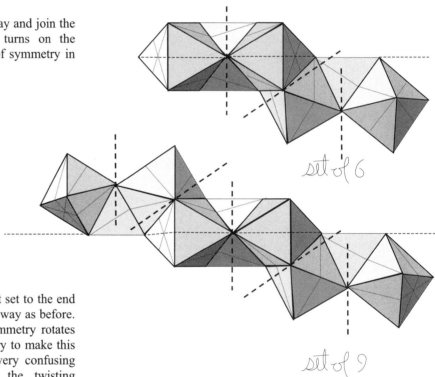

3.

set of 2

set of 3

Add one more octahedron making a set of 3. Use one of the points where 2 join as the center point in joining the third octahedron. There will be symmetry of one in the middle and 2 on either side. It will be three-fifths of a pentagon.

Make another set of 3 the same way and join the 2 set face-to-face so that one turns on the horizontal axis twisting the line of symmetry in relationship to the first set.

set of 6

Make a third set of 3 and join that set to the end face of the second set in the same way as before. Turn it so the twisting of the symmetry rotates all in the same direction. If you try to make this using one at a time it will get very confusing unless you clearly understand the twisting pattern. It is easier to develop systems using progressive sets rather than one unit at a time.

set of 9

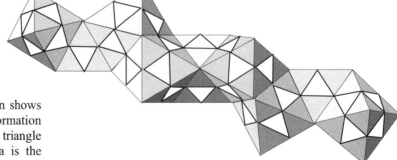

A truncating division of each octahedron shows the internal vector equilibrium helix formation where each VE is joined on congruent triangle faces. This linear twisting of octahedra is the inner helix of the tetrahelix. Connecting one half of the axial divisions of each octahedron will define the helix form of the DNA double helix design (p.139).

138

This helix arrangement of octahedra can be formed as an individual system. That does not negate that it is patterned to the tetrahedron in a linear development of octahedral connections in the tetrahelix. The context of the octahedron helix is in the two-frequency tetrahedron joined face-to-face connecting the inner spaces defined by the 6-point octahedron.

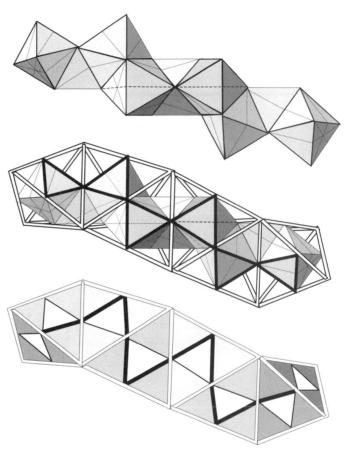

Nine octahedra in a helix pattern

The octahedron helix shown defined by the edges as it lies inside of the two-frequency tetrahelix.

The zigzag surface design on the 3 twisting planes of the two-frequency tetrahelix becomes informational to the position of the internal octahelix.

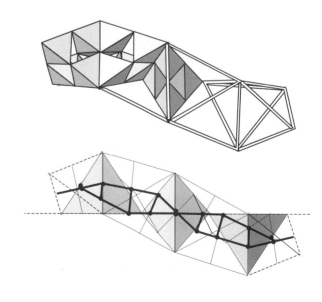

This shows 4 two-frequency tetrahedra and the edge definition or skeletal design of 5 tetrahedra without the octahedra division. It is easy to model the octahedral space by joining 2-frequency tetrahedra face-to-face in a line

Three tetrahedra joined point-to-point twisting on a central axis forming a relationship of 4 more tetrahedra. *Three tetrahedra in a helix formation are a relationship of seven tetrahedra.* Three generate four more making seven. It is the forming of triunity, basic mechanical life function showing the generative nature of structural pattern (p.69). The dark lines show one half of the connecting octahedra axis twisting in the opposite direction of the tetrahelix. This forms the DNA double helix ladder.

Octahedra stacked one onto the other are a helix formation with 6 edges in 3 sets of 3 each. The right and left hand edge lines cross each other as they twist around the in/out surface. Any anti-prism when stacked congruently on to another will form a similar balanced helix pattern to the symmetry of the top and bottom.

When the parallel twisting of edges moves out of parallel the helix breaks down and there will result a twisting out from the straight line forming. With this twisting and changing scale of individual units it moves into a spiral formation. The helix form is property of the tetrahedron (p.108). It is conditional pushing/pulling that shapes specific formations making each unique and individual without breaking continuity of pattern.

The middle drawing shows 3 black line edges moving in a counter direction to the 3 white edges, forming an intersecting tubular net. The drawings on the far right shows one of the 3 helix planes as it wraps around counter clockwise.

When a one-directional spinning is formed the other direction is usually hidden in the dynamic symmetry of directional balance, keeping to the principles of pattern formation.

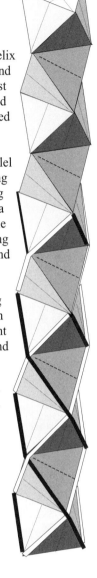
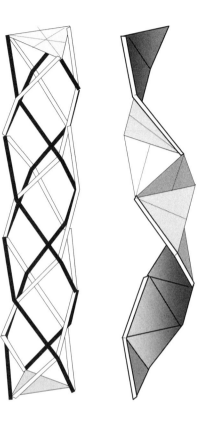

This drawing shows 6 strands, 3 sets of 2 strands each, woven alternately over and under each other into a braid. In this case the 3 sets are shown connected as 3 individual closed loops woven together. There are various combinations of strand connections that can be woven, each changing the character of the braided system.

140

OCTAHEDRON NET

Fold an octahedron and tape all the edges (p.129). Take the tape off, leaving one joined edge. Lay it flat. Each large triangle is in opposite direction sharing one half edge of the other. This is the octahedron net where all 8 faces are visible at the same time and in place to be able to reassemble it into the octahedron.

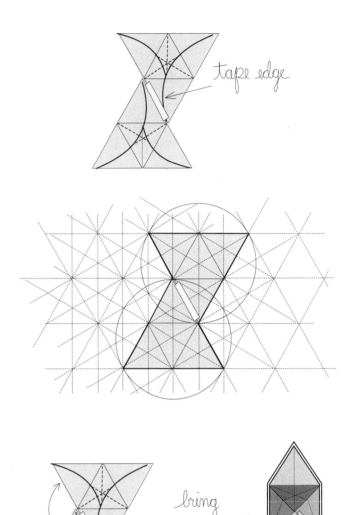

The 8 equilateral triangles in this net are part of the larger equilateral triangle, hexagon, circle matrix. The circle matrix is foundational for the individual polyhedral nets. Individual nets work because they are a part of this larger grid that is the compressed image of spherical order. Each large triangle is divided by 3 diameters/bi-sectors dividing the grid into a smaller scale of the same matrix (pp.118,121,195). By folding out the circumference flaps the relationship of the 2 circles that generate the octahedron become visible.

Join 2 adjacent angles together and tape. Do the same to the other side, making a quadrilateral opening.

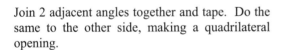

Bring one of the 2 edges of the trapezoid together with one of the square sides, either on the right or left hand side. Tape together. This gives a specific left-handed or right-handedness to the form. Folding both sides will stabilize the square opening.

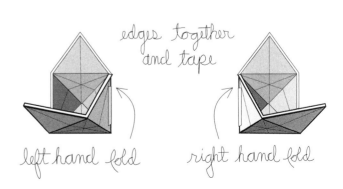

141

There are 4 stabilized reformations to the octahedron net. They are the *bi-pentacaps,* the *octahedron, tetrahelix,* and the *bi-tetrahedra,* all a reflection of the first formed 5, 4 and 3 using three diameters (p.73 #1-10). The bi-tetrahedron reflects the dual nature of the tetrahedron.

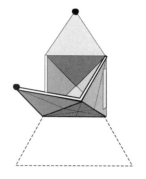

To form the bi-pentacap, bring the 2 end points from the triangle flaps together and tape, putting a piece of tape across and squeeze it in the middle. This forms 2 pentacaps that share the same open pentagon plane. Eight triangles are used to make 10 planes, 2 that are open.

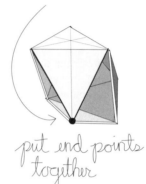

put end points together

bi-pentacap

Take the tape off and slide the 2 triangle flaps so all edges are together. Here we are back to the original octahedron that opened to the net in the first place.

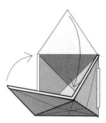

put edges together

octahedron

Open the 2 flaps just brought together and slide the edges together in the opposite directions. This forms 3 closed tetrahedra in a helix formation. This is the fundamental helix unit of 3 tetrahedra. It has no directional preference until the next unit is attached either to the right or left.

put opposite edges together

tetrahelix

142

From the tetrahelix position, push the 2 loose triangles down inside flat against the inside wall of the octahedron. This leaves a quadrilateral open cavity that is unstable.

Close the square cavity by bringing the 2 furthest-most points of the quadrilateral together. This will reform into a triangle forming bi-tetrahedra, two tetrahedra joined by a common triangle plane.

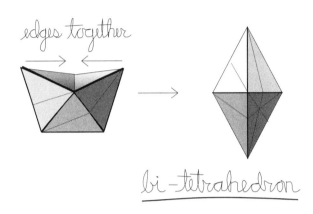

Open the bi-tetrahedra and collapse the open cavity perpendicular to the last movement into a flat trapezoid shape.

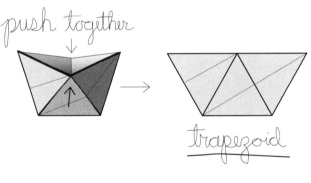

Fold one triangle over, leaving a rhomboid shape.

Fold one more triangle over in the opposite direction further reducing to an equilateral triangle, and again into a right triangle, which is the first fold of the circle in half. Compressed into the depth of this shape is all previous folding.

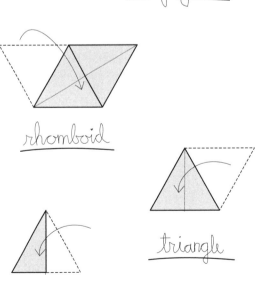

143

Open the triangle back to the full net for further exploration into the transformational process.

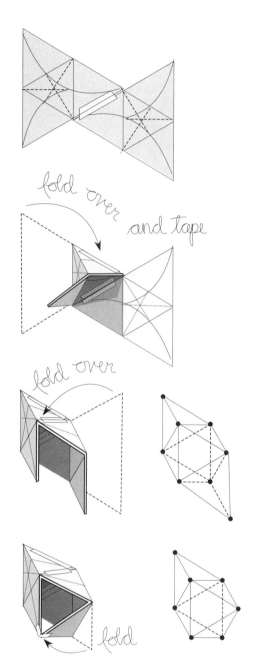

On the edge adjacent to the joined edge, fold one half of the edge to the other half, forming an open-end tetrahedron.

Then fold the adjacent one-half edge over to meet one of the open edges of the tetrahedron. Tape edges together. When all intervals lengths between points are equal to the length of the triangle it forms an arrangement of 2 tetrahedra and one octahedron, an 8-point system. (Two sets together will complete the form of the arrangement.)

Touch the 2 end points of the open triangles together. Put a piece of tape across the points and squeeze to hold them together. This forms an octahedron and tetrahedron sharing a common triangle plane with 2 open planes to the octahedron. This arrangement has 7 points.

Before taping the points together, slide one edge along the edge of the other until all edges are touching. This takes us back to the helix formation of 3 tetrahedra, an association of 6 points. One tetrahedron and one octahedron is also 3 tetrahedra.

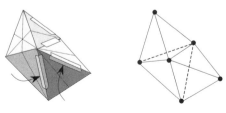

144

Again going back to the net, let's explore the square opening a bit further.

<u>Fold</u> 2 tetrahedra and tape into an octahedron net. Again notice that it is either right hand or left hand. Be consistent to orientation with multiples.

Bring 2 outside points on the long side together and tape. Do the same for the opposite side.

The square interval that is formed can be collapsed along each diagonal by bring opposite corners together.

(Make 6 of these units with a square interval and put the square edges together into a cube arrangement. There are many combinations because of the open triangles.)

Collapsing in one direction forms a tetrahelix with 2 open triangle ends. Collapsing in the opposite direction forms a tetrahelix twisting in the opposite direction with 2 open triangles on the sides.

One octahedron has been reformed into 2 tetrahedra in a helix segment of 4 tetrahedra. Joining each individual configuration in multiples will form long helix chains (p.146, *Fig.6a*)

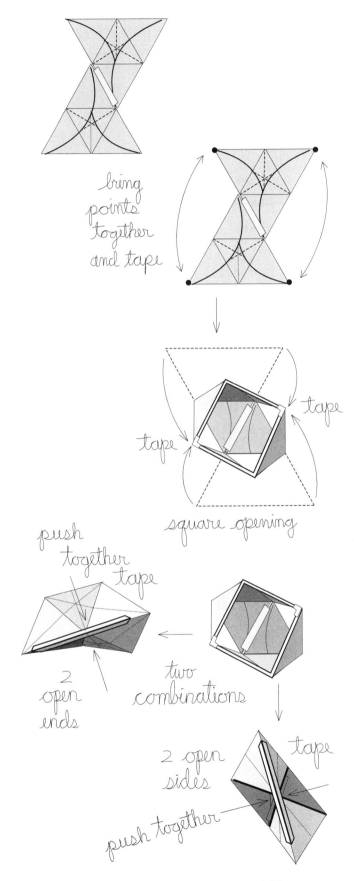

145

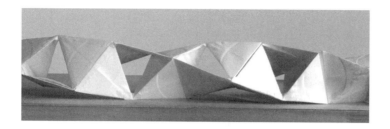

Fig.6a Three sections of reformed octahedra forms a helix (p.145). They are taped hinged between each section that follows one of the 3 twisting plains. Each section is 2 tetrahedra and 2 open tetrahedra spaces between.

Fig.6b Here the helix is opened at the joining forming a different configuration.

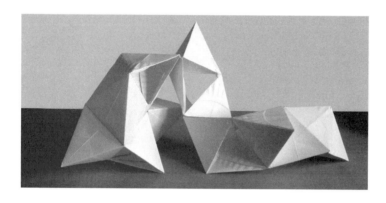

Fig.6c A further rearranging of the helix as it is partially reversed to the hinge movement.

Fig.6d. The configuration of the helix in (*Fig,6a*) is completely reversed to an inside out position. Another reformed tight twisted helix results. This shows 2 and a little more turns. Nine reformed octahedra form 18 tetrahedra which form 18 tetrahedra intervals making a full length of 36 tetrahedra.

There is a similarity to the inside out reforming of the helix shown on (P.202, *Fig.15,18*).

146

OPEN VARIATIONS

There are many ways to reform the octahedron. Here are a couple of basic variations using two tetrahedra to form the regular octahedron. These units can be used in a number of ways to develop systems of great variety.

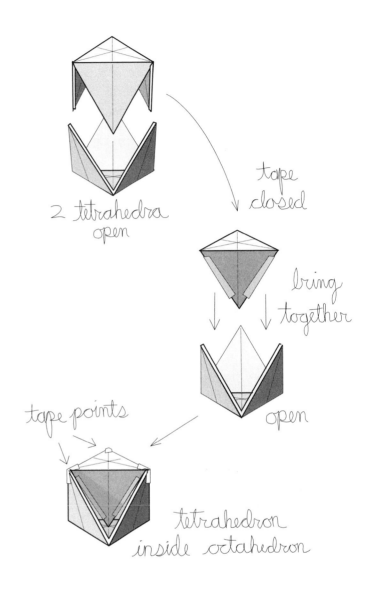

2 tetrahedra open

tape closed

bring together

open

tape points

tetrahedron inside octahedron

Fold 2 tetrahedra. Tape one tetrahedra closed, leaving one open. Put the closed tetrahedron upside down into the open tetrahedron with a point to the center of the bottom triangle. The open tetrahedron points come together with the 3 points of the upside down tetrahedron. Tape the 3 points of joining.

One closed tetrahedron fits inside the other open tetrahedron, forming a full octahedron pattern. This changes the edge joining to a point-to-point and point-to-plane joining. This is an octahedron with 3 alternate sides pushed into the center of the base triangle. When the edges are congruent, the altitude of both tetrahedron and octahedron are the same. The slope of angles of the half-open and the closed tetrahedron are complementary angles (p.85).

To tape the points together put the tape across the 2 points and squeeze in the middle so tape will stick to the underneath sides making a tight joint. (Put some glue on all points before taping if you are going to make multiples; see example on page 300, *Fig. 73*.)

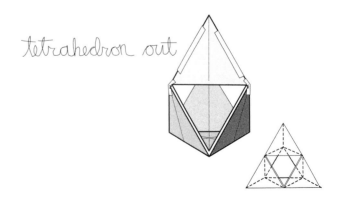

tetrahedron out

Reverse the orientation of the closed tetrahedron with the point going out instead of into the octahedron. This tetrahedron and octahedron combination opens 3 sides. Adding 3 more tetrahedra to the sides of the octahedron will form a complete 2-frequency tetrahedron with 3 sides open to the octahedron interior space.

147

Use a tetrahedron that has one side open to replace the closed tetrahedron and place it into the octahedron with the open side either to the top or to one side.

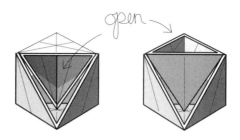

Start again with the octahedron net.

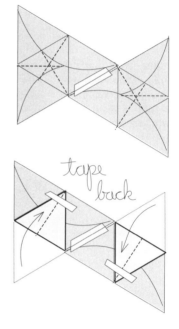

Fold over the 2 sticking out triangles leaving 6 triangles in a row. Tape the 2 down so they do not get in the way. This forms one long trapezoid.

Fold the 2 ends together and tape edge-to-edge. This forms an octahedron anti-prism band around 2 opposite and open triangle ends. This is stable even though 2 planes are open because everything is triangulated (p.132).

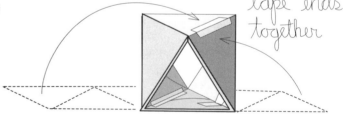

Six open octahedra edge-to-edge forming a two-frequency open octahedron (p.136). This makes it easy to see the VE division in the octahedron.

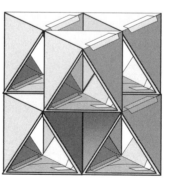

12 circles

4 ICOSAHEDRON

One circle forms one tetrahedron.
Two tetrahedra form an octahedron.
Four tetrahedra form an icosahedron.

Start with the octahedron net (p.141).

Select one two-frequency triangle as the center. Tape 2 more tetrahedra triangles to the remaining 2 sides of the center triangle, consistent in direction and position. The 3 side triangles will have moved half the edge length in the same direction around the center triangle. The two-frequency triangle is the pattern, 3 around one. This makes a left-handed or right-handed spiral depending on direction of transversal movement of the side triangles.

Join 2 parallel half sides of 2 triangle arms together. A triangle space is formed from one edge of each of the large triangles. Do this on all 3 sides. It will curve around forming a sphere with 16 triangular surfaces and 4 triangle intervals. The intervals will be located equidistant around the icosahedron. Each circle is joined to the other showing the same spiral symmetry from each center triangle that is seen in the net.

Look for 5 triangle planes coming together around each vertex; 4 surfaces and one open plane. This reflects the reformed pentagon from 3 diameters (p.74 #7) and the bi-pentacaps (p.142). (In the classroom students are told what they are looking rather than instructions. Then they figure out how to do it.)

Traditionally the icosahedron has 20 equilateral triangle faces. This is the same arrangement with 16 surfaces and 4 open triangle planes showing a tetrahedron relationship. The 4 circle centers are also in a tetrahedron relationship. Together these tetrahedron duals define the cubic pattern (p.166).

149

A variation on the icosahedron net is to unfold the triangles to the circle shapes. The circles will remain taped in the same straight-line grid only opened to the flat circle, then reformed into the icosahedron in the same way as before. The circumference flaps are to the inside and function as structurally integrated elements rather than folded under as inactive parts. This way the entire circle is structurally utilized in forming the icosahedron.

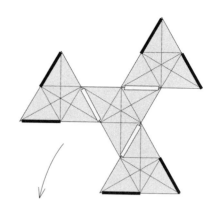

unfold circles

Open the circles, leaving the 3 taped edges. Keep track of or mark the line segments that are to be joined. Notice the diameters do not connect but continue to be perpendicular to the lines of the triangle grid matrix. This shows four primary center points to a single system.

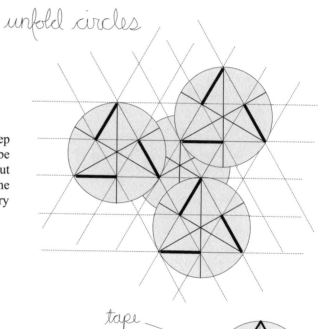

tape

Tape the circumferences to the center circle on both sides, front and back. Make sure that the folded grid lines match up accurately before taping. The dark lines show the net pattern.

Fold the dark line edges of the net together in the same spiral pattern as before. This time the circle is open and the circumferences will be folded in as you bring the creased edges together. The circumferences will form an inverted tetrahedron where there was an open triangle before. The inward-facing tetrahedra will have an irregular hexagon opening.

Using the same taped-open circles, fold the icosahedron with the circumference on the outside. This is an outside form of what you have just folded inside.

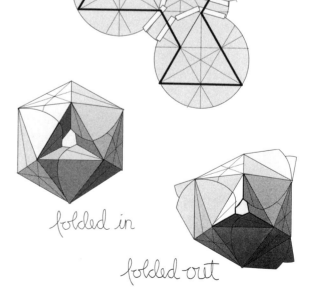

folded in

folded out

150

NET VARIATIONS

Join 3 two-frequency triangles forming a trapezoid. Attach them together putting tape only on 1/2 of the edge lengths to the smaller side of the trapezoid (tape on both sides to make it stronger).

On the long edge of the trapezoid fold the 3 untapped smaller triangles behind, forming a half hexagon interval. Tape them back against the flat surface to keep them out of the way.

Turn over with taped triangles to the back. It is now half of a hexagon shape with a smaller half hexagon interval showing 9 equilateral triangle divisions.

Make another set the same and tape these 2 sets of 9 equilateral triangles together, joining only on 1/2 of the middle edge length.

Bringing the 2 ends around touching edge-to-edge will close each open half-hexagon space. This forms 2 tetrahedra with one end point truncated to an open triangle and an open triangle bottom plane.

Pushing in on the 3 end points of the bottom triangle will force the midpoints of each side out, forming an in/out hexagon-shaped edge. Do this to both sides and bring the 2 hexagon edges together. This pushing in and together shifts the angles of the 18 triangles to form the icosahedon. Tape the 6 edges joining around the middle.

The opposite open triangles show the opposite and reverse orientation of any 2 parallel triangle planes of the icosahedon. These two open planes form a triangle anti-prism through the center of the icosahedron. This happens through all 10 planer axes (p.181).

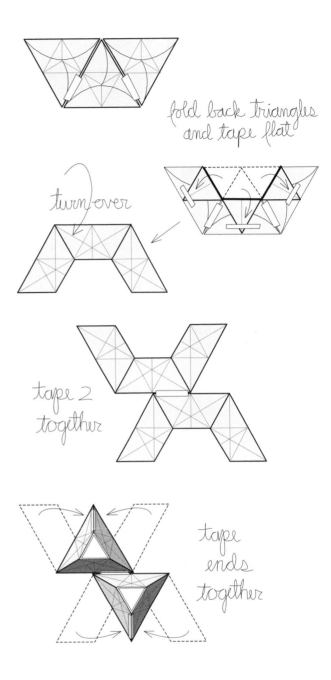

fold back triangles and tape flat

turn over

tape 2 together

tape ends together

push points in bring halves together

Another interesting connection between the tetrahedron and icosahedron is in this arrangement of triangles. Because of the 2-frequency nature of the net there are many other configurations of which the icosahedon is only one. Each net configuration reveals different movement and forms.

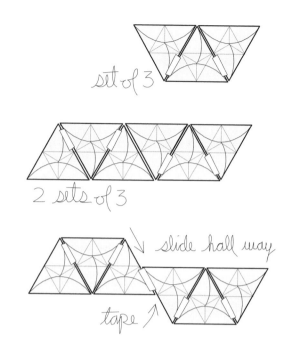

set of 3

2 sets of 3

Put the 2 sets of trapezoids together, arranging in a straight line end to end.

Slide the 2 trapezoids in opposite directions on the connecting ends. They will only touch on one half of each edge length. Tape together on both sides. Six two-frequency are 24 triangle planes.

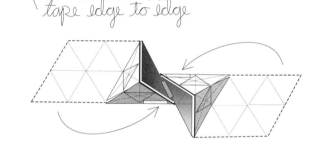

slide half way

tape

fold around and tape edge to edge

Fold each set into a tetrahedron and tape. Notice the similarity to what we have already done. Here the full tetrahedron with one open face is formed.

The open sides are unstable and don't want to stay straight. The two-frequency triangle openings are folded into 6 segmented in/out edges. Bring the zigzag edges together on the taped hinge and tape the edges together.

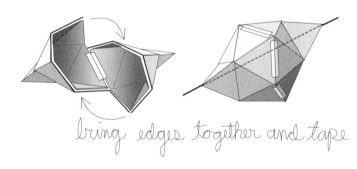

bring edges together and tape

fold over triangles and tape

Again we have an icosahedron with 2 opposite faces stellated into regular tetrahedra that can be folded flat, forming the regular icosahedron figure (p.157).

Return to the spiral net of the icosahedron (p.149). There are other ways to attach edges together reforming the icosahedron. In making the icosahedron we folded the triangular arms to form 5 triangle planes around each of the 12 corner points. This time the arms come together showing a different direction in forming. (Notice the attachments are opposite from where they are on the center triangle.)

Bring the edges together and tape, indicated by the heavy lines.

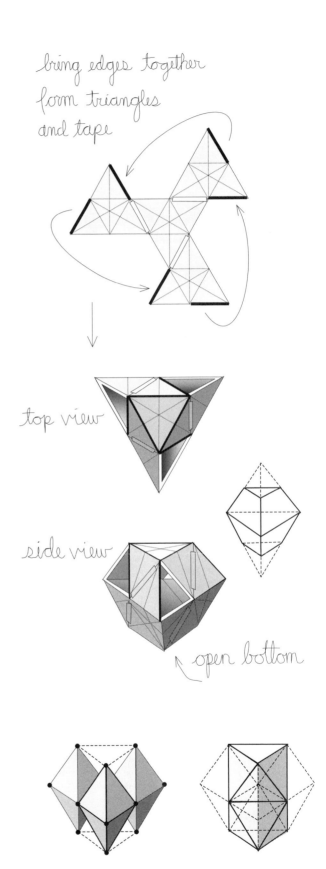

The <u>top view</u> and <u>side view</u> drawings show the reconfiguration in the form of a bi-tetrahedron where 2 opposite endpoints have been truncated. An open triangular plane joins the 2 tetrahedra. The top tetrahedron has a truncated triangle surface and 3 trapezoid planes where each trapezoid shows one rhomboid and one open triangle each. The bottom tetrahedron has one open-end triangle plane with 3 trapezoid surfaces. The direction of the open triangles reflects the direction of the net.

There are other combinations of reconfiguring in this net to be discovered. Try joining point-to-point and overlapping surfaces. They are not consistent with the edge-to-edge joining but they reveal other transitional possibilities

These 12 points show an arrangement of 3 bi-tetrahedra. Change again to a formation of 2 joined octahedra intervals forming an interior triangle prism defined by the relationship of tetrahedra. The formed and unformed are intimately interactive.

The value of this kind of exploration is in understanding something of the individualized expressions that are interrelated through movement within a single net.

Here is another way to approach folding a "solid" icosahedon with 20 surface planes. It uses 6 tetrahedron-folded circles, but only utilizes 20 of the 24 triangles.

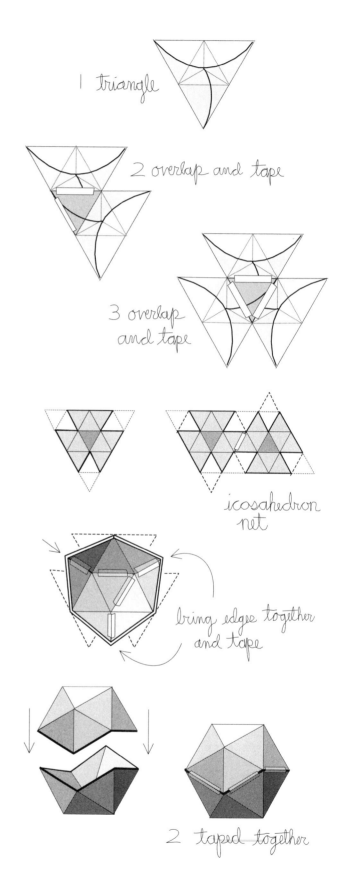

Fold 3 opened tetrahedra and overlap a corner triangle from each triangle so all 3 large triangles are sharing the same center triangle. Line up all edges and creases evenly taping them together on the front and back.

This forms 3 two-frequency edges and 3 alternating intervals. This irregular hexagon is one half of the icosahedron net.

To complete the net, make another set of 3 circles in the same way and attach one half of the 2-frequency sides together forming a center band of 10 triangles with five on top and bottom. This is a variation on the more traditional icosahedron net (p.157).

Two halves can be assembled individually or as a net. Individually bring together 2 adjacent edges that form the intervals. This makes a dish-like shape of 10 triangles. Keep the circumference to the inside as you tape the full length of the edges together.

Bring the 2 halves together edge-to-edge, and tape. The zigzag hexagon configuration of the edges fit one into the other, similar to 2 tetrahedra forming the octahedron. This completed icosahedron "solid" has 20 equilateral triangle surfaces in a spherical pattern.

Here is another variation to forming a net with 6 two-frequency triangles that will fold an enclosed icosahedron. Join three large triangle by taping only 1/2 of the joining edges on the short side of the trapezoid. Make two sets. Join the two sets by putting the short sides of the trapezoids together and sliding over 1/2 the edge length and tape on front and back.

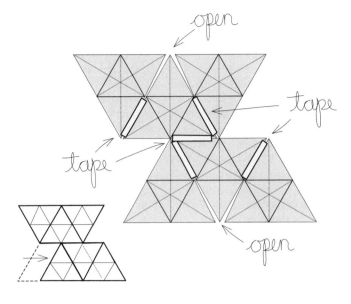

The advantage of having 24 triangles is similar to the advantage of having only 16 triangles. They both show functions of the icosahedron not seen using the traditional 20 triangle net.

The symmetry of joining the edges is seen in the net layout and much easier to do than it is looking at the flat drawing. Put the edges pointed to by the connecting arrows together, edge-to-edges, and tape. Start with the edges 1. and 2. as the drawing shows. The rest becomes obvious.

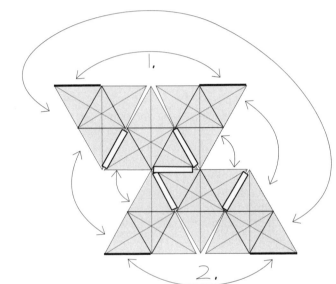

In assembling this net, the 2 sets of 3 overlapping triangles at opposite ends can be used to form a tetrahedra, folded over to a flat surface, or pushed inside leaving an inverted tetrahedron, or an open triangle plane.

The opened tetrahedra can be used as a means of joining to another icosahedra or other congruent systems. Joining open tetrahedra provides an edge-to-edge octahedron connection (p.129).

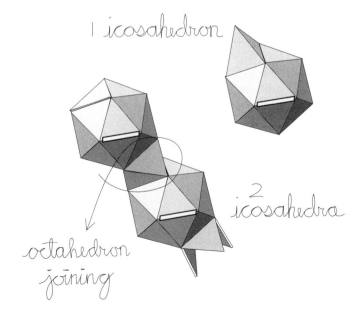

155

This icosahedron with one tetrahedron on each end demonstrates that an icosahedron cannot be formed using 20 tetrahedra pointing to the inside. (Something students seem to want to do that.) When one tetrahedron is inverted and the opposite is also pushed in the 2 will jam up in the center of the icosahedron. The edge lengths of the tetrahedra are longer than the radius of the circumscribed sphere of the icosahedron.

While many students want to put 5 tetrahedra into a pentagon, this demonstrates why it cannot be done. The angles are not congruent because of the different between 5 and 6 symmetries. Five is less than 6 and the tetrahedron is 6.

The pentagon has a smaller radius than the hexagon because 1/6th of the area has been folded in and the circumference is diminished while the outside lengths remain the same (p.74 #7). Notice the centeredness of pentagon to hexagon outline.

Buckminster Fuller would hold 13 spheres in a VE arrangement and by removing the center sphere the spheres would rearrange themselves into an icosahedron without a center, where all 12 spheres became triangulated. The 6 squares spiraled into 12 equilateral triangles. The sameness of the edge length and radius of the VE shifted to edge lengths with shorter radii that would not allow a center sphere of the same size.

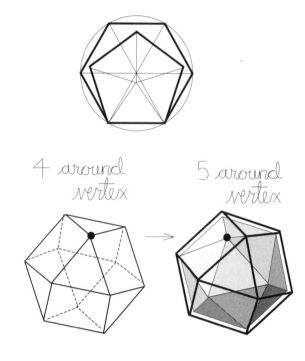

4 around vertex

5 around vertex

A center point of any positive pentacap of the icosahedron can be pushed in to create an inverted pentacap into which another icosahedron will fit. This creates a stacking of pentagonal anti-prisms that will show similar properties as the octahelix (p.140), only it will be a 5 sided helix with 2 sets of 5 opposite directions of twisting.

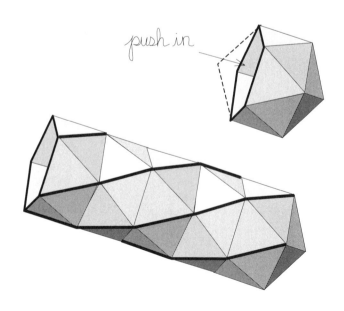

push in

The traditional icosahedron net is from the same hexagon/circle matrix, only a different arrangement of equilateral triangles. It has lateral symmetry with a "belly band" of 10 triangles with a triangle placed to the 5 edges both on the top and bottom, 20 total. The circle tetrahedron folds do not correspond to the circle grid in this configuration where the net uses overlapping hexagons. But this net can be folded from a single circle that is folded to a higher frequency grid (p.118).

This icosahedron net design shows 5 sets of the linear tetrahedron net side to side. 4x5=20. It shows an obvious 1:5 relationship between the tetrahedron and the icosahedron. It is important to explore all possible net variations. *There is still much to learn about the changes that occur during the compression and reformation process of spatial organization.*

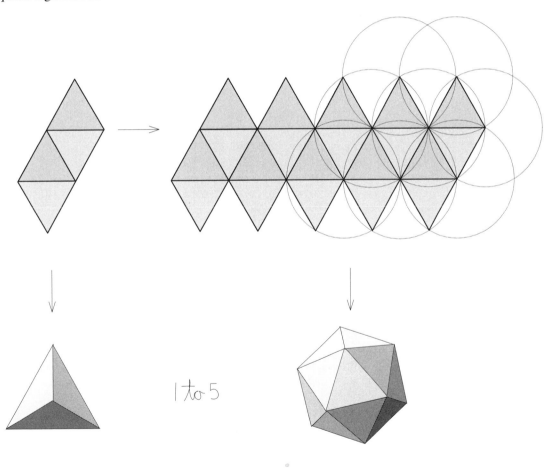

1 to 5

We have formed 2 types of icosahedron, the "solid" or enclosed and the open icosahedron. The solid is a static model; open, it becomes dynamic and can be changed without altering the basic arrangement. Each form comes directly from the tetrahedron reformation of multiple circles. There are 4 distinct forms that we can make some observations about.

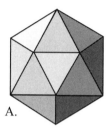 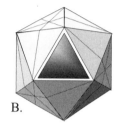 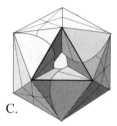 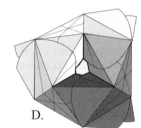

A.   B.   C.   D.

A) This is the traditional form of the icosahedon pattern showing 20 equilateral triangular surfaces. It is one of the Platonic (regular) solids. When formed from 6 circles, overlapping allows for changes with the 2 opposite faces.

B) This shows a more economical approach to modeling the icosahedron using 4 circles with only 16 triangles. There are 4 equilateral triangle openings in a tetrahedron pattern. Sixteen surfaces and 4 openings form the 20 regular triangle planes. This open shell reveals a dual tetrahedra relationship in a cubic pattern. The openness allows internal connections when joining multiples.

C) This drawing shows greater use of the 4 circles forming an inverted partially formed tetrahedron in each of the 4 openings. There is increased strength in this form.

D) The fourth drawing shows the inverted tetrahedra extending to the outside. This is an odd form of a tetrahedron relationship with an icosahedron center. This is the male interlock to the female form in fig. C without closing the open connection.

The traditional properties of the icosahedron consist of 20 congruent triangular faces, 12 end points, and 30 edges. From this we can observe a net of 31 major axes. There are 6 point-to-point, 10 face-centered and 15 edge mid-point axial connections.

 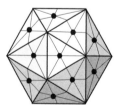  

end points   face-centered points   edge-centered points   all points

The 6 end point axes are the same 6 axes of the vector equilibrium (p.156), only having shifted to a different angulation between them. The 6 axes reflect the 6 points that define the octahedron, which are the 6 connecting points of the tetrahedron pattern in spherical order.

There are 20 face-centered points. The open icosahedron shows 4 open faces equally placed in a tetrahedron pattern. The numbers tell us that there are 5 sets of 4 triangle faces that can be located in the icosahedron pattern. The same holds true for the center points of the triangle faces, making 5 individual tetrahedra. The form of 5 tetrahedra intersecting in an icosahedron pattern is called a <u>compound</u> and looks like a star because of 20 points. While the proportions are different, the same organization is observed in the full stellated icosahedron (p.166).

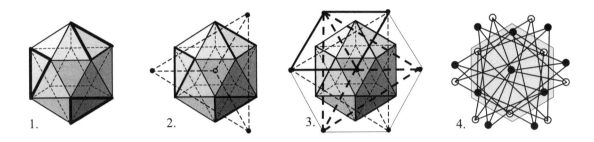

1) Four equally spaced triangles. 2) The four center points extended to show the tetrahedron connection. 3) Two tetrahedra by extending four more triangle faces equally spaced with the first four. This forms a cube pattern indicated by compression of the square to the rhomboid shape. 4) A schematic drawing shows the compound of all five tetrahedra.

<u>Below</u>. The octahedron and cube patterns are inherently part of the icosahedron. They can be found in 3 sets of parallel pairs of opposite edges. These edges are 3 perpendicular planes intersecting each other. The 6 axial points are the edge-centered points of the planes.

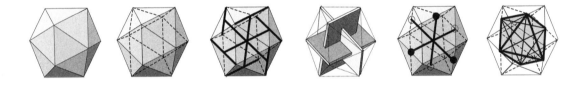

The 3 perpendicular axes of the octahedron have become 3 perpendicular axial planes that run through the icosahedron. The 3 axes remain consistent in reforming through all form changes.

Two triangles share one of each of the 6 edges; these are 12 triangles. The remaining 8 triangles form a cube pattern. This is most easily seen when the center points of the 8 triangles are stellated, (p.166). This is a different form of the same pattern in the above drawing number 3, where the tetrahedra edges are the diagonals of the square faces.

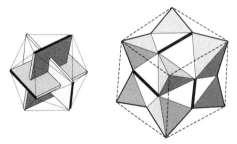

The 3 right angle intersecting planes are to a golden rectangle proportions (p.47). The division of intersects between planes follows the same proportional ratio. This is a growth ratio found basic to all 5/10-based symmetrical forms (p.167).

159

<u>Below</u>. Twelve points form the icosahedron pattern. Each point is a raised center point of twelve intersecting pentagon planes, or twelve pentacaps. Removing each of the twelve pentacaps leaves a spherical pattern of 12 preformed locations. Each stellation has been truncated, effectively removing the form from the pattern of relationship. The pattern remains pre-existent to form.

<u>Side view</u>

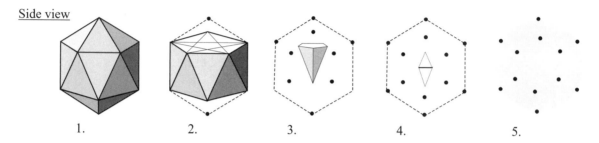

1.   2.   3.   4.   5.

1) Solid icosahedron
2) Two opposite pentacaps are removed. This reforms the icosahedron into a pentagon anti-prism.
3) Removing the bottom 5 pentacaps leaves a pentagon pyramid. Most of the upper 5 pentacaps have been eliminated by removing the bottom 5.
4) Continue to remove the remaining 5 pentacaps, which results in an invisible 10-sided center plane. Each part removes parts of the others. All parts are interrelated on primary levels of formation.
5) Twelve locations in a spherical pattern remain with the form cut away leaving a relationship of 12 pentagon planes. We see the dodecahedron pattern inherent to the icosahedron.

We must recognize pattern as preformed beyond the range of human perception within a larger context. We do not see pattern but we do see the organization of multifunctional results of duplication of pattern in energy/mass formation. Pattern is not gravity conditioned where as all physical forms and systems are.

As with so many things, appropriate tools extend our understanding beyond what does not exist in our normal range of perception. The word "invisible" directs us to look for what we do not see within what we can see. What is invisible is what is held within and surrounds what is visible. As we look closer and with greater depth of intention in a broader field we will see more.

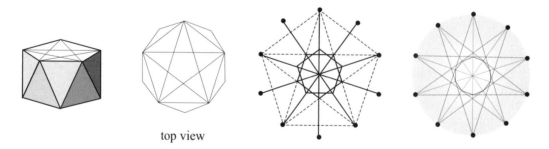

top view

Above. The pentagon plane is a 5/10-fold symmetry held within 10 perimeter points of intersections, again the number for the tetrahedron. Ten is inherent to all 5-point symmetrical division of the circle; 5 are obvious, 5 less obvious. There are 10 triangles around the center band of the icosahedron. the same number in the pentagon star surrounding the center pentagon. The flat pentagon ratio of division is the compressed pattern of spatial 5-10 symmetrical arrangement (1:2) that has origin with the first fold of the circle.

160

Of the <u>Five Platonic Solids</u>, the *tetrahedron*, *octahedron* and the *icosahedron* are the only three regular polyhedra that made with equilateral triangles. These three are primary before the other two; the *cube* formed of squares, and the *dodecahedron* made with pentagons.

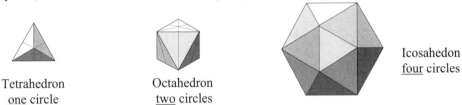

Tetrahedron
<u>one</u> circle

Octahedron
<u>two</u> circles

Icosahedon
<u>four</u> circles

These three polyhedra are from the same diameter circle. There is a very specific rate of expansion of surface and volume. The three primary polyhedra show the beginning sequence of a <u>geometric progression</u> of folded circles; 1,2,4, (1+2+4=7). These three are the only regular polyhedra possible that are absolutely consistent to the first fold, using only equilateral triangles. The individual symmetry of 3, 4 and 5 is primary to the circle. and found in each. These three polyhedra are a higher order of classification within the five regular polyhedron class.

The triangle is primary structure. It has minimum parts, minimum area, minimum perimeter, with maximum strength. It is first before all polygons. One and two is a ratio of division that forms a tri-unity in four points. By observing the different parts of each polyhedron, as a developing progression of three, structural order becomes obvious. Count the surfaces (sides), ends (points) and edges (lines) by touching the parts. Touching generates familiarity, leaving an experiential imprint not possible by only looking.

Tetrahedron	4 sides	4 points	6 edges	→	$4+4+6=14$	$(1+4)=5$	
Octahedron	8 sides	6 points	12 edges	→	$8+6+12=26$	$(2+6)=8$	$5+(2\times8)=21$
Icosahedron	20 sides	12 points	30 edges	→	$20+12+30=62$	$(6+2)=8$	$2+1=3$

The octahedron and icosahedron together is number 7 [8 + 8 = 16, (1 + 6 = 7)]. By adding the 5 of the tetrahedron we get 12. The commonality between 12 and 21 is 3. The difference (21−12) is 9 (3²). By regrouping this information we can see multiple relationships in the pattern of these polyhedra. The tetrahedron is a pattern of 5, traditionally associated with the pentagon, growth, self-replication, and generation. Two fives is ten, the dual tetrahedron, the tetrahedron number of four spheres in order. One diameter and four points make five parts in the first fold. Both the octahedron and the icosahedon are reduced to the number 8. Four is forming of pattern to itself, an octave, consistent to the first fold of the circle (p.36).

The triangles between all three are congruent, coming from the same size circles. The area increases at a ratio of 1: 2: 5. The Golden Ratio is inherent in these numbers, (1 +2 = 3, 3 + 2 = 5, 5 + 3 = 8, ...) It is about growth, and next we have 8 + 5 = 13 (the VE), which is the centered system of spherical order reflecting the 1 + 3 = 4 of the non-centered tetrahedron of spherical order.

<u>Below</u>. The congruent faces of one polyhedron placed onto another and viewed from point symmetry shows the same pattern of lines folded in the 8-frequency diameter division of the circle (p.118).

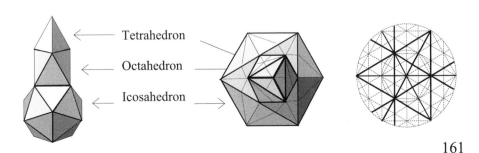

Tetrahedron

Octahedron

Icosahedron

161

5 DUAL POLYHEDRA

STELLATE & STRING TETRAHEDRON

The process of <u>stellation</u> is raising the mid point on a flat plane perpendicular from that plane, forming a pyramid relationship of points. This is first seen in the reconfiguration of three diameters (p.47) and in folding the tetrahedron (p,101). Placing two congruent, regular tetrahedra together face-to-face shows a dual stellation. The altitude of both tetrahedra is a stellation in both directions from the center point of the common triangle plane. This figure is called a *bi-tetrahedron,* two joined tetrahedra sharing a common face, a 6-sided figure with 7 equilateral triangles; a redistribution of 6 around 1 is 7.

Use one tetrahedron as center, stellate the other 3 faces by adding 3 more tetrahedra. (This shows 4 outer points and 4 from the center tetrahedron. Eight is the number of corner points on a cube, sides to an octahedron. There are 5 tetrahedra.)

Connect the 4 end points using tape (p.67). Fold tape over along its length giving greater strength, a thinner line, and no exposed sticky surface.

The *dual* function of the tetrahedron is self-forming, shown by the tape outline of the 4 stellated faces, the edges of the 6th expanded tetrahedron.

The tetrahedron self-generates on varying scales. The first tetrahedron is held within the second. The stellated end points will always be in alignment with each succeeding generation providing a continuous expanding measure.

<u>Below</u>: 1) Tetrahedron inside with end points touching the center of the triangles planes.
2) Points extending beyond planes.
3) Tetrahedra are equal size showing octahedron.
4) Cube pattern from number 3.

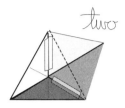

one   two

tape all three sides
when joining

five tetrahedra

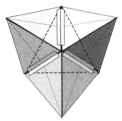

string tape between
end points

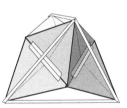

six tetrahedra

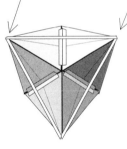

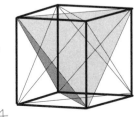

cube pattern

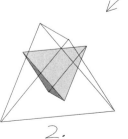

points
touch
faces

1.   2.   3.   4.

Using individual tetrahedra it takes 4 circles joined edge-to-edge to form a stellated tetrahedron. Here is a variation using the triangular net of only 3 circles.

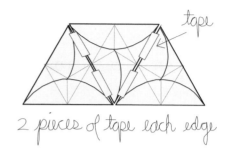

tape

2 pieces of tape each edge

Fold 3 tetrahedra and open to the triangle. Make a linear net joining edge-to-edge. Tape each half-edge length individually as shown to the right. Do not use one long piece of tape; it needs to fold at the midpoints of the joining.

fold edges together

Bring the 2 end points of the long edge of the trapezoid base together. Use two pieces of tape as before to join each half edge individually. This forms a large two-frequency tetrahedron with one open triangular plane.

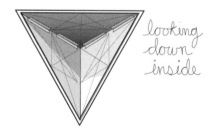

looking down inside

With open side up, bring 2 of the 3 end points together forming a smaller tetrahedron. Tape the edges together. This forms an open quadrilateral.

Move the remaining end point to the center joining, completing the 4 tetrahedra, and tape the edges together. The fifth center tetrahedron exists as the relationship of points and edges, and the 6th tetrahedron is the relationship of 4 end points.

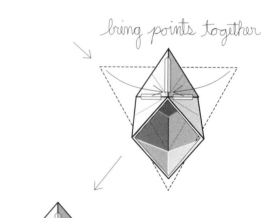

bring points together

Three is more efficient than 5 or in forming the stellated tetrahedron. Each way of modeling the stellated forms are informational to differences. In exploring ways of forming, look for the most economical ways possible using the least number of units. They often reveal transformations that formulate to larger systems.

fold point to center and tape edges together

STELLATE & STRING OCTAHEDRON

<u>Fold</u> 2 half-open tetrahedra and tape together, forming an octahedron (p.129). Make 8 more tetrahedra and tape them individually closed.

Place 4 tetrahedra equally-spaced around the 8 triangular faces of the octahedron, forming a solid two-frequency tetrahedron, reflecting 4 spheres in spherical order (p.173).

Add the remaining 4 tetrahedra, filling in around the first 4, completing the stellation of the octahedron. (This reveals 2 interpenetrating two-frequency tetrahedra sharing an octahedron center. One tetrahedron is pointing up the other pointing down in the opposite direction.) This is 14 spheres in the closest packed order, (p,93).

String all 8 points of the stellated octahedron with tape to see the cube outline. The cube is the dual function of the octahedron. Let's count parts again:

Octahedron	6 points	8 faces	12 edges
Cube	8 points	6 faces	12 edges

The center points of the faces and the number of faces are reciprocal one to the other while the edges remain the same.

The square face is a relationship of tetrahedra. The diagonals of the squares are edges of the two tetrahedra, the structural nature of the cube. This can be made without the octahedron center by joining the bottom edges of 8 tetrahedra.

Put 4 *tetrastars* (stellated octahedron) together in a tetrahedron arrangement, stringing the points.

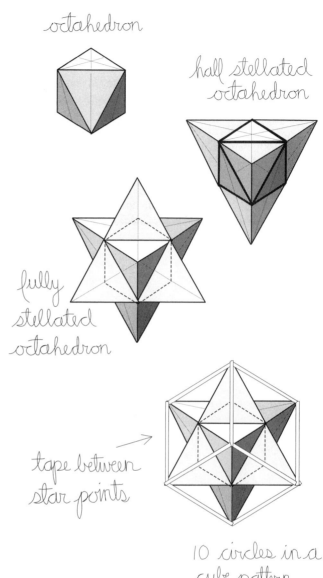

octahedron

half stellated octahedron

fully stellated octahedron

tape between star points

10 circles in a cube pattern

4 cubes in a tetrahedron pattern

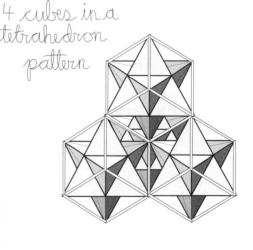

165

STELLATE & STRING ICOSAHEDRON

Stellating the icosahedron in stages reveals interconnections between polyhedra.

Make an icosahedron (p.149) and place 4 tetrahedra on faces equally spaced around the icosahedron. Tape them in place along the edges. The 4-circle icosahedron shows the open planes equally placed, as do the 4 center points of the circles. Stringing the stellated points shows the tetrahedron with an icosahedron center.

icosahedron

tetrahedron

4th tetrahedron behind

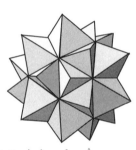

8 tetrahedra

Make 4 more tetrahedra and place them equally spaced around the first four (look for the placement of triangles between the stellated points). String the 8 points. We now have a cube with an icosahedron center (p.159).

cube with icosahedron center

Make and attach 12 more tetrahedra, stellating the remaining faces for a fully stellated icosahedron.

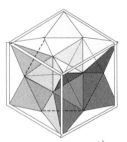

20 tetrahedron, stellated icosahedron

Using tape, string all the points adjacent to each other. The tape edges reveal 12 pentagons, the <u>dodecahedron</u> pattern, as the dual to the icosahedon, reflected in the numbers.

Icosahedron	12 points	20 faces	30 edges
Dodecahedron	20 points	12 faces	30 edges

You might want to use different color tape or colored yarn to differentiate between individual systems when stringing the stellated points. String the 5 cube systems individually to see the compound of 5 interpenetrating tetrahedra (p.167).

Using only tetrahedra, all 5 regular polyhedra, <u>The Five Plutonic Solids,</u> have been formed showing all dual relationships.

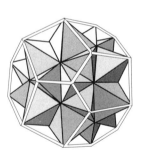

dodecahedron

dodecahedron and cube

166

When both the cube and dodecahedron patterns are strung we see each edge of the cube is a diagonal to each of the 12 pentagon planes. There are 12 pentagons and 12 edges to the cube. Each pentagon has 5 diagonals, meaning the dodecahedron also <u>compounds</u> 5 cubes. Each regular arrangement shows aspects of the others when viewed comprehensively.

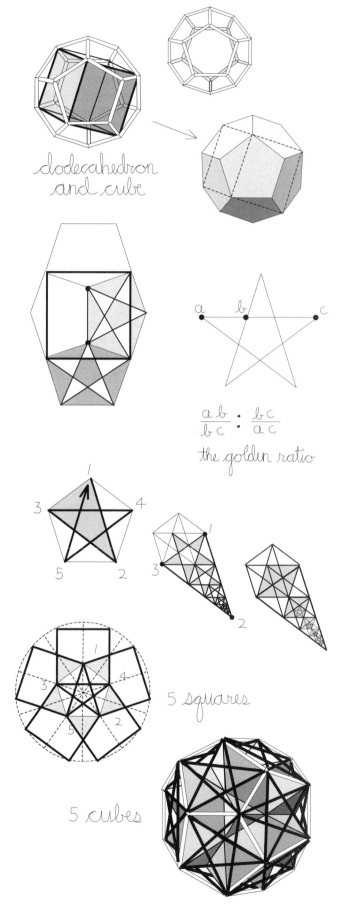

dodecahedron and cube

Each square side to the cube has 4 pentagons that move through it on 4 different planes. These planes form a 2-pointed, 4-sided kind of pyramid. The joining of partial edges of 4 pentagons forms the 2 points.

Five diagonals form the pentagon star. Each diagonal divides 2 other diagonals in a golden ratio, or golden proportion. (*ab* is to *bc* as *bc* is to *ac*) This is a self-referencing growth ratio related to the Fibonacci series (pp.46, 159).

$$\frac{ab}{bc} \cdot \frac{bc}{ac}$$

the golden ratio

The pentagon star is a closed system of 5 intersecting diagonals proportionally to each other in a division of 10 (p.160). The triangle 3, the quadrilateral 4, and the pentagon 5 are all part of the fractal development in the pentagon relationship. The spiral inherent to the tetrahedron is primary to this growth also found in the 9 creases of the tetrahedron folds. This same pentagon growth is observed in the arms of the hexagon (p.113).

The octahedron shows 3 intersecting squares; the dodecahedron shows 5 intersecting cubes. They are pictured here as squares.

5 squares

String all 5 cubes in the stellated icosahedon, each with a different colored yarn or tape. This makes compounding of proportional cubic function in the dodecahedron very clear.

5 cubes

Without triangulation of the tetrahedron in multiplicity (stellation) the cube and the dodecahedron would not exist. They are the outer formation of expansion to an inner tetrahedral movement.

167

Fig.7 Here are the primary 3 regular polyhedra; the *tetrahedron* folded from <u>1</u> circle, the *octahedron* from <u>2</u> circles, and *icosahedron* made from <u>4</u> circles. This is the beginning of geometric progression. Seven circles are used to make the 3 primary patterns of the tetrahedron.

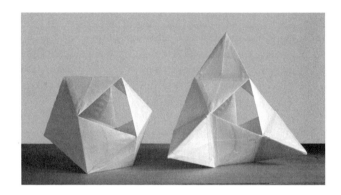

Fig.8 *Left*, The icosahedron is shown with 16 triangle surfaces and 4 open planes. *Right*, When stellated with a tetrahedron on the center triangle of each of the 4 circles a tetrahedron relationship is formed showing the icosahedron center open in 4 tetrahedral locations.

Fig.9 In this picture the 4 open spaces are covered with tetrahedra forming a cube arrangement of 8 corner points. Three intersecting rectangular planes are located by the 3 pairs of opposite edges at right angles to each other (p.159). Here we see the icosahedron positioned inside the cube. To string the 8 points will clearly show the edges of the cube.

Fig.10 The fully stellated and strung icosahedron forms a star configuration. The edge relationships of all 20 points reveal the dodecahedron as dual to the icosahedron. By stringing one of the 12 pentagon planes with yarn the flat pentagon star with all the golden ratio proportions is revealed. The pentagon is based in the icosahedron which is a specific relationship of 4 reformed tetrahedra.

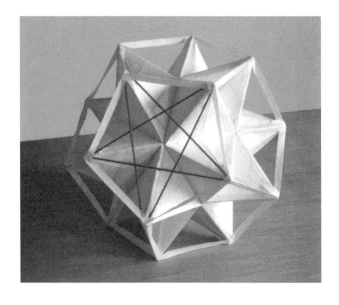

168

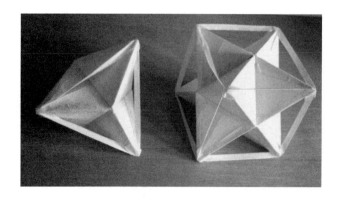

Fig. 11 Here are both the tetrahedron and the octahedron are stellated and strung with tape. They show the dual configurations of the center point and surface planes of the tetrahedron and octahedron. The points are extended perpendicular from the surface. The tetrahedron self generates and the octahedron generates a cube relationship of tetrahedra.

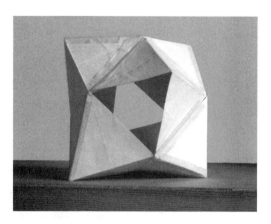

Fig. 12 This models a cube relationship stellated with 6 half octahedron pyramids made using an open tetrahedra for each. There are 2 opposite open triangular planes that are off center to the primary axis reflecting a rhombic organization (pp.234, 363).

Fig. 13 The stellated cube (*Fig.12*) when strung with tape between the 8 end points show the edges of the octahedron.

The point of modeling the Platonic solids in this way is to see the development of the interrelationships between them and how the tetrahedron is parent to them all. It is importance to understand the function of primarily pattern and the forming process of associations and transformations through movement and not just to see them as static individual polyhedron solids.

Fig.14 This model is a fractal development of the octahedron by the stellation process. It is made by folding only tetrahedra starting with 9" diameter paper plates and cutting the diameters down to about 1 ¾ " diameter circles. More are used than I care to count, but clearly this is a process without end. It was started by using 8 tetrahedra forming the stellated octahedron, (p.165) and developed out from there being consistent to the stellation process.

THE FIVE PLATONIC SOLIDS

Tetrahedron   Octahedron   Icosahedron   Cube   Dodecahedron

These pictures represent the Five Platonic Solids, the only five regular formed patterns of spatial organization that we know to exist in the universe. They are traditionally referred to as solids because they were originally modeled in solid materials and used to demonstrate a classification of absolute regularity of shapes and angles not found in any other polyhedra. These solid forms *represent* the only known patterns that are parent to formations on all scales of perceived phenomena. Pattern is not subject to change; non-responsive to gravity and conditional influences on any scale; pattern exists as plan before any formation occurs. It is the physical material formed to pattern that gets pushed, pulled, distorted, and changed.

These solids show three polyhedra made of triangles (3), one of squares (4) and one of pentagons (5). Observing the points, or vertex where planes come together, we find an interesting relationship of numbers. Three sides meeting at a vertex are found first in the tetrahedron, then with the cube, and the dodecahedron. Here two polyhedra formed without triangles, 4 and 5 have 3 inherent to vertex points. Four sides joining at a vertex happens with the octahedron with triangles. Five sides at a vertex occur only with the icosahedron, again with triangles. Four and five are relationships of three. The two polyhedra made with triangles inherently demonstrate the 4 and 5. These two show a distribution of structural pattern in two primary ways, the number of shapes and the way the shapes come together. All five are structural arrangements by virtue of all 5 displaying triangulated aspects. The 3 joining of planes is in the tetrahedron (3), the cube (4), and the dodecahedron (5). The 4 and 5 joining of planes is with the triangles (3). Again all parts show different relationships of the same pattern, each displaying individual symmetries and unique properties of form, serving as sub-patterns for further down-stepping into greater complexities of design. The tetrahedron is principle, preceding the other four.

Counting the number of sides of each they are all even numbers; 4, 6, 8, 12, 20, and. (sum is 50)
The numbers of points are all even; 4, 6, 8, 12, 20 . (sum is 50); same as the sides. The numbers suggest that points and sides (planes) are different forms of a reciprocal function.
The numbers of edges are all even; 6, 12, 30, 12, 30. (sum is 90) (6+12+12)=30 [3x30=90]
[(2x50)+90=190] 1+9+0=10, the tetrahedron; the primary pattern formation in first fold of the circle.

There are other ways to classify these five regular polyhedra (p.175), and other relationships between the numbers that represent the physical properties of each individual polyhedron (p.172). Below is the arrangement of points that corresponds to the images on the top of the page.

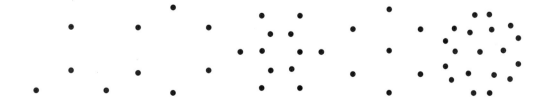

When counting the number of planes, points, and edges of each of the 5 regular polyhedra, we have seen some interesting number relationships appear. Let's go further with the numbers.

	faces	points	edges				
Tetrahedron	4	4	6	→	4 + 4 + 6 = 14	1 + 4 = 5	
Octahedron	8	6	12	→	8 + 6 + 12 = 26	2 + 6 = 8	
Icosahedron	20	12	30	→	20 + 12 + 30 = 62	6 + 2 = 8	
Cube	6	8	12	→	6 + 8 + 12 = 26	2 + 6 = 8	
Dodecahedron	12	20	30	→	12+ 20 + 30 = 62	6 + 2 = 8	

4 x 8 = 32   3 + 2 = 5

In the *reciprocal* function between the octahedron and cube, the icosahedron and dodecahedron, there is an inverse relationship between number of points and faces while the number of edges remains consistent for each set. The face-centered points are reciprocal with the end points. All points are center points and subject to changing shape and scale and the edge lines show the movement of change. The numbers of connecting edges don't change; they appear in different places defining form change. Numbers and shapes are merely symbols representing extraordinary transformations of boundaries in-forming spatial pattern.

In adding the numbers of each polyhedron down to a single digit we see another number relationship. The tetrahedron shows the number 5. The two sets of the other four display a wonderful symmetry of 26 and 62, each reducing to 8. The sum of all four 8s gives us 32. Adding 3 and 2 we get 5, the over all number quality of the group of 4 regular polyhedra. These four collectively reflect the tetrahedron from which they are derived. The reciprocal numbers of the octahedron/cube are 6+8=14, 1+4=5; and of the icosahedron/dodecahedron they are 12+20=32, 3+2=5. While showing different symmetries they are the same number pattern and, combined, reflect the 5 of the tetrahedron. The tetrahedron being dual to itself is 2 times 5; 10 being the number of the tetrahedron. The 5/10 (1:2 ratio of first fold) is the symmetry of the pentagon and associated with proportional growth. Folding circles and the numbers reveal relationships that are closer than what is acknowledged by traditional classification that tends to separate based on comparing external descriptions.

These number connections reflect the relationships of the development of multiples of the tetrahedron from which the 4 other regular polyhedra are generated. There are three made of triangles, one of squares, and one of pentagon faces. The 4 and the 5 are inherent within the 3, the tetrahedron. (3+4+5=12, 1+2=3) Without the three there is no 4 or 5. Three is structural pattern, principle and first condition.

The tetrahedron is the only polyhedron that generates itself. The dual comes in a different size and orientation. The other four polyhedra are not self-generating, they are reciprocal functions of different forms generated by tetrahedra. There are three primary patterns and three secondary. By count there are six regular polyhedra; two tetrahedra and the four other. This does not deny the five regular polyhedra classification; it only gives greater meaning by recognizing the difference in the tetrahedron dual and understanding the tetrahedron is parent to itself and therefore to the other four.

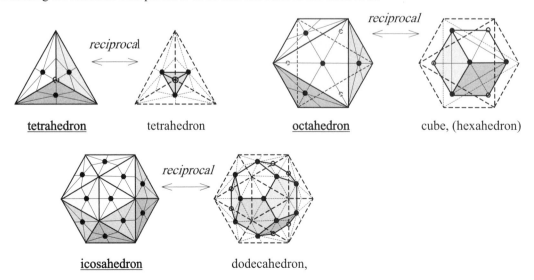

tetrahedron tetrahedron octahedron cube, (hexahedron)

icosahedron dodecahedron,

172

These regular polyhedra represent spherical locations of tetrahedral movement within spherical order. The spherical equivalent of a 2-frequency tetrahedron is, in the closest packing of spheres, 20 spheres. There are 5 tetrahedral units of 4 spheres each. There are 4 units of 4 closest packed around a center unit of 4. (4x4)+4=20 It is a non-centered arrangement of spheres.

<u>Below</u> Four spheres touching each other traditionally represent the tetrahedron pattern. This forms a minimum spherical pattern of a two-frequency tetrahedron. We tend to want to define the tetrahedron by disregarding the six points of connections and counting only the four center points of the spheres. By counting only center points we throw away more than half of the point locations and most of the information. There is no spherical equivalent to a single tetrahedron of four points.

When 4 tetrahedral units of 4 spheres are put together in the same way, to a higher frequency order the space for another tetrahedron of four spheres is formed. The midpoints of touching spheres form the octahedron pattern but the space holds exactly 4 more spheres in a tetrahedron arrangement. This fifth tetrahedron will be in opposite orientation to the other 4. The four-frequency tetrahedron (6 tetrahedra) show a corresponding spherical order of 20 spheres. This is similar in number to 16 triangles (4 circles) that form a unit of 4 intervals, making 20 planes of the icosahedron (p.145). This five-set tetrahedron packing is reflected in the number 5 by counting up the number of points, lines, and planes of the tetrahedron on the previous page. Five reflects the number of tetrahedron in a stellated tetrahedron (p.199).

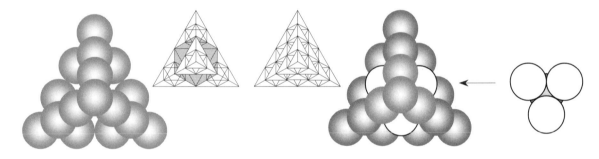

arrangement of four tetrahedra   arrangement of five tetrahedra

Polyhedral forms are linear expressions of the multifunctional nature of spherical order. The relationship of twenty spheres in the closest packing of spheres when shown in a drawing looks like a six-frequency triangle where the opposite facing triangles are the octahedral spaces. That is a scaling process where each frequency level shows a changing up and down of opposite facing triangles (p.101). There are no preformed polyhedra. It is all spherical pattern. The sphere is about the Whole and polyhedra are parts of the Whole. There is intelligence in spherical order that shows only three regular, non-centered polyhedra patterns, the tetrahedron, octahedron, and cube; and one centered system, the VE. They are all revealed by straight-line connections of center points and tangent points, all occurring within the multiplicity of the sphere. Through spherical shifting does the icosahedron come into being (pp.100, 156).

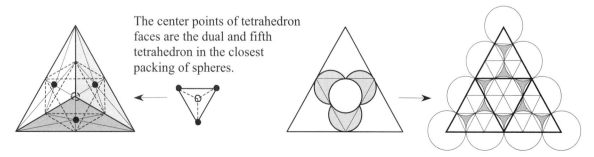

The center points of tetrahedron faces are the dual and fifth tetrahedron in the closest packing of spheres.

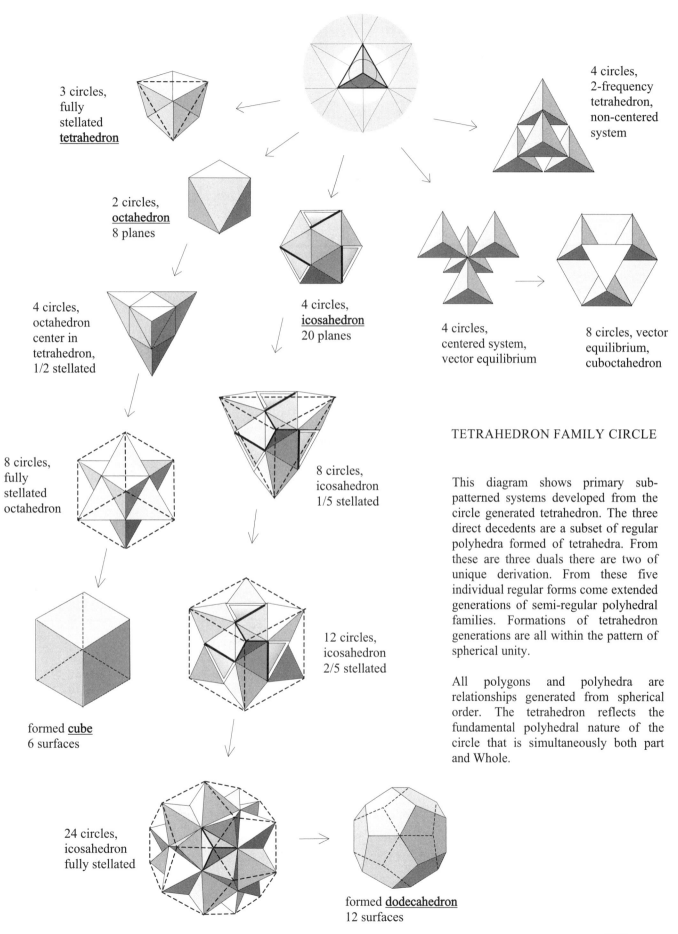

3 circles,
fully
stellated
tetrahedron

4 circles,
2-frequency
tetrahedron,
non-centered
system

2 circles,
octahedron
8 planes

4 circles,
icosahedron
20 planes

4 circles,
centered system,
vector equilibrium

8 circles, vector
equilibrium,
cuboctahedron

4 circles,
octahedron
center in
tetrahedron,
1/2 stellated

8 circles,
fully
stellated
octahedron

8 circles,
icosahedron
1/5 stellated

12 circles,
icosahedron
2/5 stellated

formed **cube**
6 surfaces

24 circles,
icosahedron
fully stellated

formed **dodecahedron**
12 surfaces

TETRAHEDRON FAMILY CIRCLE

This diagram shows primary sub-patterned systems developed from the circle generated tetrahedron. The three direct decedents are a subset of regular polyhedra formed of tetrahedra. From these are three duals there are two of unique derivation. From these five individual regular forms come extended generations of semi-regular polyhedral families. Formations of tetrahedron generations are all within the pattern of spherical unity.

All polygons and polyhedra are relationships generated from spherical order. The tetrahedron reflects the fundamental polyhedral nature of the circle that is simultaneously both part and Whole.

175

These thirteen semi-regular polyhedra are classified according to the shapes and symmetries derived from a method of <u>truncating</u>, systematic cutting apart the five Platonic Solids. Traditionally this is the second level division into the primary "solid" polyhedra. The drawings include the number and shapes of faces that describe the outer appearance. The apparent value of these figures lies within the parent source.

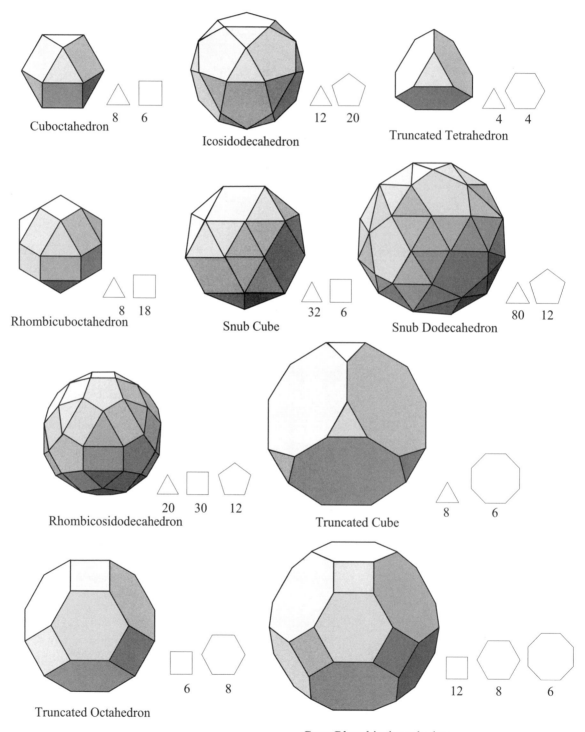

Cuboctahedron   8   6

Icosidodecahedron   12   20

Truncated Tetrahedron   4   4

Rhombicuboctahedron   8   18

Snub Cube   32   6

Snub Dodecahedron   80   12

Rhombicosidodecahedron   20   30   12

Truncated Cube   8   6

Truncated Octahedron   6   8

Great Rhombicuboctahedron   12   8   6

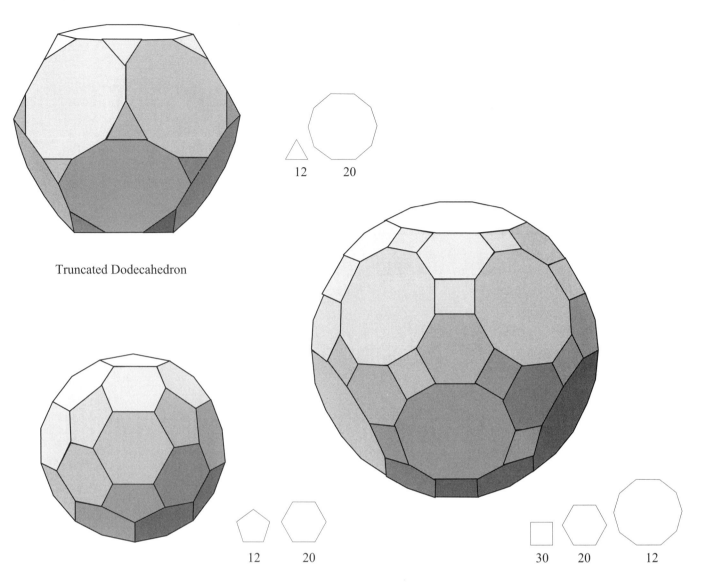

Truncated Dodecahedron

Truncated Icosahedron

Great Rhombicosidodecahedron

This size triangle has been used as the common measure for drawing these images to show the relative proportional size between them. Cutting the corner points from a triangle makes a hexagon, from 3 sides to 6. Doing the same to the square forms an octagon, from 4 to 8 sides, and from 5 to 10. Each cut decreases the edge lengths. In the same way, systematically cutting off the end points of solid polyhedra will decrease the size of what remains. Here the inverse is shown where the edge lengths stay the same, and the division yields polygons with greater perimeters that result in larger scale polyhedra proportioned to the same measure. Division is without scale and moves in both directions, in to and out from. This kind of transformational division can take place only if there is a pre-formed, unchangeable pattern to begin with. Division is a multiplication function principled to the Whole.

All numbers of faces of each polyhedron are even. There is only one polyhedron that incorporates all three primary faces; the triangle, square, and pentagon; 3, 4, 5, making it unique. The octahedron is both a truncated tetrahedron, and is regular. Some are truncated from two different polyhedra, others are not. Since these 13 polyhedra are derived from polygons formed out of the tetrahedron, then we assume there is that pattern inherent in each that relates them all beyond individual properties. When classifying we often overlook importance differences in order to fit a generalized definition. Understanding evolves and expands while classifications tend to remain static and resistant to change.

The number and shapes of polygons that make up each semi-regular polyhedra show individual symmetries consistent to the 3, 4, and 5 of the three diameter circle. There is a direct correlation between the evolving from points to planes that creates more points. This is clearly demonstrated in the reciprocal functions of relationships between the six regular polyhedra; the reciprocal function of the tetrahedron to itself, the octahedron to cube and the icosahedron to dodecahedron. This reflects the duality of the circle (2 sides). The six and eight in the truncated cube reflect properties of the cube where the six squares are six octagons; the eight points are eight triangles. This shows an increase in frequency of the square to an interval of twice the number of sides, keeping the same edge length. When looking at the great rhombicuboctahedron the eight triangles have changed to eight hexagons (another increase in frequency and not edge length) and have generated 12 squares (relationships of triangles). With the truncated octahedron the six squares remain the same but the eight triangles have opened to become eight hexagons. We see openings occur creating new intervals, adding complexities to the regularity of pattern generation. Division within the tetrahedron is expressed in outward reformation of division into all the regular solids and by extension to semi-regulars.

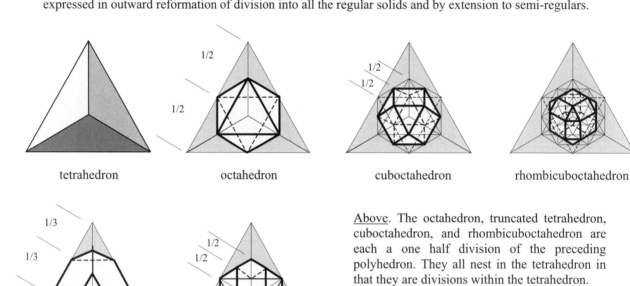

tetrahedron   octahedron   cuboctahedron   rhombicuboctahedron

truncated tetrahedron   truncated octahedron

<u>Above</u>. The octahedron, truncated tetrahedron, cuboctahedron, and rhombicuboctahedron are each a one half division of the preceding polyhedron. They all nest in the tetrahedron in that they are divisions within the tetrahedron.

<u>Left</u>. The truncated tetrahedron is a one third division of the tetrahedron edge length. The truncated octahedron is one half division.

<u>Below</u>. The same process of division is within the cube and with the five-fold symmetry. None of these individual polyhedra are separate from each other. They all have common root in the duality of the tetrahedron and share the unity of the circle/sphere.

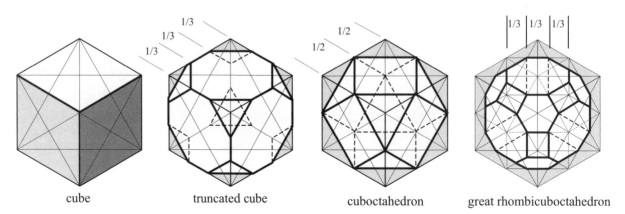

cube   truncated cube   cuboctahedron   great rhombicuboctahedron

178

Pictures of polyhedra are visually confusing. The best way to read them is through hands-on experience. Squares not parallel to the picture plane get compressed to rectangles and rhomboids; congruent edge lengths and angles change. The numbers of parts change becoming compressed and hidden behind others. The more hands-on experience we have with fundamental spatial forms, the easier it is to read the images and understand the abstract conceptualizations about them. Young students, we all, need the grounding of a physical medium to inform the mind.

It is necessary to expand the context of traditional classifications in order to develop greater meaning. Recontextualizing what is known with what is not known opens the mind towards greater truth. With it comes a higher sensitivity to beauty and an increased appreciation of derived goodness. The circle provides simplicity of known and unknown from which to expand understanding about the connections between various classifications of polyhedra. The circle forms one tetrahedron. Two tetrahedra form an octahedron. Four tetrahedra form into an icosahedron. Theses are 3 primary directives of spatial organization through circle/sphere form changing. By adding more tetrahedra from the center out, the relationships of the cube and the dodecahedron are formed. The Five Platonic solids are codified by a regularity of shape and angles. That is a good start, but there is so much more to explore.

Similar to opening in flowers, massing of bubbles, and growing crystals, all forms in nature reveal what is hidden through movement and transformation. The opening of points (circles/spheres) to planes, and planes that divide into more planes generating more points, reciprocal into and out from movements show the re-arranging of interrelated systems. What appears not to exist is pre-existent and most often hidden within our simple conceptions about it. It is an extraordinary thing to be able to simply fold the circle into a tetrahedron.

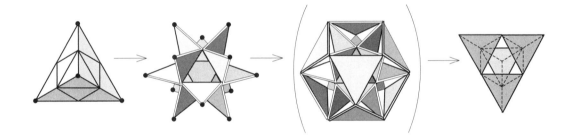

Above: The transformation of a solid 2-frequency tetraheron insideout forming an open 2-frequency tetrahedron. Four points of a 2-frequency closed tetrahedron open becoming 12 end points that form 8 triangle planes and 6 square planes of the vector equilibrium. A total of 14 octahedra and one tetrahedron have been reorganized by this movement. The non-centered tetrahedron system has opened to form the primary centered system without the center. Stability lies with the 4 triangles that form the 3 intersecting squares defining the octahedron. (In the third drawing 8 triangle planes have been added, leaving the squares open, to show a more recognizable form of the vector equilibrium. This addition would stablize the system preventing any further change.)

Without adding the 8 triangles, the opening direction of the tetrahedron now turns into itself closing and reforming into 4 tetrahedron with the octahedron center open. The entire outside surface of the 2-frequency tetrahedron has become totally hidden within the 4 tetrahedra of the 2-frequency tetrahedron in opposite orientation from where it started. This reformation goes beyond traditional transformation of individually separated polyhedra. It is possible because the pattern form of the tetrahedron creates its own dual. (see pp.206-7 for folding this model.)

SYMMETRY

The absolute sphere is absolute symmetry.

Symmetry is a self-referencing system of balanced movement. The first fold of the circle is symmetrical, a dividing movement forming a duality in the ratio of 1:2. The folded diameter is a line of symmetry between the two parts are without separation; the same in shape, size, and moving around the greatest length.

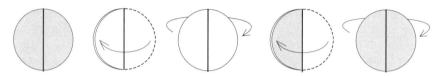

360° axial symmetry. There are 4 positions: congruent, front/ back and in/out.

Individual symmetries are useful for identifying and classifying movement patterns of congruent properties, establishing sets and subsets of interrelated similarities. In that way symmetry is a good counting tool. With one complete rotation around a given axis, the number of similar occurring positions is the designation of the symmetry of a given figure. The sphere has an unlimited number of axes with an unlimited number of similar positions. Symmetry is about movement and it is all held within the spherical Whole. The first fold in the circle shows only two similar positions around that axis, front and back.

Seven axes of the tetrahedron

The regular tetrahedron has _four axes that go from a corner point to the center of the opposite triangle face._ There are _three axes that go from the mid-point of edge to the mid-point of the opposite edge._ They all intersect through the center of the tetrahedron. These seven axes show the symmetry of the tetrahedron. This reflects the seven points of hexagon pattern, the most possible number of associations between three.

These 4 axes from point to face have three different positions of rotation each.

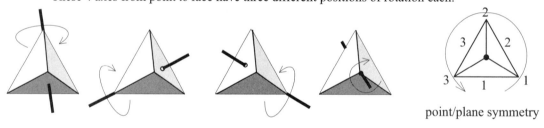

point/plane symmetry

When the tetrahedron is rotated a full turn around the axis there are six positions of symmetry; three along the edge line and three on the triangle face. In each symmetry three positions are indistinguishable one from the other. The beginning and ending of all circle forming is spherical movement.

The three axes that run through the midpoints of edges, _dihedral_ axes, can be rotated to only two positions where they both look the same. This 2-fold rotation happens two times, starting with triangles pointing up and again starting with triangles pointing down. These three axes run through the six points of the octahedron, and are shared by both tetrahedron and octahedron.

line symmetry
with axial end
point

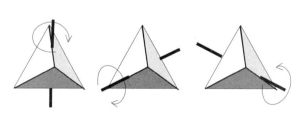

7 axes defined in 3
diameter/bisectors
of the circle/triangle

180

Thirteen axes of the octahedron

The octahedron pattern has three axes of point symmetry with four separate positions. It also has four axes through the center of each pair of triangle faces with six positions of rotation. There are six axes through the midpoints on each edge with two identical positions in rotation. There are a total of thirteen axes in the octahedron. Thirteen reflects the primary centered system of the closest packing of spherical order.

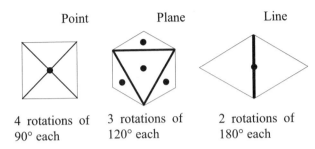

Point	Plane	Line
4 rotations of 90° each	3 rotations of 120° each	2 rotations of 180° each

Left. A picture of the square shows the point axis; in the hexagon the points in the triangles are the center axes of the planes; the rhomboid shows the midpoint axis of the edge line.

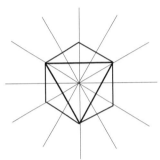

Right. All thirteen axes of the octahedron are shown compressed into six diameters showing twelve equal divisions of the hexagon. The center point is the thirteenth axis. The octahedron has a 4 axis 3-fold (2 times, pointing up and down), 3 axis 4-fold, and a 6 axis 2-fold symmetry. The octahedron has a 3, 4, 2-fold symmetry.

Thirty-one axes of the icosahedron

The icosahedron has 31 axes (notice the reversal of digits in the 13 axes of the octahedron). There are twelve points showing six end point axes, twenty triangles with ten planer axes, and thirty sides having fifteen mid-edge point axes. The icosahedron has a 6 axial 5-fold (twice, up and down), 10 axis 3-fold (twice), and a 15 axis 4-fold symmetry. The icosahedon has a 5, 4, 3-fold symmetry.

Right. The six axial points have ten different positions because of the ten alternating triangles around the middle band of the form. Start with a triangle pointing up and there are five positions in a single revolution. Start with a triangle pointing down and there are five in the same way. This happens similarly with the plane axis of the octahedron and the midpoint edge axis in the tetrahedron. The line with the tetrahedron, the plane with the octahedron and the point axis with the icosahedron show the double rotation, one with each of the three primary regular polyhedra.

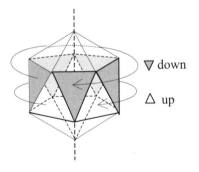

▽ down

△ up

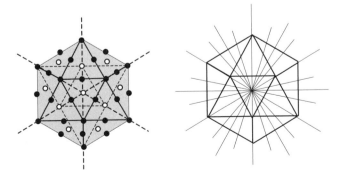

Thirty-one axes are shown to the far left as they appear in the icosahedron. The near left shows all thirteen diameters, in a two-dimensional compressed form. Again the center point is the end view of the thirteenth axis.

181

The tetrahedron reveals symmetries basic to all polyhedra. The octahedron and icosahedron can be thought of as symmetrical elaborations on the tetrahedron similar to the way the cube and the dodecahedron are reformed symmetries of the octahedron and icosahedron; all founded in the nature of triangulation.

There are as many combinations of symmetrical as there are forms of polyhedral complexities. All are enumerations of point, line, and plane axial rotation. The tetrahedron shows two axes of symmetry that reveals three rotational forms. The first axis runs through the point and opposite triangle plane. In combination we see point axis and planer axis that in the octahedron and icosahedron they are two. The second axis goes through the midpoints of opposite edges. This axis shows two positions of triangles, two facing "up" and the other two facing "down". The single rotation is two different systems of axial symmetry, equal in opposite positions. There is a dual function of the center band between the two axial poles, particularly with the planer axis of the octahedron and the point axis with the icosahedron.

The tetrahedron point/plane rotation is 120° with each turn; three times from face-to-face returns to the starting position. These three positions are indistinguishable from each other. The dihedral axis, edge midpoint, is two positions, 180° each rotation. The 360° division of the circle is the context for symmetrical calculations for all polyhedra.

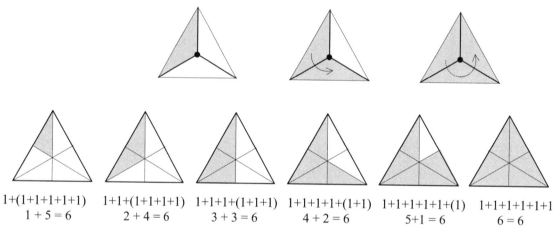

$$1+(1+1+1+1+1) \qquad 1+1+(1+1+1+1) \qquad 1+1+1+(1+1+1) \qquad 1+1+1+1+(1+1) \qquad 1+1+1+1+1+(1) \qquad 1+1+1+1+1+1$$
$$1 + 5 = 6 \qquad\qquad 2 + 4 = 6 \qquad\qquad 3 + 3 = 6 \qquad\qquad 4 + 2 = 6 \qquad\qquad 5+1 = 6 \qquad\qquad 6 = 6$$

Above. When counting one right-angle triangle as a sub set and adding in sequential rotation there is a symmetry that develops in the accumulative shifting that forms a reciprocal function of complimentary angles.

The first triangle can be given in different expressions of relationship of one right triangle to five others within the equilateral triangle. Each of the progressive relationships above can be described in differing set combinations of small right triangles. The equilateral triangle is a single set that shows various grouping, or sub-sets of six parts.

Below. There are other sub set combinations of both right hand and left hand right triangles contained within the divisions of the equilateral triangle. It is often in the exploration of varying combinations of sub sets that design motifs are discovered and further developed.

Point, line, and plane symmetrical movement is often represented in images and algebraic graphing functions using *rotation*, *reflection* and *translation* to refer to movement from one location to another. This is easily demonstrated using the 2-frequency equilateral triangle where any one of the corner triangles can move to any other triangle location through three specific movements of symmetry and combinations.

Point rotation
turn

Line reflection
flip

Plane translation
slide

These are virtual movements. The real movement is in shifting our minds' focus of location. Nothing of the divisions on the paper moves. This movement can happen in various combinations. The rotation position of one triangle to another position can be a double flip, or a turn and flip, or flip and turn, or a slide movement. The line position can be a slide and flip, or a turn, or a turn and flip. The plane position can be a turn, it can also be a double flip, or a flip and turn, depending on the path you want to take. These three movements are fundamental to the development of planer tessellations because they reference basic patterns of movement in space. They can be useful in understanding the division and design of the surfaces of symmetrical polyhedra that look asymmetrical. This is important for understanding transformational functions in graphing movement and change. The value is in understanding the movement process, not the location, because there are many ways of getting from here to there, and anywhere else. (Physical moving triangles can be done by folding circles into the flat shapes and playing with them, as we do pattern blocks, to demonstrate the various combinations of planar movement functions).

The first fold of the circle is a reflect movement. The next two folds to divide the folded circle into thirds is a combination of rotation and reflection. All folding and moving of the circle is reflective and rotating movement, there is no transversal movement. The sliding of planes from one position to another requires multiple parts on a surface plane. The movement of multiple circles, folded to at least three diameters, as a part of a larger matrix of circles can well demonstrate translation by moving folded circles from one location to another, keeping to the creased grid (p.70).

<u>Right</u>. *Bilateral symmetry* is the first fold in the circle. It is a division of the circle plane where the diameter is also the bisector. Each semi-circle is congruent to the other, yet opposite to each other when opened flat. This reflective movement happens three times in the hexagon where the three diameters serve as multi bilateral bisectors for the equilateral triangles and for each other.

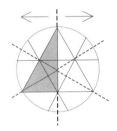

Inverse symmetry usually refers to two lines going through the same point of intersection. This is another function of three diameters in the circle. When an image moves through a focus point and emerges reflected upside down on the opposite side, it is a spherical function. The inverse can also be seen as a rotational movement on the circle plane, but not as a single flip or slide.

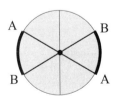

The inverse movement is modeled in the VE where any two opposite tetrahedra spaces are connected through the center point defined by three of six diameters. The tetrahedron can turn, flip, and slide to all other positions in the VE (p.89). This is all regulated by spherical order compressed into the circle where symmetry is revealed through polyhedral reformations of divisions by folding.

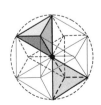

183

SPIRAL

The properties of a circle do not change with size. By drawing *concentric* circles space is generated between boundaries of differently measured, parallel circumferences. The proportional difference is critical to discerning specific movement between boundaries. Any point can be center to the outward expanding boundaries of circles (pp.51, 106-7). The concentric movement out is at right angle to the direction of compression. Changing scale is movement going into or out from circle defined boundaries. To move through concentric locations creates a spiral movment which is always relative to local boundaries.

The three axis division of the sphere is the six division of the circle. This provides proportional continuity through all scaling of compression. In drawing concentric circles we can imagine a regular curved path drawn between each circumference that corresponds to the movement from one hexagon section to another. This forms a continuous spiral path between the relative boundaries of the concentric circle compression. The spiral becomes an open circumference connecting through scale of endless movement in both directions from any location. The openness and closedness of the spiral path depend on how many sixths of the circle are traveled before reaching the next circumference boundary. In this way the specific angle of opening between diameters of the spiral is determined by the path length between measured boundary segments and the distance between boundaries.

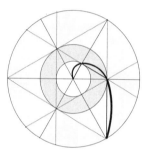

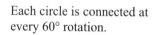

Each circle is connected at every 60° rotation.

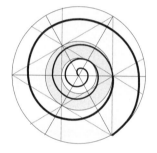

Each circle is connected at every 420° rotation, a full circle plus sixty degrees; seven intervals.

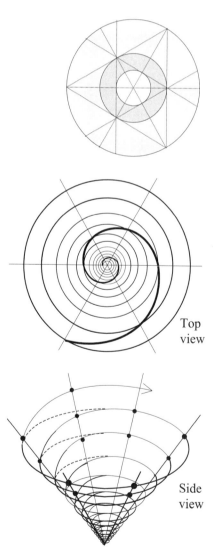

Top view

Side view

We can plot this spiral movement in space by showing each intersection between the diameters and circles as a transition point. Each point can be given a number. This is similar to giving the end points of the first folded diameter a letter A and B (p.80). The spiral is infinite, but for us it has a specific starting point at any given intersection and then the points are numbered consecutively out in both directions. (See following pages.) We will use zero to reference the starting point in both in and out, positive and negative. Everything is in the circle of compressed space, which makes point-by-point location a logical way to reference the spiral movement through this space. The movement is linear and numbers are a good method of tracking the circle movement through a spiral cone-shaped path.

From zero point —located at any radial/circle intersection— we number the intersections in sequence around the circle moving out each ring at a time. The numbers going in towards the center we call negative; those going out from the center we called positive. Absolute zero is the balance, the unity between them is our starting location. There is nothing inherent in the numbers or position that make them positive or negative. It is only in our agreement to relative direction that we find a common logic. Similarly the outside surface of a folded circle is only in relationship to the surface folded in, which is equally reversible. The point of origin is always the reference to differentiate between relative positions in what is without origin.

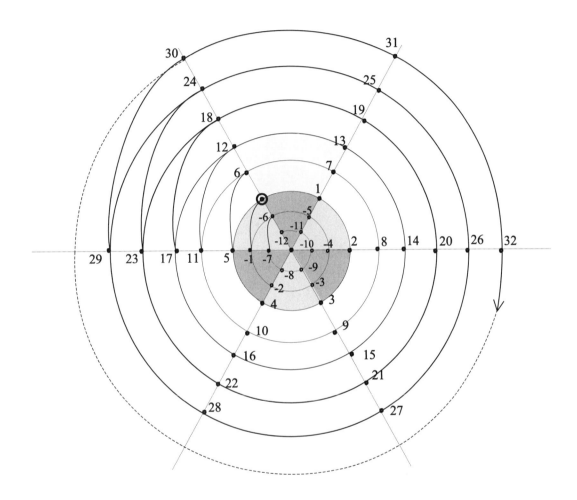

This circle with three diameters gives us a base 6 system for counting. That is reflected on each of the radii where each number changes by intervals of six. If we did it with four diameters it would show a base eight number system, or with five diameters it would be a base ten system. There are direct correlations of these different number based systems to the three, four, and five symmetries found within the hexagon nature of the circle/sphere. The circle can be divided into various base systems of positive and negative numbers (p.119).

This all seems abstract to the actuality of folding three diameters into the circle and reforming the surface by decreasing the angle less than 360°, forming a spatial configuration that grows in multiples by increasing, or decreasing sequentially each circle diameter. These drawings express a generalized form of concept about the spiral movement in the same way the folded circle is the individual expression of the same process. Both describing similar functions in different forms (pp.277, *Fig.*34-5; 278, *Fig*37; 282, *Fig.*49).

Count <u>OUT</u> the numbers starting from zero around the open circumference. Add the digits of each double-digit number. A repeating pattern in sets of 3 is formed. The same numbers are repeated on the opposite half of the diameter but in a difference sequence. Do the same on each of the other two diameters. Add the three numbers in each set finding different relationship of numbers on each diameter. The opposite half of each diameter has the same numbers with each diameter having its own number sequence; 3, 6, and 9. These sets of three numbers reduce further to 3, 6, and 9, which, with further compression, leaves of course 9. Not coincidently we have started with the circle and end up with the number nine.

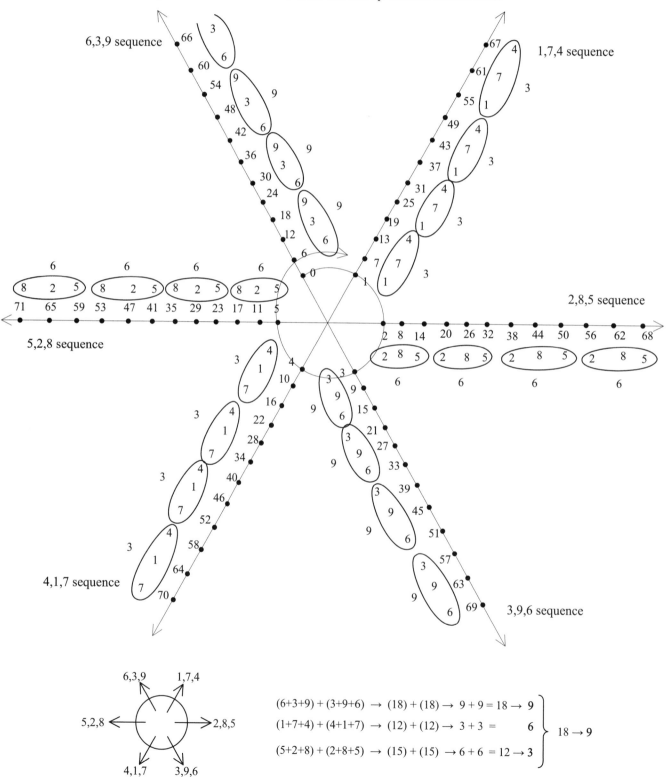

$$(6+3+9) + (3+9+6) \rightarrow (18) + (18) \rightarrow 9 + 9 = 18 \rightarrow \mathbf{9}$$
$$(1+7+4) + (4+1+7) \rightarrow (12) + (12) \rightarrow 3 + 3 = \quad\;\; \mathbf{6}$$
$$(5+2+8) + (2+8+5) \rightarrow (15) + (15) \rightarrow 6 + 6 = 12 \rightarrow \mathbf{3}$$

$$18 \rightarrow \mathbf{9}$$

186

Count **IN** starting from zero point anywhere on the continuum. Add numbers and group, in sets, the same as in counting out. Observe the differences and similarities.

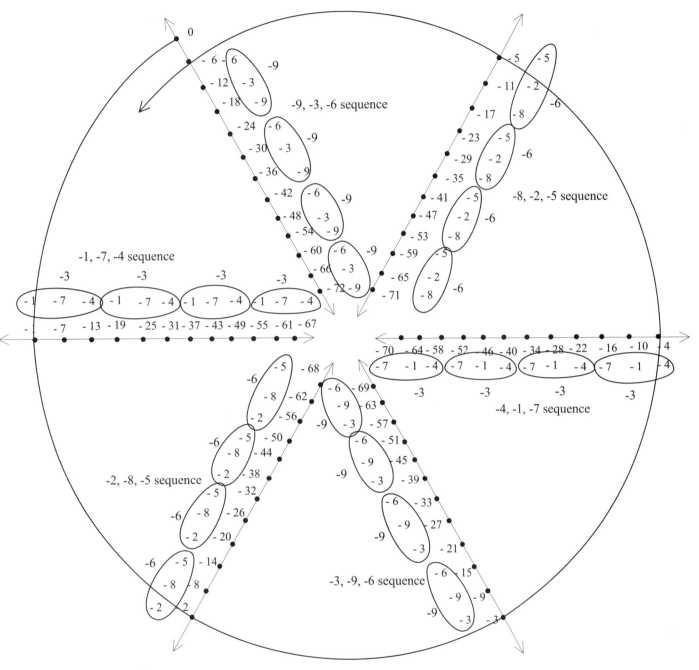

Each diameter shows the same numbers but in different relationship to each other depending on which side of the circle they represent and local point of origin.

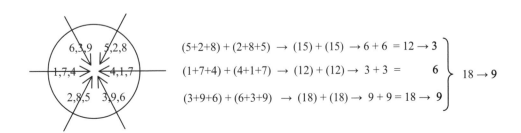

$(5+2+8) + (2+8+5) \rightarrow (15) + (15) \rightarrow 6 + 6 = 12 \rightarrow \textbf{3}$

$(1+7+4) + (4+1+7) \rightarrow (12) + (12) \rightarrow 3 + 3 = \qquad \textbf{6}$

$(3+9+6) + (6+3+9) \rightarrow (18) + (18) \rightarrow 9 + 9 = 18 \rightarrow \textbf{9}$

$18 \rightarrow \textbf{9}$

187

Going in and going out are the same numbers, the same combinations. All numbers being reduced to a single digit turn out to be the last number in the primary single digit sequence. Starting from a circle it ends at nine. The nine of the spiral is the same nine that numbers the creases to form the tetrahedron which itself is spiral in nature (p.108). It is the only polyhedra dual that does not change going both into and out from itself through all scaling of measure. The tetrahedron generates an opposite orientation with each rescaled dual that moves consistently in a spiral path. Through infinite scaling the hexagon pattern of three diameters is a function of spherical pattern.

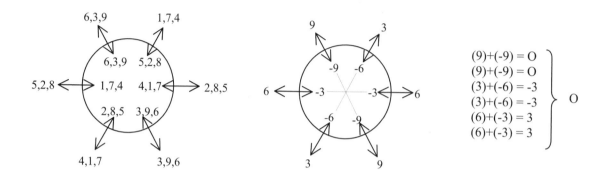

It all adds up to where we started. The fullness of the circle is the compression of the Wholeness of the sphere. IN-moving is towards unqualified Wholeness and OUT-moving is towards infinite expression of unqualified potential. Duality is not the expression of separation; rather it is the symmetry of structural progression towards unqualified realization.

Spiral paths, both 2 and 3-D, are seen with many different degrees of opening in circular movement, often calculated in angulation of radial movement. For some the angle is consistent and the length or size of segments changes. In others there is a progressive increase/decrease in the angle opening. Spirals happen in relationship to the complexities of their context and the conditional constraints, they are not separate from the conditions that form them. The form of any spiral is in the dynamics of a moving balance between tension and compression which is a natural function of expanding and contracting growth (pp..304, *Fig.87a-c*, 383, *Fig.132a-d*).

The spiral moves or grows through different scales in different forms where movement is the stability. The spiral is a patterned movement that can take any form or combinations of forms from any location.

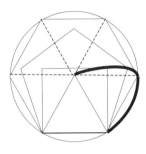

The triangle, square, and pentagon, when sharing the same edge length of the hexagon inscribed in the circle, reveal an interesting curving path from the circle into the center location. The curve connecting end points is irregular and reflects the difference between 3, 4, 5, and 6. More sides to the polygon, 7, 8, 9, etc. having the same measure will fall outside of the circle, moving the center, and the spiral will then fall towards the base line.

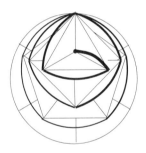

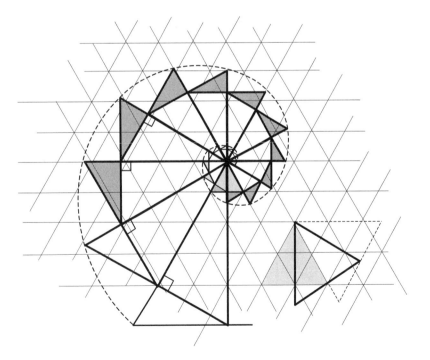

Above. In folding 3 diameters through 6, 5, 4, to 3, the circumference moves in as the center moves off of the plane. (p.74 #7) This is a spiral function of 3 diameters showing right angle movement connecting 3, 4, 5, within the circle 6. It relates to stellation and cone functions.

Above. The hexagonal grid holds compressed spiral information. The triangle diagonal extends to form the side of another triangle where the side of the first triangle is extended, becoming the diagonal to the second triangle. This right angle movement out is the same movement in but differently formed. The smaller extended right triangle is the measure to follow inward. The angle sequencing is based on 12, the rotation of 3 diameters on 6. The out-facing base lines of the triangles are all parallel to the lines on the grid folded into the circle. Each progressive equilateral triangle has a different internal divisional organization reflecting the same grid.

The tetrahedron viewed from the edge shows a square that generates another proportional spiral. The square also shows a right angle triangle off of the diagonal. It functions in this four-fold symmetry in the same way we see in the equilateral triangle. There is a square root proportional growth ratio between the changing length of the *hypotenuse* and the fixed measure of the leg of the right angle.

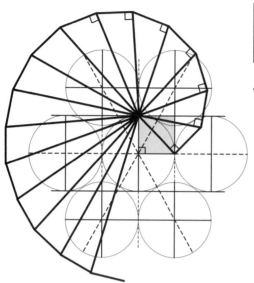

  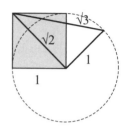

The square is a flat tetrahedron

<u>Left</u>. The right angle grid is a function of the hexagon pattern and the spiral out of the edge length and diameter relationship is the same for the square as it is for the triangle. Three comes first, sets the pattern, then four and five. What happens in space will always be reflected in the information abstracted from the compressed images. Most often we work backward using abstract constructed overlays trying to understand what is spatially observable.

<u>Below</u>. The more commonly shown square spiral is where each diagonal becomes the edge length of the next square in both directions. The perpendicular movement from the diagonal of the square out starts at the midpoint of the diagonal, whereas in the above example the perpendicular movement starts from the corner.

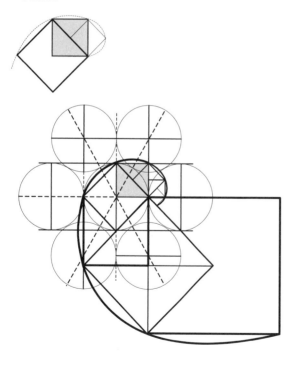

<u>Below</u>. The golden rectangle is simply a variation of the diagonal-edge relationship. It is the square divided in half, where the diagonal or hypotenuse of half the square is the new diameter to the proportional expansion of the circle. Otherwise stated: from the center of the circle the diameter increases from ½ radius at right angles to the full radius. That is the diagonal of half the square as located in the circle matrix (p.87).

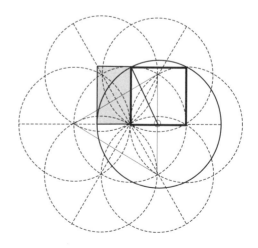

Drawings are abstractions, images composed of points and lines that form areas. There are two kinds of lines; straight and curved; both are of the circle. There are three interactions to making a line; a starting point, directional movement, and a stopping point. In space lines do not exist by themselves. They are descriptions of individual movement within systems. Polyhedra are associations of a minimum of 4 points in space connected by six straight lines that define enclosed planes forming a closed system of movement. Points, lines, and areas are not separate things. In separating them there is a problem of losing the context, loosing connections, and thus the meaning for the parts in the first place. Formulas are short cuts, they leave much of the process out, with a bias towards product.

All polyhedra can be modeled as a spatial net of lines, points of intersections, and planes. A Swiss mathematician from the first half of the 1700s, Leonard Euler, saw repeating relationships consistent to all polyhedra. Comparing the different properties of the five Platonic solids, he found a number constant to all of them. The sum of the number of points (vertices) and the number of surface areas (faces) equals the number of lines (edges), plus the number 2. They all had the number 2 in common, which was the constant factor in balancing the relationship of number of points to areas and lines. Euler wrote this formula: *number Vertices + number Faces = number Edges + 2*, or, nV+ nF = nE+2 or V+F=E+2.

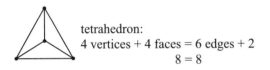

tetrahedron:
4 vertices + 4 faces = 6 edges + 2
8 = 8

icosahedron:
12 vertices + 20 faces = 30 edges + 2
32 = 32

There are variations of how to rearrange this formula while still keeping the balance. Subtract the number 2 from each side: (nV+ nF)–2 = nE. The number of edges can be subtracted from each side: (nV+ nF)– nE = 2 or V+F– E = 2. Or subtract E+2 from each side, then (V+F)– (E+2)= 0. Apply this formula to the outer description of any polyhedra or any 3-dimensional network of points, lines and surfaces, and it will work. One line uses the end points of two other lines in a closed network, leaving one line sharing the end points of two others. By adding the number 2 the balance of the whole system is restored to the abstraction of counting parts. The 2 is duality, principle to symmetry of movement.

Euler found a similar relationship in the networks between points, lines and the polygons they form. He wrote another formula that applies to 2-D nets: *number Points+ number Areas = number Lines + 1*. nP+nA=nL+1 or P+A=L+1. This formula can be rearranged as the formula above to give a variety of balanced relationships between the three primary parts of a drawn image.

3 P + 1 A = 3 L + 1
3 + 1 = 3 + 1
4 = 4

6 P + 1 A = 6 L + 1
6 + 1 = 6 + 1
7 = 7
(again we see the hexagon as number 7)

The drawn line has two points, three parts. To make two points equal to one line segment a number 1 needs to be added to the line side to make a balanced equation. It does not matter how many line segments are added together or in what arrangement; the formula will always work because it is about the triangulated nature of generating a line and the interaction between three parts as lines are joined.

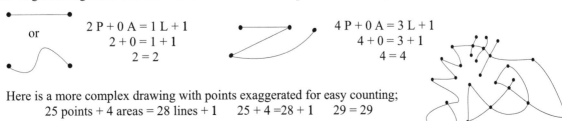

or

2 P + 0 A = 1 L + 1
2 + 0 = 1 + 1
2 = 2

4 P + 0 A = 3 L + 1
4 + 0 = 3 + 1
4 = 4

Here is a more complex drawing with points exaggerated for easy counting;
25 points + 4 areas = 28 lines + 1   25 + 4 =28 + 1   29 = 29

<u>Below</u>. A demonstration of the net development in which a line starts from a point, it stops at a point. Each point is a circle. As the circle net develops through various stages of complexity the one thing that holds constant is the point and line relationship. Another way of saying that is, *the relationship between the line (part) and the Whole is the point* (p.50).

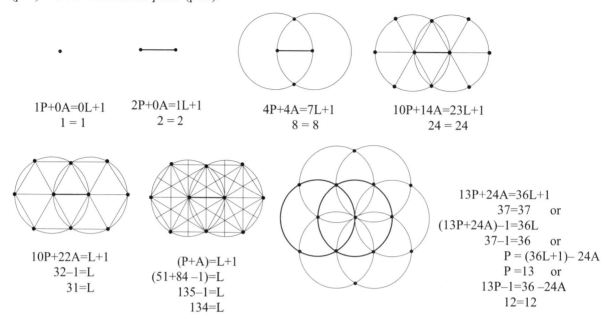

When one line is added to another a point gets absorbed at the point of joining, and another line segment is formed. Through division, one line becomes two with three points, one point is shared by two lines. By adding one (point) to the line side, the balance of one line and two points is retained. This has nothing to do with meaning, context, intention, or value of symbols; it is about the nature of, and the mechanics of the three parts of a line as a simple or rudimentary triangular system.

SETS

Euler's formula works the same way with multiple groups, or sets of nets. By adjusting the plus one to the number of sets, the formula remains consistent. We can do that because one line is two sets, one set of one line and one set of 2 points.

Let S represent number of sets in the formula. $nP + nA = nL + nS$ or <u>P+A=L+S</u>

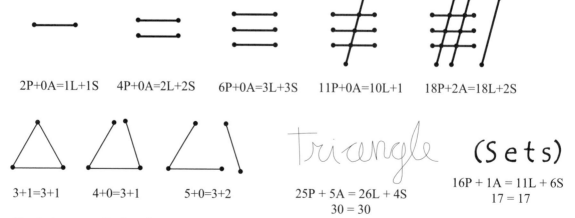

The 1 changes to 2 when there are two sets.

Looking at the sequential development of relationships between points and lines we can see a progression of numbers that shows the difference between them.

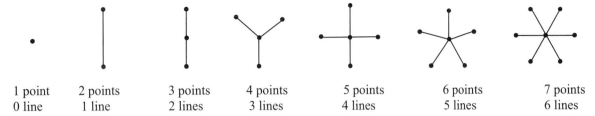

| 1 point | 2 points | 3 points | 4 points | 5 points | 6 points | 7 points |
| 0 line | 1 line | 2 lines | 3 lines | 4 lines | 5 lines | 6 lines |

Without areas the relationships between points and lines is clear. By adding one to any of the lines there will be a balance of lines to points. A circular arrangement of points shows a different relationship. There is a balance with the circle reflecting the Wholeness of the circle/sphere unity. There is nothing to be added.

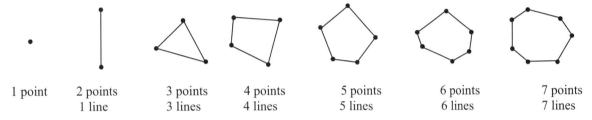

| 1 point | 2 points | 3 points | 4 points | 5 points | 6 points | 7 points |
| | 1 line | 3 lines | 4 lines | 5 lines | 6 lines | 7 lines |

Starting from a point there are two basic moves that can be made.

1 From a point we can move away and return without retracing or crossing over the line to finish at the same point. That gives us one starting/ending point, one line of direction, and one boundary area. One movement yields a balance of three parts.

2 Starting from that point again we can move away and stop without returning or crossing over. This shows two points, one line and no area; again three parts. There is separation showing two points and a line of movement without boundary. The points are doubled because of incomplete action. Needless to say, much of our thinking tends to be linear and formulistic in straight lines; incomplete. If we can understand that each point is a circle then we have greatly increased the information and re-established connection. Every line is a radius that has two intersecting circles (p87), except in the case of number 1 above, which supercedes number 2.

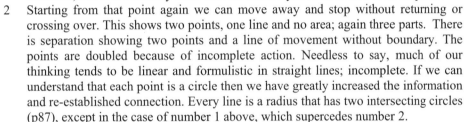

One point is a circle, is a compressed sphere, is Whole. The line is only an expression of one part of the circle movement. In folding the circle in half the longest line possible in that circle is generated. There is no separation between two points, three lines, and two areas. (For those of you counting, that makes 7.) 2P+2A=3L+1. The difference between the parts and the whole is represented by adding one point/circle whole. Points and circles differ only in scale, and the properties of numbers are not affected by scale, it is then necessary to add the circle/point context to give balance to a line of movement.

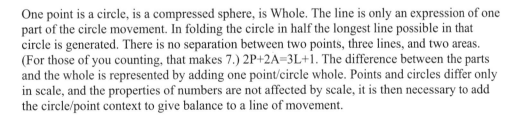

0 1 2 3 4 5
By Euler's formula (19+2=15+6) the zero is both point and circle and can be counted three times, once as a point, one line and one area. This is in line with drawing a circle, starting with a point and coming back to the same place. This is the only time drawing one line shows one point.

The issue of the Whole and parts helps clarify a pattern of movement that underlies both 2-D and 3-D point and line nets. Another interesting demonstration of point and line development is branching, lines of connection starting from a single point location (p.54).

The point and line description of the arrangement of the closest packing of the same size spheres is called the isotropic vector matrix. This matrix is a totally interconnected network of points and lines where all spheres have been reduced to points and all tangent relationships reduced to lines connecting the points. This point line matrix forms an infinite multiplicity of triangles, hexagons and squares. It is sometimes called the tetra/octa truss (tetrahedron and octahedron) when it is used as an open beam system in construction because of its inherent structural nature and minimum use of material. Given its useful nature it is an abstract reduction, a generalization about the relationship of spheres. Often left out of this point line definition are the divisional points of spherical connection. That is the mid point on each edge that makes a triangle two-frequency. This usually comes in later on as an add on divisional function. The points of connections are divisional within the spheres as much as connecting points when adding on spheres (pp.39,41,90-4). The matrix is tetrahedral and to reduce it to triangles as the basic unit is to misinterpret the information. The triangle is a planer concept derived from spherical compression. The sustainability of the triangle is in the nature of the sphere/circle relationships represented by this matrix of tetrahedra.

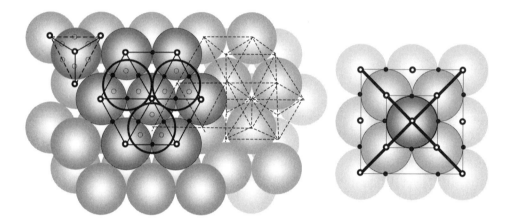

Above. This packing of spheres represents the closest possible packing of spheres of the same size. Left, each sphere fits exactly into the depression formed by sets of three spheres of each other, making all four an interconnected tetrahedron set. Right, the matrix is viewed from the side where we see each sphere nestles with four other spheres, showing a square symmetry. Both the square (octahedron) and the triangle (tetrahedron) symmetries are aspects of the same order of spheres. It is a matter of the symmetrical angle of viewing. All the lines fall inside of the spheres. The white centered rings are the center of spheres, the black dots are the points of connections and the smaller rings are hidden points of connections behind the spheres. The straight lines are the connectors. (p.31).

Spatial pattern is preexistent to image. The triangle is a symbol, a way of imaging triangulation. Triangulation is the pattern of arrangement reflected in the first movement of the sphere. The reality of triunity as a concept has become abstracted into a triangle removal from any context. The word "three", the symbol 3, or the triangle are not real, they represent an experiential concept of the first unified association, a threefold touching through movement of the Whole. Three spheres in space is a planar arrangement, four spheres is minimum spatial formation and maximum forming potential of the triunity. The tetrahedron pattern of four spheres is simply reformation of one sphere, nothing has been added or taken away. The tetrahedron/octahedron association is the isotropic vector matrix where both are part of each other; even though they can be formed and function individually in our imagined and real worlds of perceived separation between triangles and squares.

Symbols only represent selective impressions about an event, never the event experienced. We must continually reevaluate and expand our understanding, otherwise we get trapped by definitions and the meanings we give symbols and the experience losses meaning. Distortion is inherent in the process and accommodations are made to balance the perceived differences. Euler's formulas are an example (p.91). Importance is placed on spatial models because our local point of origin is spatial. The outer experience is what conditions us for the inner growth which reshapes our understanding of the outer experience. We understand the circle as a drawing until we fold it, that changes how we understand what a circle is.

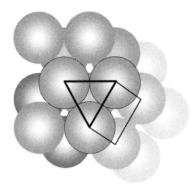

The nature of the individual triangle and square can only be fully understood in the context of the circle/sphere. The triangle is unique; principle before all else. The square is triangulation in formation, without separate (p.69). There are four triangles to every square and three squares to every triangle. The triangle embodies structural generation, allowing continuity of pattern in the evolution of formation and change.

Below. 1) A symbol for the triangle with the three edge lines extended. 2) Add three more lines each parallel to the three sides of the triangle positioned by the points of intersection. 3) With each point of intersection new triangle intervals are generated. This process draws an endless triangular grid of great fractal complexity. It is based on parallel reflection determined by position of point locations that are triangulated. This represents only one plane of four equally-spaced planar directions evidenced in the tetrahedron unit and extended to the isotropic vector matrix. (see Vector Equilibrium, p.99)

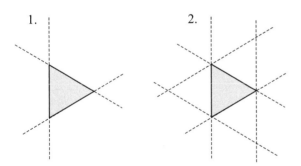

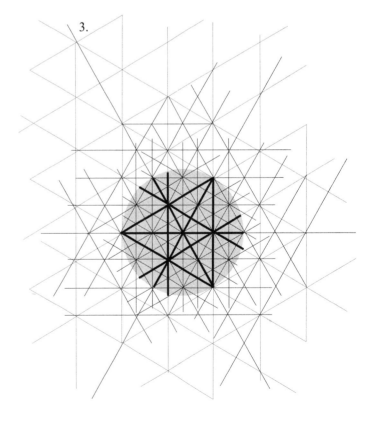

Three lines that have been extended and multiplied in parallel will form unlimited points of triangular intersection on an infinite planar grid. The primary angles of the flat grid are 360°, 180°, 120°, 90°, 60°, and 30°. Each newly generated point is a center point of endless connections.

This image represents one of four planes that extend equally from every vertex point making every point a center point to the VE. Our growing understanding in bioelectrical generation and genetic functioning is substantiates this underlying grid matrix is spherically foundational.

6 TRANSFORMATIONAL SYSTEMS

Systems are associations of individualized parts organized with intention towards a greater plan. Systems are dynamic and never complete within themselves, always changing and evolving. The interactions of systems form communities that combine with other communities, forming greater states of organizations that are ordered to greater purpose of meaning. Through compression the sphere moves from undifferentiated unity to differentiation; two circular planes with boundary and space between them. The circle is a simple system. Through movement the circle generates a tetrahedron pattern that is a complex of individual parts; *points*, *lines*, and *planes*. There is an inside and outside with minimum definition and maximum boundary that can be formed in multiplies creating interactions. These simple systems are reformed into more complex systems of discernable ordered parts giving form to the tetrahedron, which can be reformed into families of interconnections that support generations of individual development (p.175).

The tetrahedron is the first fully-enclosed polyhedral system intimate to the nine-line process of folding the circle. The tetrahedron is reformed in multiples to form the octahedron and icosahedron. These three regular forms are structurally patterned sub-systems of the tetrahedron pattern. Out of these new systems are formed the cube and dodecahedron. They are all tetrahedral by nature of origin, designed to different symmetries. The numbers of parts increase as systems develop and combine, becoming more complex. The five regular polyhedra are the first formed, differentiated systems basic to the order of spatial organization. Interaction takes three forms; point (spherical), line, and planar. These modalities define the symmetries inherent to all systems of formation.

Systems exist because they are generated to support the context in which they exist. The larger context finds expression in unification of the development and organization of parts. Systems usually serve in multiple capacities to support individual parts even while performing multiple functions to the larger context. Classifying systems is simply a way to keep track of individualized similarities of appearance and function and to differentiate in an overwhelming complexity of interrelated parts. Because this often creates separation, we lose sight of the context. Movement is a system of transformation, transportation, and communication showing patterned interaction of constituent parts and expresses coordinated intention within the greater whole. This speaks to the gravity of beauty, the economics of truth, and the practicality of goodness.

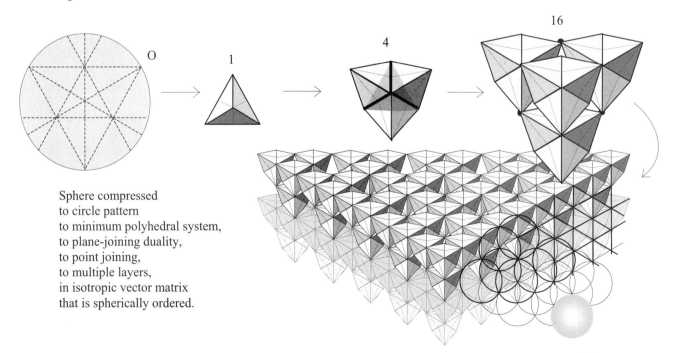

Sphere compressed
to circle pattern
to minimum polyhedral system,
to plane-joining duality,
to point joining,
to multiple layers,
in isotropic vector matrix
that is spherically ordered.

TETERHEDRON RING

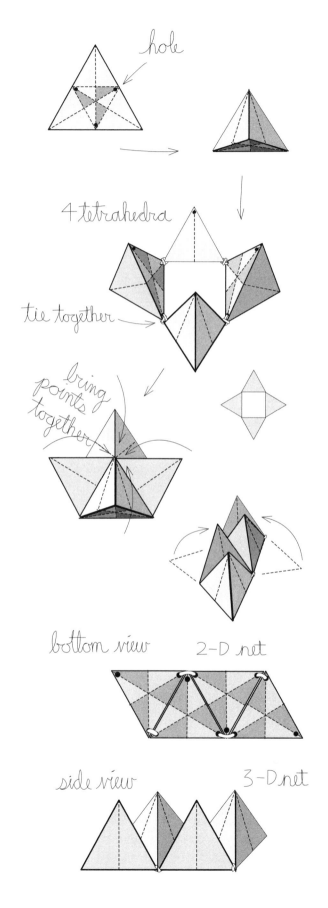

Fold 4 tetrahedra.
Open them into the equilateral triangle.
Punch a hole in from the mid-point on each edge.
Fold and tape all 4 tetrahedra individually.

Option; color in some way the center triangle of
each circle to keep track of the bottom triangle
faces of the tetrahedra as they move.

Arrange them into a square with 2 opposite
tetrahedra sitting on the colored center triangle.
The other 2 opposite center triangles are placed
so they face to the inside of the square interval.

Tie the 4 corners together, forming a necklace of
4 tetrahedra. String gives the greatest range of
movement. 4 punched holes are left unused.

Bring the 4 top points together. This forms one
half of the 8 tetrahedra that form the vector
equilibrium (p.231). (Putting 2 halves together
with square intervals joined, form the stellated
octahedron/cube arrangement (p.233).)

Open the ring and put 2 opposite tetrahedra flat
to the inside of the square. The square interval
will change to a rhomboid permitting all
tetrahedra to sit flat in a line edge-to-edge.

Looking at this line of tetrahedra with all the
center triangles facing up on one plane, forms the
net of triangles for one tetrahedron. The flat net
shows only 6 points. The other 4 are on the
opposite side. which you wouldn't know from
the image to the right.

 The full count shows a 10-point system, which
includes the 4 points of the tetrahedra. Ten is
recognizable in the two-frequency tetrahedron.

From the "net" configuration move all the tetrahedra so that all 4 center triangles are facing in, forming a stellated tetrahedron (p.163). All tetrahedra are joined edge-to-edge. Unlike starting from a single tetrahedron where we generate the dual by stellating the faces, here we observe the dual form in a single system where all parts are physically connected.

The dual tetrahedra of different sizes are at right angle to each other. One is the center tetrahedron, defined by bottom faces; the other is the relationship of stellated points. The points of the smaller penetrate through the faces of the larger triangles.

Rearrange the tetrahedra into a two-frequency tetrahedron where all 4 center triangles are on the inside leaving the octahedron an open space. The 4 tied corners now define one of the 3 intersecting square planes of the octahedron.

Dropping the top tetrahedron into the octahedron space, one quarter of the tetrahedron is truncated, in this case moved to the inside. (Tetrahedron inside octahedron, p.147).

Go back to the net position where all 4 tetrahedra are in a line.

This time move all the stellated points together putting all the center triangles on the outside. This forms a tetrahelix arrangement (p.201)

By flipping one of the end tetrahedra, the helix rotation will reverse and go in the opposite direction. The same happens when 2 as a unit are flipped to the other 2 tetrahedra.

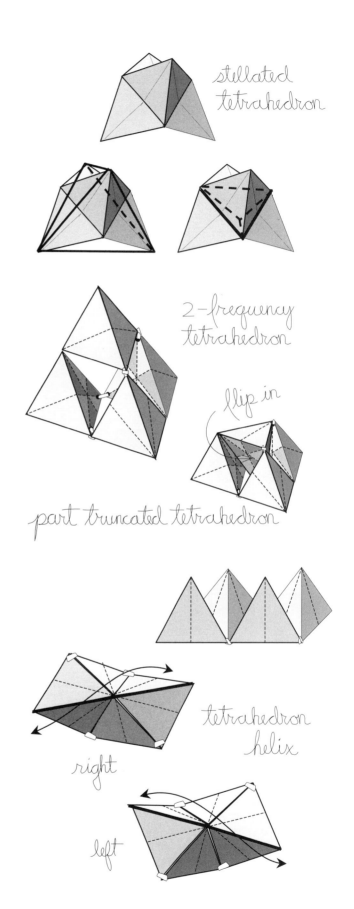

stellated tetrahedron

2-frequency tetrahedron

flip in

part truncated tetrahedron

tetrahedron helix

right

left

From the helix formation flip both ends to form an <u>irregular</u> pentagon. All 4 tetrahedra will be in a face-to-face arrangement where the fifth section of the pentagon is the spatial interval. The interval length between the two end points is longer than the edge length of the tetrahedra making this dual pentacap irregular (p.156).

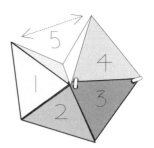

All these arrangements of coherent static positions are only the identifiable forms in the movement of 4 tetrahedra connected to each other. Look for other in-between arrangements that have a coherent form but are without stability.

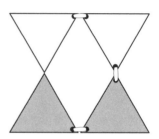

There are many levels of complexity to explore in what appears to be a very simple system. Each position is the result of rotation, reflection, and surface-to-surface movement. Use the centered triangle (marked if you colored them) of each tetrahedron as they are moved from one position to another, to register changes in arrangements and orientation. Different orientations can add other levels of information without changing the form of arrangement (p.329).

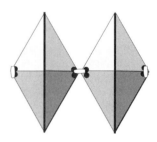

This drawing shows the base net line moving half way to forming the stellated tetrahedron where the 2 end tetrahedra join point-to-point. This forms 2 more tetrahedra intervals, making 6 tetrahedra total before moving into the 5 tetrahedra of the dual form. It is also the half way position moving into the two-frequency tetrahedron. There is much to observe about symmetry, balance, and change. Move it from one coherent position to another, going from an open loose arrangement of parts to closed "solid" forms that open spatial intervals between them. The unity between connections is never broken.

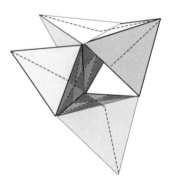

TETRAHELIX

Fold a circle into a tetrahedron.

Make 2 tetrahedra and join them edge to edge using the hinge joint (p.66).

Make 2 sets of 2 hinged tetrahedra. Join the 2 sets edge-to-edge in a line. The hinge joining will be perpendicular to each other in a chain, joining on opposite edges of tetrahedra. You now have a set of 4 tetrahedra.

Make another set of 4 tetrahedra as before. Join the 2 sets of 4 tetrahedra keeping to the same joining pattern where the movement pattern shows the axis of the hinge at right angles to each other. The development has been O, 1, 2, 4, 8 tetrahedra, a *geometric progression*.
Working in sets is easier to keep
track of when developing
in progressively
larger sets
(p.89)

This string of 8 tetrahedra will twist down to a helix formation. It doesn't make any difference which direction you twist, one direction is the inside out of the other and both go in the same direction. An easy way to see this is to color or mark the three twisting planes different colors when in one of the helix positions. When twisting it to the opposite direction, the colors will disappear to the inside while the direction of spin remains the same.

To change direction of spin, reform the helix in sets of 2 tetrahedra alternating every 2 triangle sides. Two side out followed by two sides in. With two of one color in, then 2 of one color out, alternately on each of 3 twisting planes, the direction of spin changes in a ratio of 1:2.

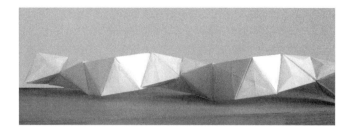

Fig.15 This is the most fundamental helix form of tetrahedra joining surface-to-surface. It was formed using sets of 3 tetrahedra end-to-end (p.142). Changing the set numbers between hinging allows different movement (F*ig.16, Fig 17a*).

Fig.16 This helix is a reformation of the one above only using sets of 2 tetrahedra joined into sets of 4 (2-4 reforms differently than sets of 3-6). It moves differently than the helix in (*Fig.15*).

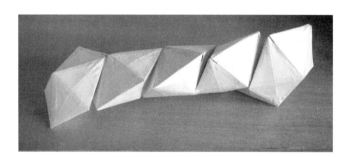

Fig. 17a This helix is the inside out rearrangement of *fig.15* above, hinged in sets of 3 units each. It is a tighter twisting than *Fig. 16*. One side of each tetrahedron continues to be hidden to the inside.

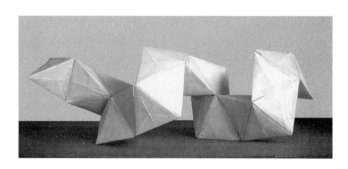

Fig. 17b The above helix (*Fig.17a*) has been hinged between each set of 6. This shows the helix turned inside out twisting at every sixth joining.

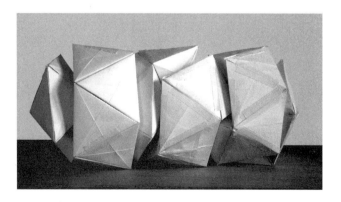

Fig.18 Here is another variation of the tetrahelix with a very tight winding of tetrahedra in sets of 3-6. This segment contains 54 individual joined tetrahedra. It is similar to the one on (p.146, *Fig.6d*).

TETRATORUS

Start with the chain of 8 tetrahedra (p.201). Bring one end around to join up with the opposite end, edge to edge, forming a circle pattern. Use the same tape hinge to join ends together.

This ring of tetrahedra is called a torus (p. 212, 234, 369-70). It is one continuous surface that moves through its own center. (To understand a torus put your finger through a ball of clay making a hole similar to a donut. What has happened to the spherical surface of the ball as the hole appears?)

The tetrahelix surface is formed by 4 triangular planes; one of the four is always hidden on the inside showing only 3 at time. In the torus form all 4 sides are visible and differently configured when rotated through the center. When colored to the divisions of the planes the rotational movement of the tetratorus becomes somewhat like a 3-dimensional kaleidoscope.

Eight is the minimum number of regular tetrahedra that will form a torus circle. To make a rotating torus with fewer tetrahedra means the tetrahedra have been adjusted to some irregularity. (Unity of circle is duality; duality twice is four. Formation in duality is eight; duality to the third power, 2^3.) (p.89) Try 16, 32, or more tetrahedra units. Observe how more units move differently. Explore the kinds of knots possible with longer rings of tetrahedra.

diagram of torus with 8 tetrahedra

attach one end to the other end in a circle

open center

center closed

203

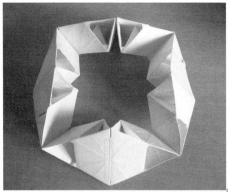

Fig.19a The torus open showing all 8 truncated tetrahedra in sets of two.

Fig.19a-e shows positions in moving and the tetratorus through the center space (p.203). Here the tetrahedra are truncated on one corner and some triangle surfaces opened to the inside. This model was made using 8 four-frequency diameter folded units. It can be made using similar truncated tetrahedra units (pp.128, bottom of 222).

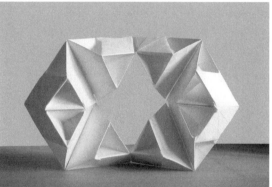

Fig.19b This open position of the torus takes a form that reflects the pattern of 2 hexagons sharing the same radius.

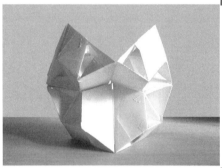

Fig. 19c A closed position from fig.19b.

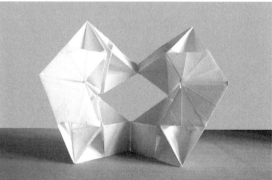

Fig. 19e Another closed position.

Fig.19d Here angles are changed showing two pentagons in reflective symmetry. They both share a congruent but unformed edge length relationship.

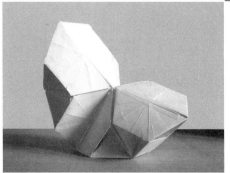

204

TETRATORUS MOBIUS

Going back to duality as principle we can see that the torus ring is 4 sets of 2 each. From that starting place we can add another set of 2 tetrahedra consistent to the hinge joining expanding to 5 sets of 2.

Open the torus ring (p.203) and add to one end 2 new tetrahedra hinged joined on edges.

The number of tetrahedra is now 10. Unlike with 8 tetrahedra, we now have 2 options to attaching the open ends together. By keeping the continuity of an inside and outside straight edge joining as with 8 tetrahedra, or we can twist the chain 180° and then attach with a hinge joining.

By twisting 180° and hinge taping edges the torus changes into a Mobius form (p.369).

Notice the regularity of the 2 edged joined triangles that form rhombus shapes. They form inside and outside surfaces when move through the center. When the chain is turned 180° and taped, you can then trace one continuous line of torus movement that incorporates 2 sides of each tetrahedron. (see drawing to right). The other 2 sides when turned through the center will trace through another continuous surface in the same direction. A torus with 4 sides now has only 2 continuous sides.

This tetratorus-Mobius system is the same 180° function that is discussed in the Mobius section. You have to turn it, through the movement you will see the Mobius surface. This is not a static model. It moves well when the turning is done from two opposite positions. It takes feeling the movement

<u>Fold</u> 4 tetrahedra and open them to the two-frequency equilateral triangle.

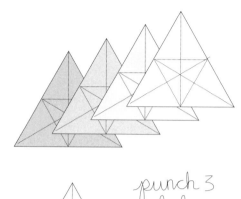

Use a hole punch to make a hole in from the edge at mid-point of each 3 sides of each triangle (p.67).

Arrange the 4 triangles into a tetrahedron net. This can be done in one of two ways; 1) four triangles in a linear grid, 2) four triangles in a triangle grid. Join these 4 triangles together tying at the mid points of each triangle, using small twisty ties or string. Either way there are only 3 places of joining.

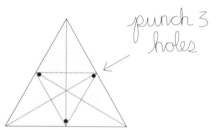

(To make a very clean joining, instead of punching holes, sew the points together with needle and thread. This is a tighter point-to-point joining with good rotational movement.)

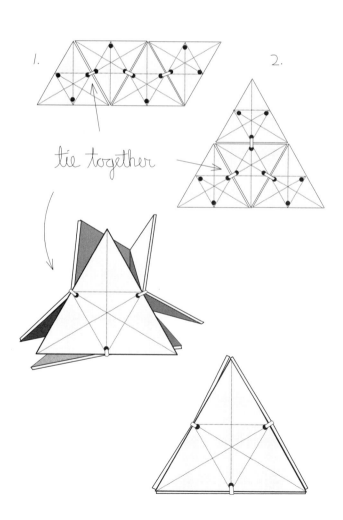

Bring the other 6 holes together in 3 more tying locations joining edge-to-edge. All 6 tied locations are the octahedron, a stable 6 point triangulated system.

The 12 points of the triangle ends come together forming the 4 end points of the two-frequency tetrahedron.

1) Explore the movement of the tetrahedron by pushing in one triangle flap. This will form a concave tetrahedron. The triangle plane remains but the surface shape is a trapezoid.

2) Fold over the 2 remaining triangles on top of the first folded-in triangle. One corner of the tetrahedron has now been "truncated". There are 3 trapezoids and 2 triangle faces. The 3 triangles can be pushed down to an inverted tetrahedron.

3) One by one, fold each of the other 3-corner tetrahedron flat. The tetrahedron is now fully truncated being reformed into the octahedron. Traditional cutting off is replaced by a folding-in process where nothing is separated or discarded. There is no waste in transformation, just change.

4) Open 3 triangles folded over on one side of the octahedron and bring them together forming a tetrahedron and octahedron combination.

5) Open out the rest of the folded-over triangles so they form the original tetrahedron with an open side and the center triangle as a tetrahedron.

6) Continue to fold the triangles over to form 3 more concave tetrahedra as above. There are other configurations to be explored that show interesting combinations of various open and closed parts of the tetrahedron and octahedron pattern.

7) Fold the 4 tetrahedra of the two-frequency tetrahedron leaving the octahedron as an open space. Here the outside of the large single tetrahedron is folded completely to the inside forming 4 smaller tetrahedra. The inside is outside, forming the same outer configurations to a totally different formed tetrahedron. It goes from a single unit to a unit of 4. Much like one sphere becoming 4 without separation (p.30).

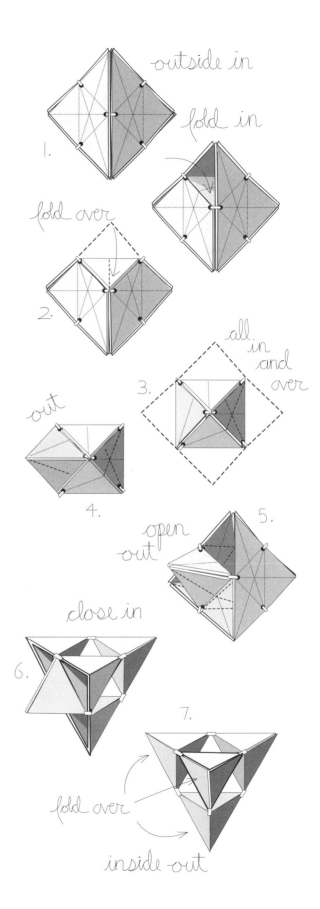

8) Open the 4 tetrahedra half-way so the edges make a cross at the 6 midpoints that form the octahedron. The 12 end points form 3 intersecting squares. The 12 edges are diagonals to *6 squares,* planes defined by the end points. There are also *8 triangle* planes formed by the 12 points.

8a) This shows the relationship that is the vector equilibrium.

8b) This drawing shows 4 tetrahedra opened to form 5 octahedra in a tetrahedron pattern, a reflection of 5 tetrahedra in spherical order (p.173). This is also the open tetrahedron unit that can be used in the reformation of the benzene matrix (p.211). Three and the center octahedra are visible, with one hidden to the back.

9) Close one set of 2 diagonals bringing 2 triangles together, forming one edge of the original tetrahedron. Do the same thing to the edges crossing on the opposite side perpendicular to the first. The remaining 4 triangles sticking out at the proper angles will form the 4 points of another partially formed tetrahedron which intersects with the first more fully formed two-frequency tetrahedron (p.165).

9a) This configuration shows the 8 corner points forming the cube relationship. This is not stable for there is nothing to hold the open triangles. The octahedron is the stable fulcrum to all the open transformations that have occurred within this system.

The nature of forming pattern is in the economy of parts used to transform through maximum number of combinations of arrangements. Each configuration represents only part expression of the repositioning and reforming possibilities with these 4 circles.

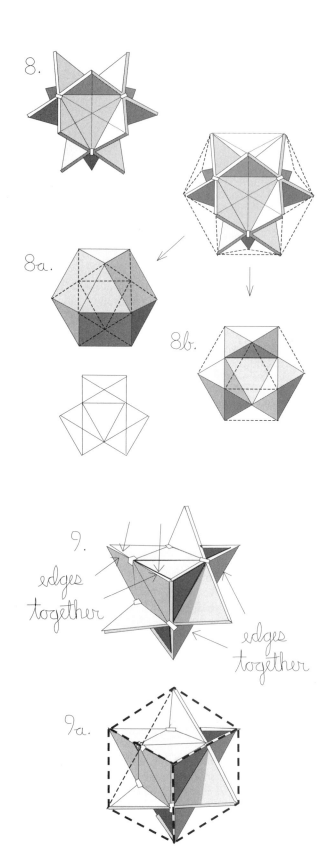

One benefit of the transformational nature of dynamic systems is the ease of partial disassembly and reassembling. This suggests a practical adaptability to a changing environment and affords a convenient means of accommodation.

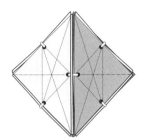

Similar to the vector equilibrium sphere (p.95), relieving the triangulation that holds the square plane by removing two ties, allows the square relationship of tetrahedra to collapse.

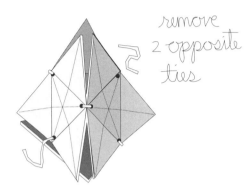

remove 2 opposite ties

Remove the two ties from opposite ends of the octahedron pattern. 6 minus 2 equals the instability of the square (4). It goes from structural to non-structural. Hold the system up and look at the square after having removed the ties. Push 2 opposite corners together and it will find its own symmetry by going flat. From this flattened pile fold the two opposite triangles diagonal to each other under to form a rhomboid.

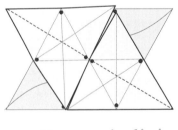

collapesed flat

Below. The first and last images represent 2 views of the tetrahedron. Removing the diagonals from the square it has no stability, it collapses on the rotational movement of each corner. It collapses to a rhomboid shape. Through transformation the square becomes the triangle. With diagonals removed the square is no longer structural until it moves off of the flat plane back into a tetrahedral formation. The compressed image distorts the edge length and diagonal change of scale. This is abstract compared to simply holding the tetrahedron, which the drawing represents, and turning it from one position to another, without change of scale or edge lengths.

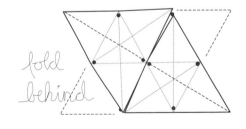

fold behind

  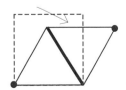 

209

Now open the two-frequency tetrahedron to the vector equilibrium positions. Punch a hole at the end of each 12 points.

Fold 4 more tetrahedra and open to the triangles. Punch a hole at the mid-point on each edge as with the first 4 (p.206). Place a tetrahedron on each of the triangular planes formed by the open 12 points of the VE. Tie the center triangle to the end points. The end points of the newly attached tetrahedra can be closed forming a tetrahedron, opened to join with another open tetrahedron to form an octahedron, or opened further, and by adding 3 more open tetrahedron to each, will form 4 icosahedra (p.149).

Observe the formation of 4 icosahedra made by adding 3 more open tetrahedra to each of the intervals created by the first 4 tetrahedra. The 4 icosahedra are in a tetrahedron pattern.

Open the remaining 4 tetrahedra and complete 4 more icosahedra, making 8 icosahedra in all. The two-frequency tetrahedron has been opened to the VE that has then generated eight icosahedra, all touching point-to-point, clustered to an octahedron relationship revealing the dual tetrahedron. The icosahedra are shown without the 4 open triangle faces. This is one direction of expanding the tetrahedron that started with 4 circles.

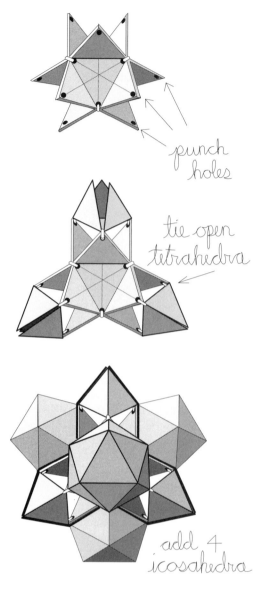

punch holes

tie open tetrahedra

add 4 icosahedra

add 4 more

BENZENE RING

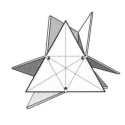

Fold 4 tetrahedra, open to triangle, punch holes at the midpoints of each edge forming the two-frequency tetrahedron and tie (p.206). Keep the tetrahedra open.

Make 6 of these in the same way.

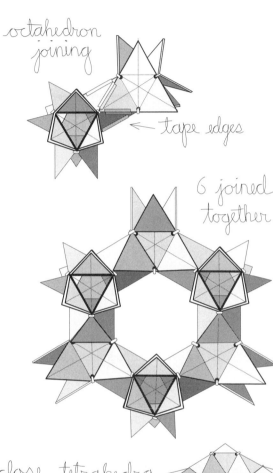

octahedron joining

← tape edges

Join 2 open-end tetrahedra in an octahedron joining (p.129). Tape them together edge-to-edge.

6 joined together

Join all 6 together in the same way in a circle. There is an alternate orientation of tetrahedra around the circle. This forms 6 octahedra connections between the 6 tetrahedra, which are now reformed to octahedra, 12 octahedra in all. The ring has 12 open tetrahedra where 12 more tetrahedral units can be joined, expanding the octahedron form of the matrix indefinitely. There is again the option of joining icosahedra.

close tetrahedra

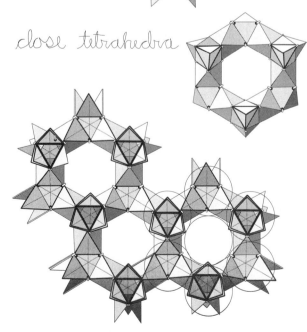

By closing the tetrahedra it becomes a solid ring.

This is one layer of how rings are connected. As the matrix develops, the hexagon openings appear in infinite parallel planes in 4 different directions to the same angles of the isotropic vector matrix, or the closest packing of spheres.

By opening the end points of the tetrahedra and joining them to form an octahedron connection, we are in effect joining tetrahedra point-to-point without rotational function. The form of the tetrahedron is altered when making the octahedron connections, but the pattern is not. With this octahedral joining of tetrahedra they can only be in opposite orientation to each other. This is a function of the dual nature of the tetrahedron and the point to opposite plane symmetry. It is the same octahedron/tetrahedron relationship found in the isotropic vector matrix (p.98). It reflects opposite nesting in the closest packed order of spheres. This is easily seen in the VE form of spheres and in the two opposite inscribed triangles in the hexagon.

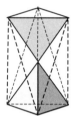

anti-prism seen from the side   anti-prism seen from the top

Six tetrahedra when joined point-to-point in a ring shows every other tetrahedron will be positioned in the opposite orientation. This models a similar arrangement in the chemical compound benzene.

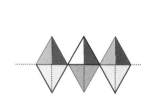
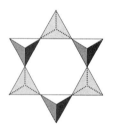
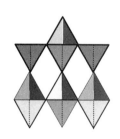
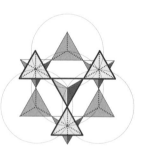

side view of one layer   … and top view   side view of two layers   … and top view

Above. When joining two layers of rings there is an off-center overlapping. The second layer is not a complete ring. The center tetrahedron on the top layer is the joining of three incomplete rings indicated by the circles. The overlapping of rings is a function of the reverse position of every other tetrahedron and is in line with the circle matrix that shows overlapping to be different layers (p.31, 120). It reflects a sphere fitting into a three-sphere depression (p.98).

Right. The benzene ring is in the form of 6 partially truncated tetrahedra resulting from opening the tetrahedra to octahedron connections. By closing the open triangle flaps 12 smaller tetrahedra are formed revealing the 6 two-frequency tetrahedra. Folding down the three sides of the 12 tetrahedra can further collapse the tetrahedra to show a ring of 12 octahedra. An endless matrix of interlocking tetrahedra can be reformed into face-joining octahedra. Multiples of this octahedron ring are interesting to explore; similar to (p.336).

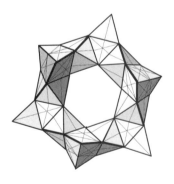

<u>Below</u>. By adding a second layer of 4 tetrahedra in a centered system to the non-centered system of 6 tetrahedra, a pattern of the truncated tetrahedron is formed from 10 tetrahedra. The two-layers combined form an interval relationship that describes a truncated tetrahedron in the open space. This arrangement of 10 tetrahedra can be identified as the same 12 points that form the truncated tetrahedron.

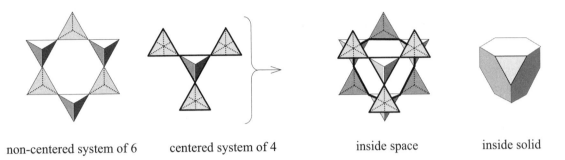

non-centered system of 6   centered system of 4   inside space   inside solid

The space as a solid removed from the ten tetrahedra shows the two middle layers of the four layers of individual tetrahedra that have been filled in. The forming potential of systems is in the associative qualities of sub-patterns and the interactions of parts to any given form.

<u>Left</u>. Divisions defined by the relationship of the four corner tetrahedra.

<u>Right</u>. Opening and folding over the 10 tetrahedra can reform the benzene system into a relationship where the truncated tetrahedron is twisted to the center of the four corner tetrahedra.

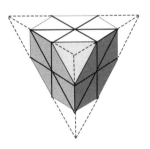

All these triangular surfaces are congruent and can be stellated by raising the center point which is a function of folding and unfolding the tetrahedron in different ways. Separating joined edges and opening surfaces can multiply points as well as creating new triangle planes. There are many combinations to reforming the surfaces by folding into and out from various point locations. With model in hand these drawings will help clarify the underlying system of what you have assembled and point the way for further explorations.

If you get this far, open up the exposed tetrahedra and continue to add more tetrahedra, developing the benzene matrix to see how far you can go, and what else will emerge in the reforming process. As you explore this pattern further other ways of forming this same arrangement will emerge, and it will get bigger.

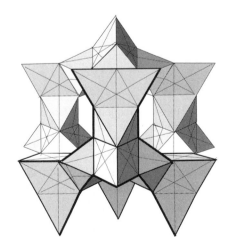

OPEN STELLATED TETRAHEDRON

Fold 2 tetrahedra. Join them half open, one edge of center triangle to the edge of the other center triangle. Secure with bobby pins, one on each end.

Make another set of 2 the same way. Join the 2 sets together using the 4 center triangles to form a tetrahedron. Use more bobby pins.

The 4 tetrahedra each open to an octahedron relationship of points forming 6 open tetrahedra spaces around one formed tetrahedron. Each open octahedron will form 4 intersecting triangle planes. This forms a pattern of tetrahedra and octahedra that is a planer replication of spherical order. Four octahedra forming a tetrahedron center is reciprocal to 4 tetrahedra forming an octahedron center (p.122). This shows a similarity with a different forming of parts. Closing the 4 octahedra forms the stellated tetrahedron.

Fold and tape closed 4 more tetrahedra. Tape, or tie, one of the 3 points of each open octahedron. This expands the center tetrahedron into a three-frequency tetrahedron revealing the tetrahedron and octahedron matrix of 5 individually formed tetrahedra, 6 tetrahedra intervals and 4 octahedra. Four tetrahedra around one tetrahedron reflect what is observed with the spherical packing of 5 tetrahedra (p.41).

Make 4 of these 3-frequency tetrahedra. It will take 32 circles. Put them together in a tetrahedron pattern, making a 6-frequency tetrahedron. The ends can be attached by tape or with punched holes and tied. It will take 128 circles to expand to a 12-frequency tetrahedron.

214

tetrahedron open half way

bobby pin 2 together

4 intersecting triangle planes

open

closed

add 4 tetrahedra

the fifth tetrahedron is hidden behind

6-frequency tetrahedron

Fold 8 tetrahedra. Follow the same joining procedure on pervious page to form the open stellated tetrahedron. Use the bobby pins to hold the center triangles together when forming the center octahedron. This time the closed tetrahedra will form the stellated octahedron.

Each open tetrahedron forms the octahedron relationship. It will take twenty bobby pins.

open all tetrahedra

The stellated octahedron with 8 points will open to make 8 octahedra intervals around the octahedron center. This is a system of 8 around one, 9 octahedra. The tetrahedron in-between spaces are much narrowed.

tetrahedra pointing out

Make 8 more tetrahedra and individually tape them closed. Tape or tie one closed tetrahedron to the 3 points of each of the 8 octahedra spaces. The tetrahedra can be placed with the point going out or going in towards the center. Either way is a beautiful and unique form variation on the cube pattern.

tetrahedra pointing in

There is a lot of movement between the open triangle sides of the octahedra until tetrahedra are added to stabilize the squares of the octahedron spaces.

215

IN/OUT TETRAHEDRON

The in/out star pattern formed in alternately folding 3 diameters of the circle (p.75) is used with subsequent folds to reveal some interesting variations in forming the tetrahedron.

<u>Fold</u> the triangle.

<u>Left</u>. Push in on the short side of the diameters, from the midpoint of each edge to the center of the triangle.
<u>Right</u>. Push in on the long part of the diameters from end point to center. This is similar to the first folding of the 3-armed star.

Differently proportioned stars are formed when you reverse the pushing in and folding the short side. Both will be developed side by side to observe the difference. (Fold the inscribed hexagon to make folding easier (p. 344) or crease really well with fingers when folding 3 arms.)

Tape the open edges together.

The long arm star is a tetrahedron relationship of points. The short arm star is a bi-tetrahedron. Opening and pushing in on the center axis of the short arm star before folding it will change it to a long-pointed 3-armed star.

Each of these 3 reconfigurations can be joined in multiples to different designs. Join 4 of each of these variations into individual systems of the two-frequency tetrahedron. Here we can begin to see the potential of the movement of the unit in the potential of the larger system. As this system develops into 4 then 8-frequency the collapsibility will allow a great deal of flexibility in the growing system. While the individual unit remains limited in movement, the accumulation of units increases the movement that is not possible without this multiple growth.

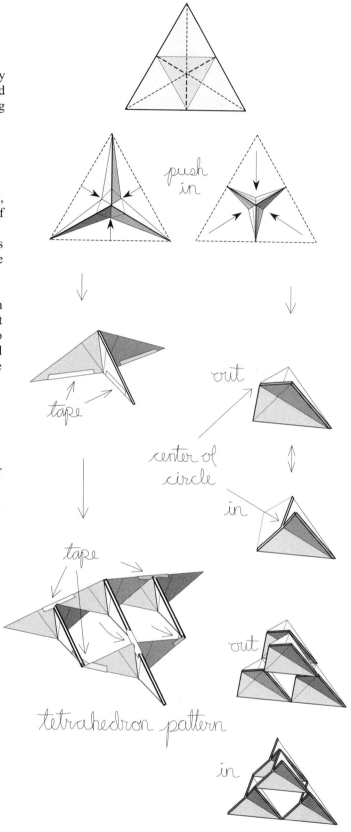

216

Looking at a one-point view distorts the image and limits our understanding. Two points of viewing often confuse because we do not see the in-between changing of perspectives. Viewing from two different symmetries is often like seeing images of two different objects. Looking at images gives little information or indication of movement.

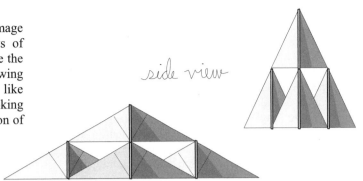

side view

top view

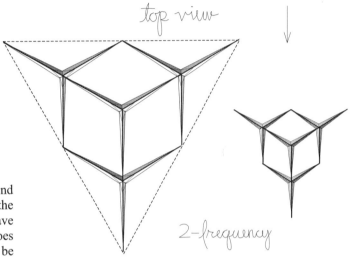

2-frequency

When these two systems are viewed from an end point the proportions remain the same but the form becomes linear; the scale appears to have changed. This view suggests movement but does not go beyond simple collapsibility. This must be experienced in hand.

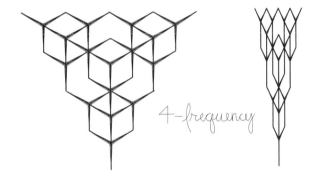

4-frequency

8-frequency

The inherent collapsibility of the tetrahedron pattern (first fold of the circle) is in this star-like configuration. Increase of frequency generates greater flexibility and fluidity in reforming the system. Use this collapsible form to further explore other systems by joining on edges and planes as well as point-to-point joining.

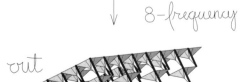

out

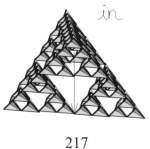

in

217

COLLAPSIBLE CUBIC

The cube is a relationship of the dual tetrahedron; it is fourteen spheres in the closest packed order. The properties of the cube are fundamental to all space-filling systems (p.253). The 90° angles reflect the first movement of the circle in half. The edges of the tetrahedron are the diagonals of the square faces of the cube. Without some form of tetrahedral stability, the diagonals, the 6 squares of the cube will collapse.

<u>Fold</u> the large triangle. Fold the short segments of the bisectors in, the longer segments out, making a long armed, three-pointed star (p.216).

Use bobby pins to hold the in-folds together. The triangle has been divided into 3 arms that can move in any combination of complimentary angles.

Fold and bobby pin another circle in the same way. Join 2 points of each together, point-to-point using tape.

Make another set of 2 units. Join the 2 sets of 2 together so that all corners have three edges coming together. This forms an irregular cube. It is a dual tetrahedron relationship (p.163). The 4 inner points with the 4 outside corner points form a stellated tetrahedron space, shown in the line drawing.

As you hold the cube by opposite points it will spiral down and collapse to a flat rhomboid shape, the intersections of two circles one radius.

Further development is seen on (p.354, *Fig.103*).

218

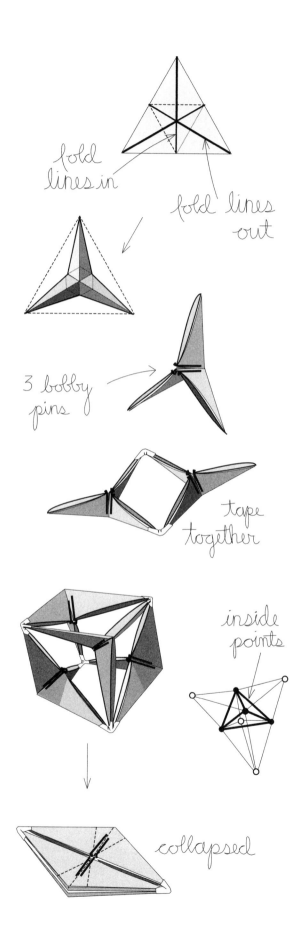

This cubic system has an irregular form; the corners are not right angles.

Make another irregular cube and join them together face-to-face, taping along the edges. Two will collapse in the same way as one. There are 2 directions to develop; add one again or double with each increase.

With doubling there are 3 directions: 1) continue in a linear mode by adding end-to-end; 2) make it planer by adding them together side-to-side; and 3) take the cube to higher frequencies in a spherical form. Anything less than the full spatial development of the cube form is only part of a higher frequency cube in a linear or planer modality.

By doubling the planer set of 4 a two-frequency cube of 8 units is formed in a geometric progression; 2 sets of 4. This is a centered system and has direct relationship to the VE.

Eight sets of 4 cubic arranged tetrahedra will fit together. The in/out edges must coincide, fitting one to the other. The irregularity gives a directional bias to the orientation when joining. There are no obvious diagonals to this system but stability is achieved through the triangular pattern of each unit by developing a center in common. (Use string or twisty ties to join units together.) This shows a divisional center to the cube that is unformed in the spherical packing (p.93,253). The edges extend beyond spherical boundaries to the space between.

There are three face designs in this two-frequency system. There are 4 sides of the first, and one each of the other two opposite faces.

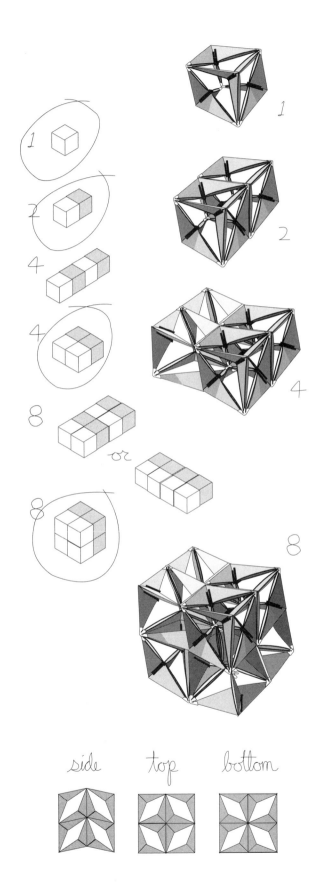

side   top   bottom

Looking at simplified images of the designs of the square planes of this irregular cube, we can see how the higher frequency develops. The growth of spatial complexity is not indicated. The outside of the single unit becomes the inside of a higher frequency system. This is simply the fractal nature of triangulation. The third arm is hidden behind the 2 arms that are shown.

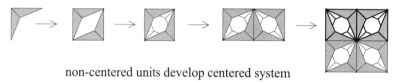

non-centered units develop centered system

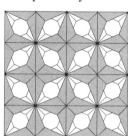

expanded system

This drawing represents the progression from the single circle unit through a four-frequency development as seen from a top and bottom view. This progression and the one below are two views of the same dimensional unit. It develops in the same way, forming different designs because of the difference in angles of position. The squares are configured with one center in and one center out, which fit one into the other. This is convient for stacking. It reflects the dual nature of pattern formation.

Two different centered designs of a single system

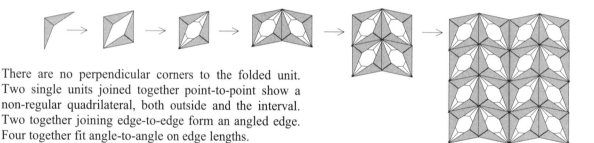

There are no perpendicular corners to the folded unit. Two single units joined together point-to-point show a non-regular quadrilateral, both outside and the interval. Two together joining edge-to-edge form an angled edge. Four together fit angle-to-angle on edge lengths.

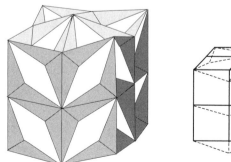

As flat images these 3 designs will not fit together in the same way 6 squares fit together forming a cube pattern. Spatially they are parts of each other generated from the same cubic matrix (p.253). When the image is compressed and isolated from spatial context the original relationships get lost.

220

TRANSFORMING TRIANGLES

This is a simple transformational system made by joining 6 equilateral triangles point-to-point in 2 sets of 3.

<u>Fold</u>   6 equilateral triangles. It is not necessary to fold full tetrahedron. Tuck the circumference flaps over/under and tape down, or glue them to the underside. Punch holes at all 3 end points of each triangle.

Join 3 triangles together point-to-point in a two-frequency triangle. Use twisty-ties or string to hold them together. (Leave enough slack in tying so each can move easily.) Make 2 sets of 3.

Put one set on top of the other and join them by tying the two points together at the 3 ends. Again leave enough room in tying for rotational movement.

Now let's explore this system.

In moving the 3 points off the flat plane towards each other an octahedron pattern is formed using only 3 of the 8 surfaces showing 6 points.

Bringing the points together 2 tetrahedra are formed, one cupped inside of the other, both with an open bottom plane (p.73 #3).

221

From the flat triangle position hold midpoints together on the bottom edge and spread the triangles at the top point. This takes the arrangement of one octahedron and 2 tetrahedra. The 2 open rhomboids without diagonals allow movement.

Push the top point all the way down to the midpoint of the bottom edge and the 6 triangles will flatten to a hexagon. This system will move through both stable and unstable configurations.

Open the hexagon to show the 3 open quadrilaterals forming 3 open triangles on the top and bottom. When the 2 triangles on each side are on the same plane it will be a triangle prism relationship. There are 6 triangle surfaces, 2 open triangle planes, and 3 open quadrilaterials.

Squeeze the three joints around the middle together, forming two tetrahedra joined at a common point. It remains a triangular prism or anti-prism as long as the top and bottom planes are parallel (p.92,231).

Move one tetrahedron to the same orientation as the other so that all corresponding edges are parallel. This forms a square interval, similar to the pattern above, only forming one half an octahedron and two tetrahedra. It is one half of a two-frequency tetrahedron.

By making another unit of 6 triangles the same and joining them, squaring the interval, the two-frequency tetrahedron is formed. Twist and lower the top unit forming a partially truncated tetrahedron.

222

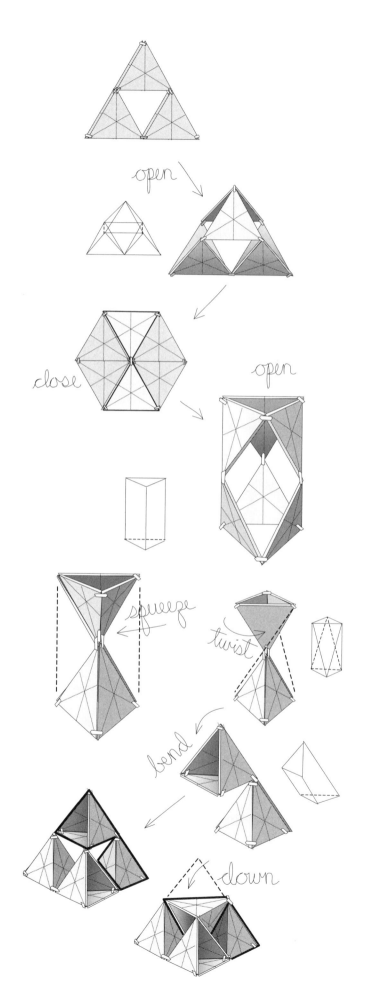

From the 1/2 tetrahedron configuration bring the tetrahedra together edge-to-edge where the space inbetween is a triangle length apart. This shows the arrangement of the tetrahelix.

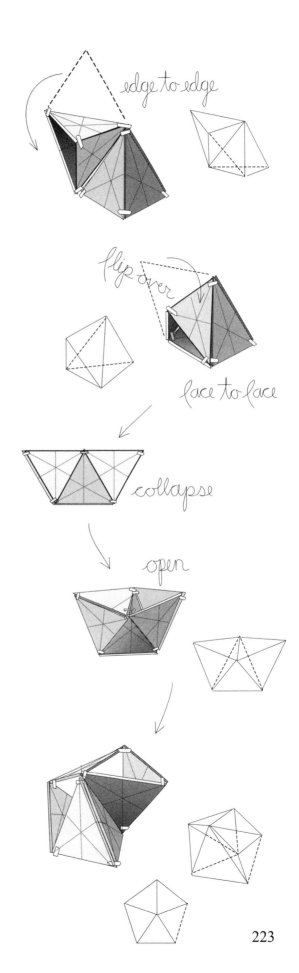

Flip one tetrahedron around the edge to the other tetrahedron, joining surface-to-surface forming a bi-tetrahedron.

Hold the joined surfaces together and open at the other end to flatten out to a trapezoid, one half of a hexagon.

Open the half hexagon and reform into the helix pattern unit of 3 tetrahedra joined face-to-face; again a different forming of the helix.

With a little reforming the 3 tetrahedra will go into a pentacap pattern of 5 tetrahedra.

As with most arrangements only the points are necessary references. From point locations all the spatial intervals and unseen connections can be determined. Reformation then becomes a matter of rearranging points by the planes and/or edges that are formed, and the movement of what is left unformed.

223

TRANSFORMING OCTAHEDRON / VE

Transforming the octahedron to the vector equilibrium and back again is a clear demonstration of truncation as a movement *into* and, stellation, a movement *out from,* a rearranging between 6 and 13 points. 6+1+3=10, the tetrahedron.

Fold 8 tetrahedra and open to the large triangle. Punch holes in from each edge at the midpoints of each triangle.

Arrange them into another design of the octahedron net as shown. Use twisty ties or string to hold the triangles together, attaching them at the midpoints of each edge, shown by the arrows.

This will form a two-frequency octahedron. Each vertex will have 4 unconnected triangles coming together. The points tied together form the 12 points of the truncated octahedron arrangement. The 6 points of the octahedron are the stellated square planes of the VE.

Push the end points to the center. Fold each triangle on the creased lines that divide the triangle faces forming the 6 square planes.

The VE is a point-centered relationship of 8 tetrahedra and 12 outside points. The two frequency octahedron has 8 triangle sides, 12 edges, one midpoint per edge. The numbers show that there is a correlation of parts. The 6 points of the octahedron when pushed to the center reveal the 12 midpoints edges of the truncated octahedron or cuboctahedron (pp.176, 178). The octahedron's 6 points become one center point successfully transforming the octahedron into the VE having 13 points. The distinction between the traditional cuboctahedron (non-centered) and the VE (centered) is one point, a point not to be missed.

Find the pentagons in the drawing to the left. They are a function of the hexagon (p.113).

224

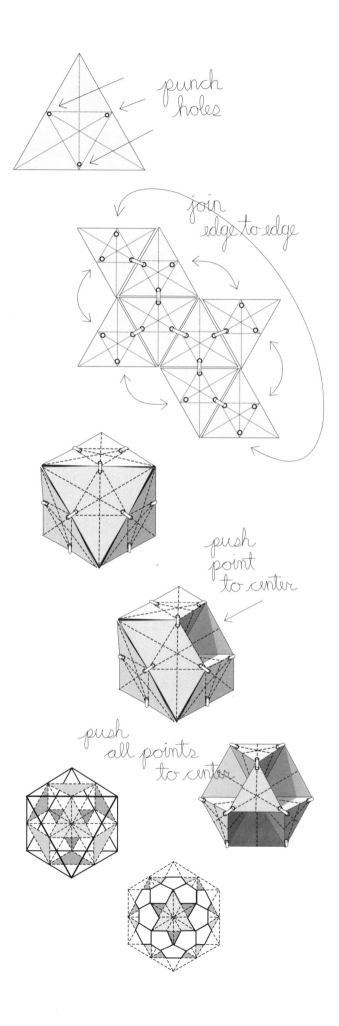

TRANSFORMING ICOSIDODECAHEDRON

Fold 2 tetrahedra opened to the equilateral triangle.

Tape the 2 triangles together edge-to-edge.

Push in the taped edges forming the pattern of 2 octahedra joined face-to-face.

Join 3 more open tetrahedra to the already joined 2 in the same way, completing a concave pentagon. The 5 taped edges go to the center of the pentagon and the center triangles of each circle form a pentagon star.

Push the open adjacent edges together and tape. This will form 5 incomplete pentagons, the dots showing the centers.

Tape more triangle circles, completing each pentagon until the sphere is completed forming the icosidodecahedron. Twenty tetrahedra to an open octahedron position and assembled to the pentagon will form the icosadodecahedron (p.176). This model can be easily transformed into a dodecahedron.

Here are two views of the icosidodecahedron; one from the triangle symmetry and the other from the pentagon symmetry. As solid plane figures they would show only 20 triangles and 12 pentagons (p.176). This arrangement is made from 20 tetrahedra reformed in an octahedron pattern. The center triangle shows the octahedron in the dark-lined hexagon. The five and 6 are not very far apart.

Locate each of the 12 triangle planes on the sphere. They are bisected by 3 diameters showing the center points to the circles. Find the long bisectors from the center point of each triangle to the end point in the center of each pentagon. Squeeze in on those creased lines. The dark dotted lines are the creases to push in.

This pushes up the short part of the bisectors, forming the thirty edges of 12 pentagons. The 20 triangles have been reduced down to 20 points. Points and planes are inter-transformable. We know this because every point is just a small circle and all polygons are parts within the circle. Transformation is what geometry does.

To make the dodecahedron, start with folding 20 tetrahedra. Push in on the long section of the bisectors. Fold them before they are taped together.

Then tape the long edges together into pentagons.

This is more direct for making the dodecahedron. It is still transformable but more difficult to transform back to the icosidodecahedron. It is easier going the other way.

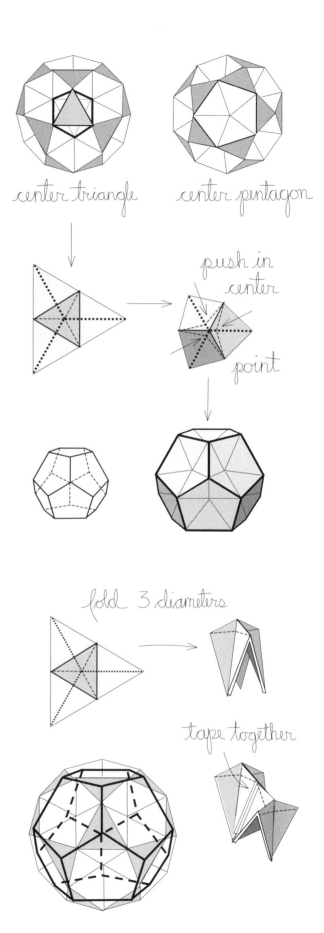

center triangle   center pentagon

push in center

point

fold 3 diameters

tape together

226

Fold the unit from the previous page with each ribs flattened to the others so they are well re-creased.

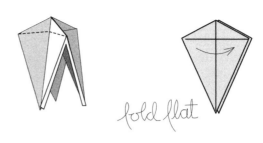

fold flat

Open 4 creased triangles and join them together, tapping edge-to-edge around a center point. The dark lines are the creases that will form the edge of a concave square.

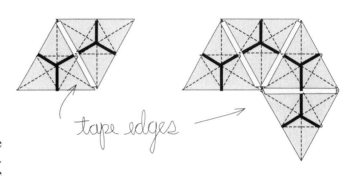

tape edges

Fold the two adjacent edges together and tape closed. Push the square in as it is taped together. This is not unlike what we just did with the pentagon joining.

tape

Make 6 of these sets of 4 and put them together following the pattern established in putting the 4 units together. Each unit of 4 circles is one-third of 4 surrounding hexagons. This is one of a number of different forming of the truncated octahedra pattern.

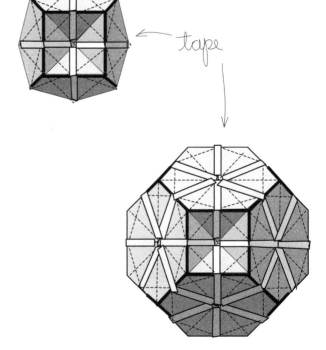

227

CIRCUMFERENCE FOLDED OUT

The circumference can be folded to the outside as easily as the inside. Cutting the circumference off from polygons and polyhedra is an amputation leaving us unnecessarily handicapped. The advantage of the circle having a circumference and leaving it where it belongs is that there are options otherwise not available.

<u>Fold</u>   a tetrahedron. Open it flat and refold the 3 end points together without folding the circumference in.

Everything is the same except the circumference is folded outside and forms three folded-over flaps adjacent to 3 edges of the tetrahedron.

Use 3 bobby pins along the creases to hold the edges and points together.

The straight edges are at right angle, 90° to the edges of the tetrahedron. The second crease division in the flap forms complementary angles of 60°s and 30°s.

Open the flaps on the 2 creases second from the end so the two 30° angles make a 60° angle. With the other two 60° angles an inverted tetrahedron is formed. There are 3 tetrahedra intervals formed by the circumference plus the edges of the formed tetrahedron, 4 tetrahedra total.

Push in on the middle crease where the two 30° angles come together closing the tetrahedral space to a flattened flap. The straight end of the flap now lies at 60° to the adjacent edge of the tetrahedron.   There are 3 distinct positions for these flaps in relationship to the tetrahedron, 90° flat, tetrahedron 60°, and a flat 60°.

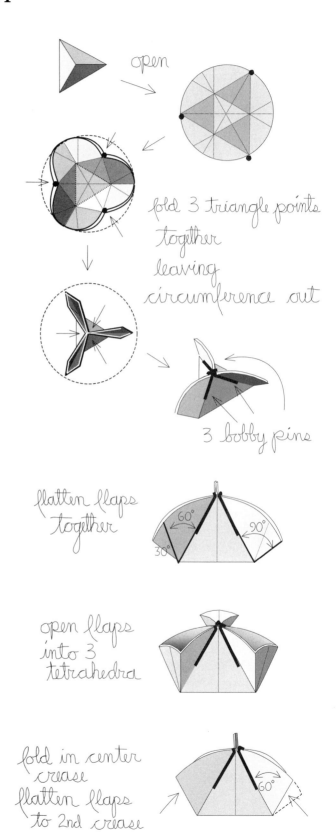

open

fold 3 triangle points together leaving circumference out

3 bobby pins

flatten flaps together

open flaps into 3 tetrahedra

fold in center crease flatten flaps to 2nd crease

There are 3 combinations of tetrahedra intervals folded to the outside; 1/3, 2/3 and the entire circumference. The tetrahedron is a relationship of the circumference folded to itself evidenced in the first fold of the circle, whether it is folded to inside or to the outside.

The potential of the structural nature of the tetrahedron is in the capacity to change form, the process of reformation. Folding the circumference out reveals 3 tetrahedra we didn't even know were there. Counting the points with circumference out shows a full number count of 10. The center point is located behind the plane that is shown. Here again we see how one is 4.

The 2 kinds of intervals formed by the circumference are the tetrahedron and octahedron spaces. The open flaps form tetrahedron spaces and the spaces between the tetrahedra spaces are octahedron spaces. This retains the tetrahedra/octahedra arrangement of spherical order. Fill in the 3 tetrahedra and the 3 octahedra intervals in this single out-folded tetrahedron and you have a 7-polyhedron system of tetrahedra and octahedra.

layer of a three-frequency tetrahedron (p.133). There are 7 polygon planes, one side is a hexagon surface and the opposite face is a two-frequency triangle, where the sides are triangles and trapezoids. This is a partially truncated octahedron, and can be see as the bottom

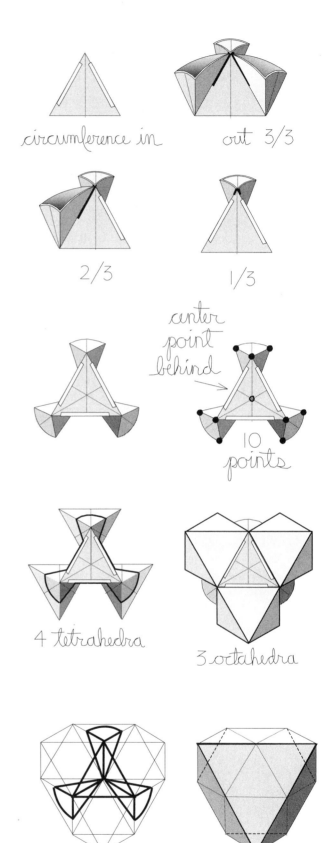

circumference in   out 3/3

2/3   1/3

center point behind   10 points

4 tetrahedra   3 octahedra

<u>Fold</u> tetrahedron flaps out opening to the tetrahedral intervals.

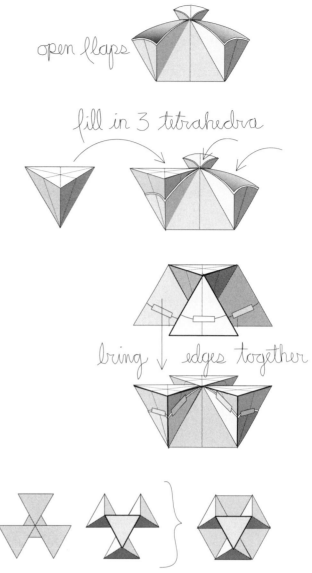

Make 3 more tetrahedra with the circumference folded inside and put them into the opened flaps. (The more accurate the folding the better the fit.)

Tape them on overlapping edges. This forms a hexagon pattern of 4 tetrahedra, 3 in the same direction and the center tetrahedron in the opposite orientation. The 3 tetrahedra pockets hold the three full tetrahedra in a regular hexagon relationship.

Make another set of 4 tetrahedra in the same way. Rotate one 60° to the other, combining alternate orientation of hexagon planes for an edge-to-edge joining. Tape where edges meet.

Two sets of 4 tetrahedra connected together edge-to-edge form a VE arrangement of triangles, squares, hexagons and 13 points (p.98, 224). There are 4 sets of opposite tetrahedra revealing 4 anti-prisms intersecting through the center (p.222).

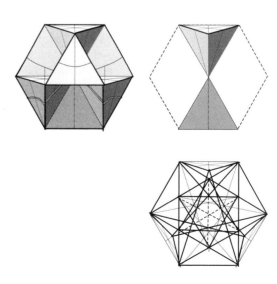

231

SPHERICAL RHOMBIDODECAHEDRON

The two-frequency tetrahedron formed with 4 tetrahedra with circumference folded inside.

<u>Fold</u> 4 tetrahedra with circumference on the outside (p.229). The right angle flaps will connect edge-to-edge perpendicular to the tetrahedron edge length. The 6 joining points of the flaps extend the midpoints (points of the octahedron) away from the edges of the tetrahedron. This is the two-frequency tetrahedron with the circumference folded outside.

Make 4 more inside-out tetrahedra. Put one on each side of the two-frequency tetrahedron with the triangle of each covering the triangle opening. The straight edge flaps join, making a 4-direction vertex at the 6 raised midpoints. Secure with bobby pins. One bobby pin will bridge the one already there holding the 2 flaps on each side. This is the stellated octahedron with the circumference out (p.236, *Fig.20a*).

The circumference can be folded over between the vertexes, indicating 12 rhombic planes. Look at the inscribed hexagon (p.344).

Fourteen outside corner points reveal the *rhombidodecahedron* reflecting the cubic arrangement of 14 spheres. There are 14 planes to the VE. The 8 vertexes of 3 edges joining show the stellated points of the octahedron (p.165). The 6 four-directional vertices reflect the 6 squares of the VE, the 6 sides of the cube. The numbers reveal an intimate relationship between the octahedron, cube, tetrahedron, vector equilibrium, and rhombidodecahedron.

This system has 12 rhomboids. Dodecahedron means a polyhedra with 12 sides, thus the name.

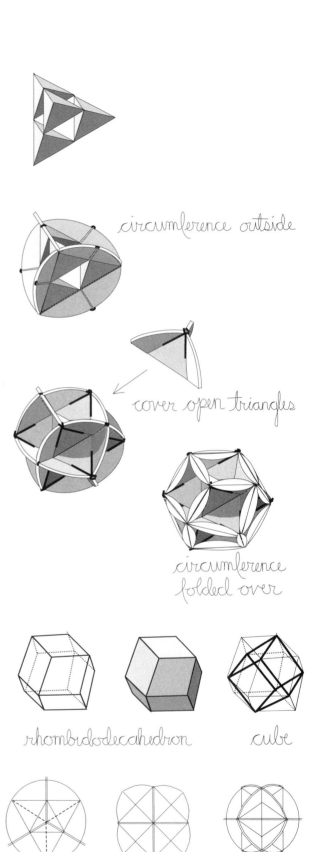

circumference outside

cover open triangles

circumference folded over

rhombidodecahedron

cube

3 views of symmetry

The rhombidodecahedron is inherent to the cube or stellated octahedron. The difference between them is a proportional movement of parts regulated by the circumference folded to the outside. When the inside becomes outside new information is always revealed.

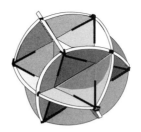

circumference outside

circumference inside

Traditionally the circumference is missing and we are limited to polygon formed models. We then miss all the good stuff that happens between the straight lines and the circumference.

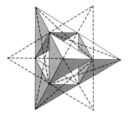

The star octahedron/cube configuration is the dual tetrahedron where all the individual tetrahedra are in equilibrium (p.163).

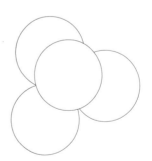

Going back further, the dual tetrahedron is 4 spheres, the minimum number in spherical order.

Four spheres is reformation of a single sphere.

233

RHOMBIC TORUS

The octahedron is an underlying system for the rhombidodecahedron. It is the pattern of interior space. It is located by the 6 right-angle crossing of edges.

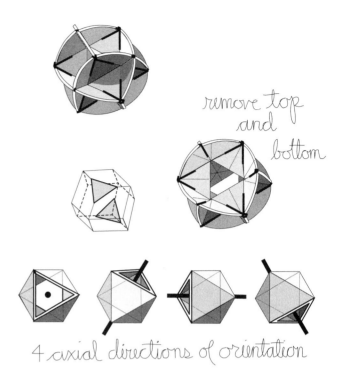

remove top and bottom

By removing any 2 tetrahedra that are opposite on the rhombidodecahedron sphere a hole is opened through the middle forming a torus pattern of 6 rhomboids with an open triangle at each end (pp.169, *Fig.12*; 236, *Fig.20b*). There are 4 sets of 2 opposite end stellations of the octahedron. (This can also be formed by stellating an open octahedron (p.148) with inside-out tetrahedra.) Each individual torus formation in the octahedron is a different orientation that aligns with the 4 axes.

4 axial directions of orientation

Fold 2 tetrahedra and open to the two-frequency triangle. Fold one smaller triangle over onto the center triangle forming a trapezoid. Do this to both circles.

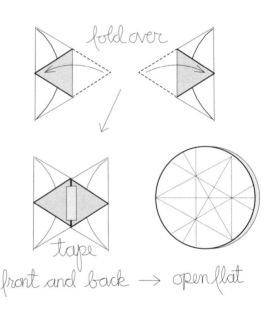

fold over

tape

front and back → open flat

Join folded edges together. Tape on both sides. Open both circles leaving the taped edges between them.

Refold each circle into a tetrahedron with the circumference on the outside. Use bobby pins to hold tetrahedra closed. The taped joint holds tetrahedra together on their base edges. Bring the 2 edges together on each end of the taped joint and hold them with bobby pins.

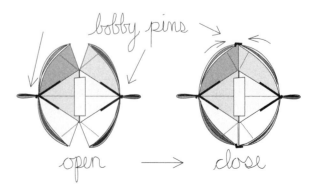

bobby pins

open → close

234

Make 3 sets of 2 tetrahedra each.

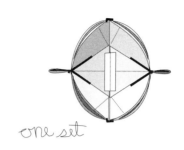

one set

Join 2 sets edge-to-edge using bobby pins or tape to hold them together.

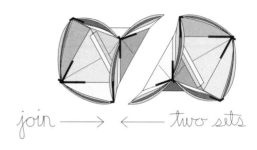

join ⟶ ⟵ two sets

Join the third set in the same way joining edge-to-edge. This forms an open triangle on each end, making the rhomboid torus through the octahedron inside space (p.148).

three sets together

To the right is an image of 3 sets of 6 tetrahedra stacked one onto the other, elongating the torus into a tube. It works best to tape the torus units together. Depending on where in line a tetrahedral section is removed the line can change directions or split off forming branches.

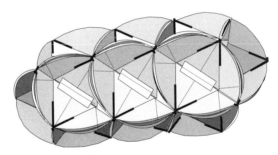

three sets of three

After putting a number of these middle bands of the rhombidodecahedron together, any number of forms with planes congruent to the system can be attached. As with any circumferences the curved edge can be opened out and folded over between the diameter end points, giving a straight edge and variable angles to the curved sections. There are many possibilities for variations including opening tetrahedra to introduce octahedron and icosahedron joining.

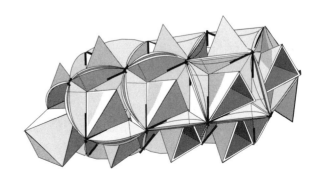

235

Fig.20a The rhombidodecahedron sphere made with 8 folded tetrahedra with the circumference folded out (p.232).

Fig. 20b *Left*; the sphere with one tetrahedron removed. *Right*; the sphere has 2 tetrahedra removed, one from each side. A wheel is formed with an open octahedron center.

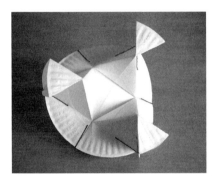

Fig.21a 4 tetrahedron are joined in a 2-frequency triangle, the fourth tetrahedra is attached to the opposite side.

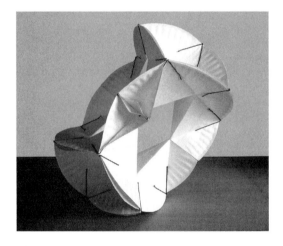

Fig. 21b Two sets of 4 shown in *Fig.21a* are joined together on the 3 congruent triangle faces of each group. This forms a bi-octahedron center space.

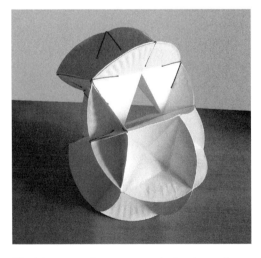

Fig.21c Here 3 more tetrahedra have been added to fill in the lower half of the system.

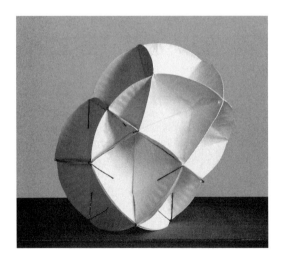

Fig.21d Again 3 more tetrahedra have been added to fill in the entire system. The top half is glued and the bottom half is held together by hair pins.

INSIDE OUT OCTAHEDRON

<u>Fold</u> a tetrahedron with the circumference out.

Open half-way as when making an octahedron (p.129). No hair pins.

Make another, and open the same way. Locate the 6 points of the open tetrahedron on both. Put the 2 together making an octahedron. It will be difficult with the curved flaps folded out rather than when folded inside. Two sets of 3 points will come together setting one into the other.

When the corner points of each half octahedron are brought together with the other there is no easy way to tape or bobby pin them. The curved edges get in the way. There is tension from the curves pulling against each other. I have found the easiest way is to sew the points together with some thread or string. That holds them very well. Punching holes and twisty tying will also work for this joining. The sewing looks much better.

Make 4 of these octahedron units with the circumference on the outside. Put them together in a tetrahedron arrangement where the curved edges join. Tape the curved edges together. Joining these 4 octahedra is a little difficult. Again sewing will make the joining easier and permanent. You can also tape and glue the edges.

Both the cube and the truncated tetrahedron combine in an interesting polyhedral and curvilinear blend.

237

OPEN CLOSE TETRAHEDRON

Fold 4 inside/out tetrahedra bobby pinned together in a two-frequency tetrahedron (p.232).

Remove 12 bobby pins that hold the tetrahedron points together, 3 around each of the 4 tetrahedra points.

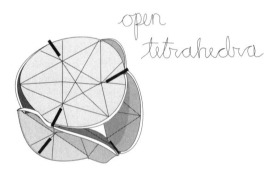

There are 6 bobby pins left holding the 4 circles together. Open the tetrahedra to circles so the 3 points of the creased triangles move towards each other, forming an enclosed two-frequency tetrahedron.

Replace the 12 bobby pins, only this time in different positions. Use 3 bobby pins to hold 3 circles together at the triangle points. Do this at the tetrahedron's 4 end points. The open two-frequency tetrahedron has just been transformed into a solid two-frequency tetrahedron. This is the pattern of an open two-frequency tetrahedron transforming into a single closed tetrahedron; a variation with the circumference on the outside (p.207).

238

One form of the tetrahedron has changed into another form through rearranging parts without adding and subtracting. The system has not changed it has simply been reorganized through movement. Changing directions within each in-between movement allows potential intersection with other systems. It is through coming together and touching that relationships change and new ones are generated. Touching in one direction means letting go in another. Each change has the capacity to bring forth multiple directions of growth, offering still greater potential as systems are developed. It is through small movements that an entire system can be changed.

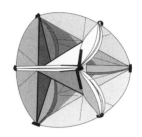

Without going into a description of each change shown in the images to the right, they represent a few possible variations. By reforming a few folds, interactions between them change and the form is changed. The pattern of the two-frequency tetrahedron remains consistent as the in/out combinations of folded surfaces open and close. Look for configurations that have congruent planes, edges, angles and spaces; they are the means to transforming systems.

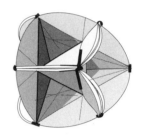

"Stem cells" have been discovered to have the potential to generate any specialized cell function in the body. It is a replication cell, first in cellular development, it carries the most generalized pre-formed individual functions. All subsequent cellular development carries this spherical pattern within individualized functions throughout all local systems. In the same way the circle is the greatest generalized pattern that carries the potential for endless individualized spatial modeling. As the folded grid is formed, each movement takes the circle further into special case or local expression, without losing the circle. In all cases any model can be reformed into any other, simply by reconfiguring the creases and multiplying or decreasing the number of circles. Priority is always given to one set of relationships over others. Any given set of creases can become dominant to a specific configuration. They all work together with bias (memory) to what is and has been. Sometimes it takes re-creasing lines to more easily move to new positions.

<u>Fold</u>   and assemble a rhombidodecahedron using 8 tetrahedron with the circumference folded to the outside (p.232).

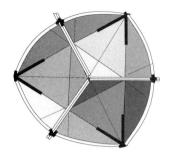

Remove the bobby pins, holding one right angle intersection where the 4 edges come together. Open the folded flat edge of the circle to make the tetrahedron interval spaces, 60° angles, using the already folded creases.

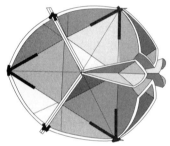

remove 2 pins and open

Make strong angles with the creases in each flap so a square interval is formed with the corners touching each other. Use 4 bobby pins to hold the corners together. The proportions of the rhomboids start to change as the point of intersection opens to form a square, with tetrahedron intervals off of each side.

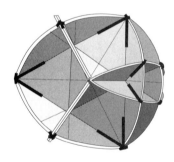

replace with 4 pins

Open and reform all remaining 4-crossing vertexes in the same way. By opening the vertexes the entire design of the sphere has changed. Where before there were only 12 rhomboids, there are now 6 squares and 24 triangles as well as the 12 rhomboids.

replace all 4 vector pins

240

Here is another approach to making this sphere without having to go through forming the rhombidodecahedron.

<u>Fold</u> the inside/out tetrahedron and open the flaps to form 3 tetrahedra intervals. Push the 90° edge of the flap in and re-crease on the 60° creases.

Join 2 inside/out tetrahedra edge-to-edge using bobby pins at the intersections forming two 180° crossings. The intervals that are formed are 60° and 120°. This joining of 2 units shows the same angles and intervals reflected in the VE. (You do not have to know the angles to do the folding.)

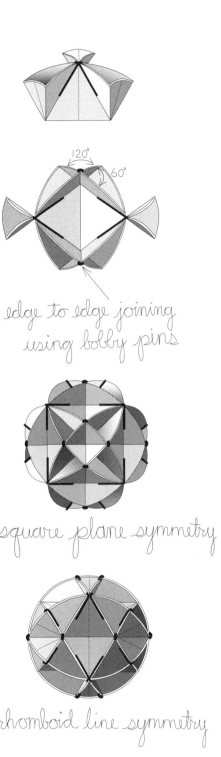

edge to edge joining using bobby pins

Make 2 sets of 2 tetrahedra joined as above. Put them together so the opened flap ends form a square interval. This will leave 4 half-square intervals on each of the 4 sides. It will be unstable until the other half is pinned in place.

square plane symmetry

Make another set of 4 tetrahedra the same way. Join the 2 sets of 4 tetrahedra together edge-to-edge, completing the remaining square intervals. This will form the complete sphere. It is very strong.

rhomboid line symmetry

To the right are drawings showing the 3 views of symmetry. The first is viewed from the square open plane, an octahedron interval. The second shows the sphere directly from the rhomboid plane along the longest axis, formed by 4 points of circle intersections. This octahedral interval has tetrahedral intervals on each side. The bottom image shows a point symmetry. This is the same alternate triangle and square relationship as the VE (p.244), only with increased proportional rhythmic openings. The openings allow you to expand out by adding tetrahedra and octahedra into the intervals.

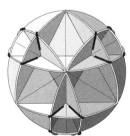

tetrahedron point symmetry

241

Going back to the rhombidodecahedron sphere there are other variations that can be explored, as with all circle configurations (p.363-5).

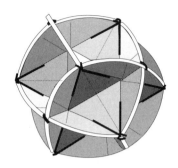

Eight in/out tetrahedra have been used to form the rhombidodecahedron. There are 8 end points plus the 6 points where the circumferences are bobby pinned together. The cube is the basis for the rhombidodecahedron in the form of the stellated octahedron (p.367). Numbers can provide connections not visually observable that provide directions to explore further.

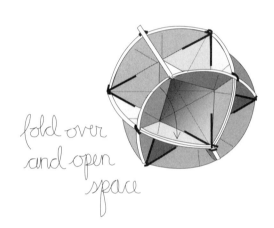

With this sphere all the flaps are doubled, allowing any of the 14 points to be moved without changing the arrangement. The flaps can be moved back and forth opening and closing spaces. There are many options and combinations in moving triangle planes. These small movement differences may have little consequence to a static model, but in relationship to functioning of biological organisms these kinds of subtle differences can be the basis for survival.

fold over and open space

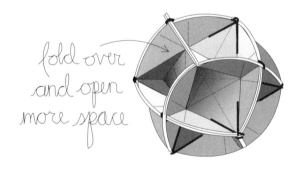

fold over and open more space

As a separate spherical model one variation is no more important than any other, there are many combinations for opening and, closing surfaces. These combinations are all interdependent on one another. That one should periodically predominate is a matter of developmental choice. It does seem clear that without the consistency of pattern these changes will not happen. This process models systemic interactions that support individual change that is purposeful towards expression of the entire system of circles.

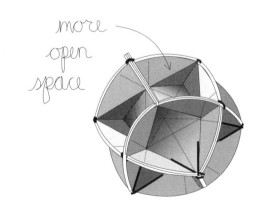

more open space

TETRAHEDRON PROGRESSION

Folding circles demonstrates a simplified process that is observable everywhere in nature. Individual static models are not continuous or interactive. The interaction of continuous inside/outside, open/closing intervals, spherical and polyhedral forming and blended combinations, along with pulsing and rotation movement are some of the qualities inherent observable in folding and joining multiple circle tetrahedra.

Four open tetrahedra form the two-frequency tetrahedron with circumference folded inside.

The same tetrahedra reformed with the circumference to the outside gives direction to evolve through various spherical possibilities.

Opening 4 points reforms the dual tetrahedron revealing 4 individual tetrahedra that open the inside octahedron space out.

The 4 open octahedron planes extend to 4 individual tetrahedra with circumference out, expanding spherical information in intervals of tetrahedra and octahedra. All arranged to the dual tetrahedron, stellated octahedron, cube and rhombidodecahedron. There is no separation in spherical order.

Both points and lines open to form planes and interval spaces of octahedra and tetrahedra. Six points open to 6 square intervals further compounding spherical complexity The shifting of circles does not disturb the matrix of spherical order. These changing forms are all the same tetrahedron realization of single sphere unity.

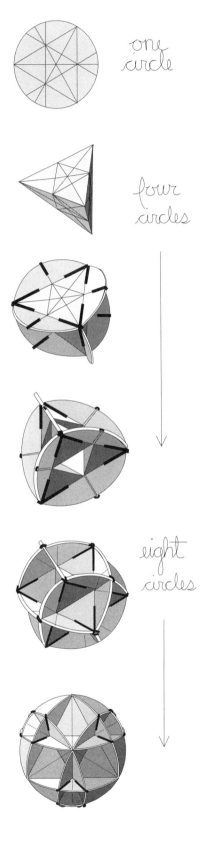

one circle

four circles

eight circles

243

DISCUSSION: TETRAHEDRON PROGRESSION

There are no great circles in this sphere. It is formed by 8 congruent lessor circle planes. There are 4 sets of 2 circles parallel to each other. Imagine a center circle inbetween each set of parallel circles and you will have the 4 great circles of the VE. (p.89).

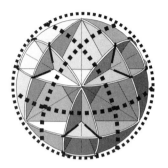

These 4 dotted lines are great circles of the VE. Think of them as 4 broad bands where the edges of each are the 4 pairs of lesser circles that find intersection with each other. This is broadband information in which there is potential for many tuning positions.

4 sets of paired circles and 8 points of the cubic pattern compressed into 6 outside points around the center seven

The 8 points of 3 circle intersections are the end points of the stellated octahedron or cube relationship of 2 two-frequency tetrahedra. There are 12 rhomboid intervals in the rhombidodecahedron relationship. The rhomboids are connected at the 8 points. Each spherical cavity holds either an octahedron or regular tetrahedron.

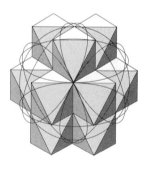

12 octahedra

24 tetrahedra

6 octahedra, one behind center one.

Above. The rhomboid openings hold 12 octahedra. There are 24 irregular triangle intervals that hold regular tetrahedra. The 6 square octahedral intervals are at right angle to each other and hold octahedra. Total fill-ins are 24 tetrahedra and 18 octahedra of the same spherical measure.

The parts we choose to move and what view we draw make a difference in what we see. Here are 24 tetrahedra seen from their edge line. The triangle face symmetry is pictured above. Opening the circumference of tetrahedra will change the form and relationships of parts, but it will not change the pattern.

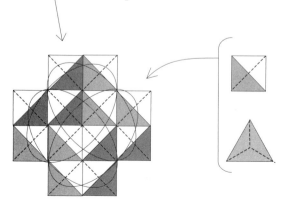

244

As a compressed image within the flat division of the lesser circles, we see 7 individual parts, a combination of intersections from the other folded circles that make up any one individual circle.

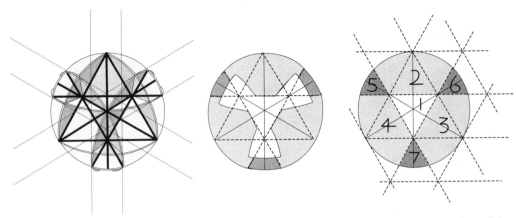

Above. The flattened divisions of combining 8 individually-reconfigured circles is the same as that of the folded creases in forming the tetrahedron itself. The 8 circles together reflect the same divisional organization folded into each circle individually. The numbers reflect the primary hexagon pattern of 7 points in the number of areas of flat circle division, reflective of the 3, 4, and 6 in the VE. These 8 circles are the broad band VE and carry more information than the narrow band of a single line. Expanding the bandwidth increases in-formation capacity.

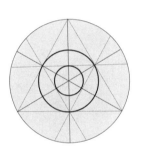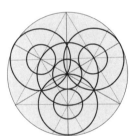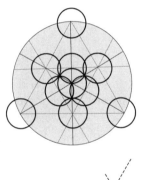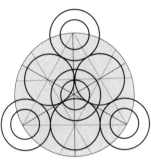

Above. All points of intersection on a plane are center points. Looking at the creased pattern we can see that every point functions as a circle (p.120). The circle has no size but when fixed to a radial measure it becomes fixed in relationship to other measured circles. All sizes of circles will always reflect the same triangular matrix of spherical order. That is the nature of the Wholeness of the circle. As circles intersect they create more center points that expand concentrically in countless potential circles that expand into and out from given local intersections (p.106-7). The digitization of information tends towards straight-line relationships as most efficient, but not practical in a gravity filled spherical environment. The spherical and polyhedral may well be a blend of the real and ideal that we have yet to realized.

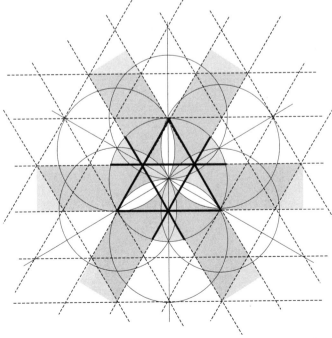

TRUNCATED TETRAHEDRON

The truncated tetrahedron is classified traditionally as a semi-regular solid polyhedron. It is derived by systematically cutting away each corner from the tetrahedron, 1/3rd in on each edge. This leaves 4 triangle ends and 4 hexagon faces (pp.178, 213, 248). As a partial form truncated tetrahedron can be derived in many ways. The following is an interesting variation that is less complicated than it looks.

<u>Fold</u> 2 inside out tetrahedra and spread the 3 flaps forming 3 tetrahedron intervals (p.229).

Join these 2 tetrahedra face-to-face on the flat ends of the open flaps. Attach with bobby pins.

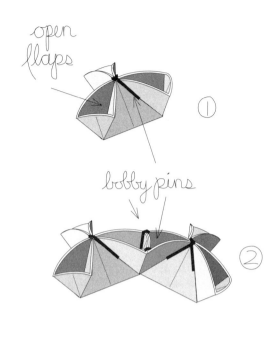

Join a third tetrahedron to the first 2 in the same way on the end flaps.

Bring the flat ends of opposite tetrahedron intervals around and join together forming a triangle. Again using bobby pins to hold the surfaces together.

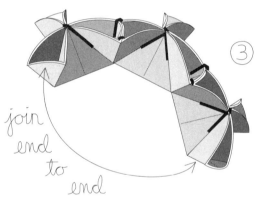

A two-frequency, 3-dimensional, planer, and triangular unit is formed.

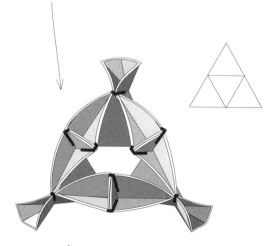

2-frequency triangle

Make 4 triangle sets of 3 tetrahedra each.

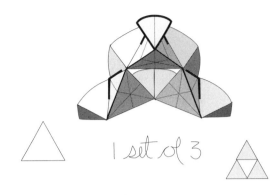

1 set of 3

Put 3 triangle sets of 3 together into a triangle. The individual tetrahedron's straight edges, where the flaps are joined, form a hexagon space rather than the triangle interval. The 3 exposed triangular end flaps facing up (white with heavy black outline) are where the fourth set of 3 tetrahedra will be connected.

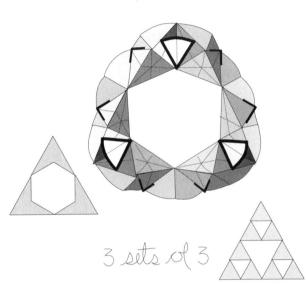

3 sets of 3

In attaching the fourth set, notice how stable and strong the system becomes. Joining 4 triangular sets of 3 tetrahedra into a tetrahedron pattern forms 4 hexagon spaces. This takes 12 folded in/out tetrahedra to assemble.

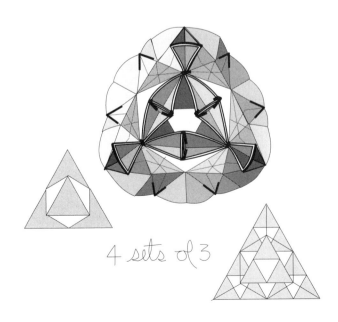

4 sets of 3

247

There are 2 primary truncations of the tetrahedron, 1/2 and 1/3 divisions. Systematic truncation increases end points, generates planes and develops complexity. A one third division reveals a combination of 4 triangular planes and 4 hexagon sides. It does not matter how irregular the form, this kind of division will always generate the same number of triangles and hexagons, albeit in different proportions. When cut at mid-length there are 8 triangles forming the octahedron. We have seen other ways to approach truncation without cutting the tetrahedron apart.

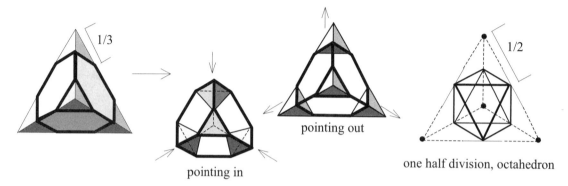

pointing in

pointing out

one half division, octahedron

Cutting corners from a static model reduces what is there, producing a loss. The Wholeness of formation neither cuts off nor appends as a growth process, it moves into and out from, forever transforming and changing. Points moving into and out from the center can be seen as the same movement in different directions.

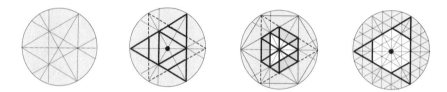

Above. Observing from a point/circle symmetry, the divisions of truncation change scale, revealing layers of information that change the shapes of intersection and vertexes. The change of form from point, triangle to hexagon is a function of moving through a high frequency grid pattern. There is no definable boundary since scale is relative and changing shapes provides continuity. Any point of the tetrahedron moving towards the opposite plane becomes an increasingly larger triangle. In process, it changes boundaries as it interacts with the triangle first going in and then out in the opposite direction. There are many ways to read this information. All parts are multifunctional and cross connect with a diverse number of systems.

Of the tetrahedron, movement is proportional in and out, very much like breathing. We only see the effect, not the movement. Solid is perceived as non-movement. Change moves through perceptual boundaries. In understanding the inseparability of the triangle and hexagon we see how much is lost in static concepts.

A higher frequency bandwidth carries more information than a single band.

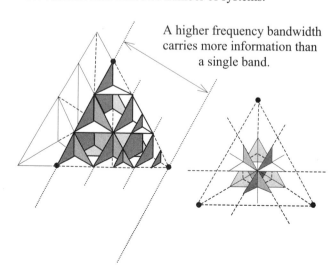

248

TRUNCATED CUBE

Fold 2 in/out tetrahedra. Flatten the circumference flaps forming the 90° angle with the tetrahedron edges. Do that to both tetrahedra.

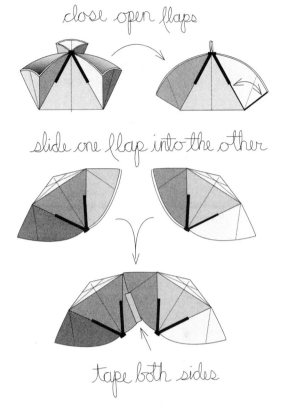

close open flaps

slide one flap into the other

Slide one flap into the other so the edges coincide, keeping the tetrahedra edges parallel and perpendicular to the slide-over edge. Tape the curved edges where they overlap, on both sides. A rectangular plane is formed between the tetrahedra with one short side missing. This reflective joining of tetrahedra allows movement in the unit in the same way that two tetrahedra are joined edge to edge. The line has become a rectangular plane. There are various combinations to explore in this sliding-flap method of joining.

tape both sides

Join 2 sets of 2 tetrahedra, flap-into-flap, 4 times to form a square where each tetrahedron is a corner square seen from the edge view. The tetrahedral relationship is hard to see in this flat one-point viewing, but easy to understand when directly in hand.

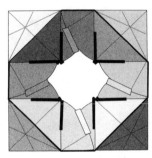

4 tetrahedra

square symmetry

Make another set of 4 tetrahedra. Put them together using the same sliding joint. This forms a truncated cubic pattern where the 8 corners are 8 triangles and there are 6 octagon open planes. This is a very different form of the truncated cube; there is movement because of the reflective joints. As a solid this polyhedron is rigid.

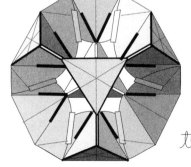

8 tetrahedra

point/triangle symmetry

As with the tetrahedron there are 2 primary ways to truncate a cube. The 1/2 division shows the vector equilibrium where the points move into a triangle plane and the square plane gets reduced to a square half the size. The 1/3 division changes the square into an octagon and the corner points into a triangle plane of different proportions. On the outside there are 8 triangles and 6 squares and the other shows 8 triangles and 6 octagons. The number 8 in both polyhedra tell us there is a direct correlation between the two. We observe 8 tetrahedra in both systems, which are in similar relationship and orientation. The octagon has 2 times the number of sides as the square. The square interval (4) has opened to an octagon interval (8) by a division of thirds. Both polyhedra are derived from a tetrahedral movement out as much as from the reduction through truncation of a solid cube.

When looking at the orientation of the 8 tetrahedra there is a reciprocal relationship between the 4 tetrahedra pointing down and the 4 pointing up. Only the intervals between them have changed.

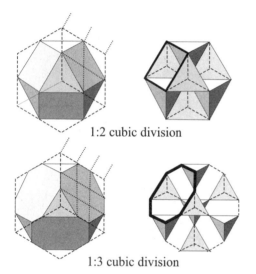

1:2 cubic division

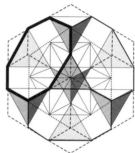

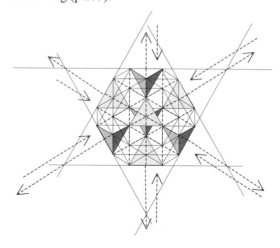

The VE shows tetrahedra joined edge-to-edge. When space is formed between the edges the rectangle planes take a truncated form (p.248).

1:3 cubic division

When the cube octahedron is understood to be only 8 tetrahedra the relationship between triangles, hexagons, squares and octagons becomes apparent. Truncating as an outside function does not predict the internal dynamics. Change one thing and all parts shifts.

While these polyhedra can be seen as separate and individually formed we must recognize that each is an expansion/contraction of multiple tetrahedra, a patterned movement as we have seen with the triangle and hexagon (p.99). Going back to the first fold of the circle 1:2 ratio expression of the tetrahedron has the inherent ratio of 1:3. (1+2)+(1+3) =7) How could it not be? The tetrahedron spherical arrangement defines an octahedron space showing three 1:2 ratio edge line division, reflected in the 1:2 division of the three diameters. It is this division into what is undivided that triunity is formed to a tetrahedron pattern. This is not an aggregation of parts. The truncated cube reside within the tetrahedral net. The vector equilibrium is compression/expansion of spherical order, a tetrahedral/cubic lung (p.253).

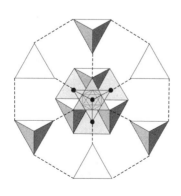

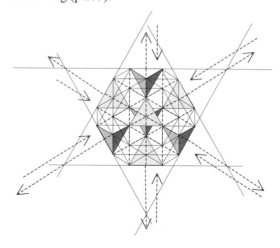

Cubes of the same measure will, when endlessly arranged one next to another, fill all space. The idea that space is empty and finite, to be filled with a solid anything, is not congruent to our experience, nor logical to our mind, and denies the spirit of creation. Cubic packing leaves no space, except maybe the contained space separated into each cube package. That makes no sense. The cube is a relationship of spheres in the same way the square is part of the circle (p.53). To understand the nature of the cube we must look to the octahedron, tetrahedron, and beyond to pattern in the form of spherical packing (pp. 90-94, 194).

<u>Below</u>. Eight corners of a cube truncated in thirds, or in half, when reassembled, will form an octahedron. The inside right angle edges come together forming the 3 axial divisions of the octahedron (p.130). These axes represent the inside out of 12 edges of a cube. Eight right angle tetrahedra are cube corners that are the axial division of the octahedron. Truncation is not just about cutting off parts.

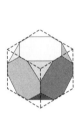 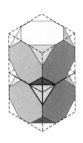 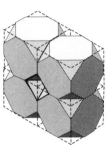 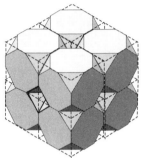

1 truncated corner, 1/8 of octahedron

2 truncated corners, 1/4 of an octahedron

4 truncated corners, 1/2 of octahedron

8 truncated corners form a full octahedron center.

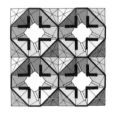

<u>Right</u>. Joining 4 truncated cubes (p.249), in a square arrangement generates 1/2 an octahedron interval in the middle. Four 1/2 truncated cube spaces are generated at the same time. Space is defined by the forms that are used, the principles of order do not change. Again we see the triangle/square and the tetrahedron/octahedron as primary directives to the reformation of spatial configurations.

<u>Below</u>. Add another set of 4 truncated cubes to the square arrangement, making a complete cubic relationship of 8 truncated units. How many truncated cube spaces and octahedron spaces have been formed? There is an octahedron space in the center of the cube, not unlike the 2-frequency tetrahedron. There is half an octahedron space centered on each of the 6 sides, one quarter of an octahedron (12 tetrahedra) on the edges, and a triangle surface on each corner; 27 octahedral locations. There are 3 layers of octahedra. How does this form correspond to the spherical packing on the next page?

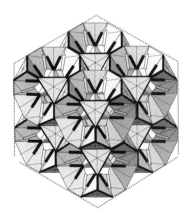

These are the same size. What are the similarities and differences? One image shows considerably less surface than the other. Calculate the implied surface area of each and compare. What is the surface area of the number of circles minus the folded-in parts used to make the one on the left? Compare the two. When would one form be better used over the other?

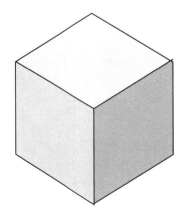

Geometry models describe the order and arrangements abstracted from the natural world. There is a correlation between the arrangement of these 27 octahedra and the development of atoms, molecules and compounds. The primary arrangement of 14 octahedra (corner points and centers of each side) reflects the closest packing of spheres. This arrangement is similar to the atomic lattice for gold. Salt is another arrangement of the cube pattern. By identifying underlying relationships of similar forms the geometry of spatial organization of physical matter is discovered. We also begin to see on a fundamental level how connected everything is to the same spherical origin of pattern direction.

Within these 27 octahedral locations there are 12 midpoints of each edge line and the center location, showing a number pattern of 13 points (the vector equilibrium). From the truncated cube (27) subtract the VE (13) and the 14 spheres of the cube are left. The numbers work in formation of spherical order. Subtracting 14 from 27 creates a separation showing both polyhedra as sub-systems of the same spherical order, one 13, the other 14 (p.93).

<u>Below</u>. The cubic form of 14 spheres is shown cut from the spherical order. The spatial arrangement compressed to a flat image reflects the same pattern in a hexagon form. Observe the truncated cube relationship. We don't see or experience this spatial interaction with the "solid" form of the cube or with solid spherical packing. The cube form is surface centered to each square but not spatially centered.

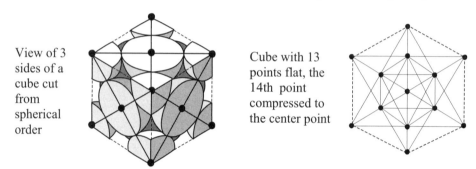

View of 3 sides of a cube cut from spherical order

Cube with 13 points flat, the 14th point compressed to the center point

The 14-sphere cubic pattern is the same arrangement as the stellated octahedron, the tetra star (p.165). All edges of the tetrahedra and octahedron lie inside the spheres' connecting center points. The cube form is a separating out from spherical context and structural dynamics. When rearranged, this volumetric arrangement of 6 half spheres and eight 1/8 sections of a sphere will form 4 complete spheres, the tetrahedron. What is the ratio of the volume in a solid cube to the volume of the four spheres of that cube arrangement? What is the ratio of cubic to spherical? There are many interesting questions to be asked if we understand something of pattern and the proportional interplay through various formed relationships.

Beauty is the unification and harmonizing of everything as part of everything else. There is an extraordinary rhythmic interplay of spherical unity reflected throughout, forming endless individualization. The magnificence of this in-forming intelligence is overwhelmingly observed everywhere.

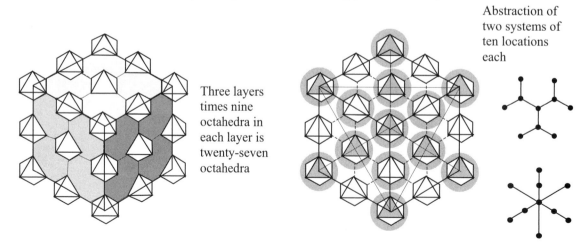

Three layers times nine octahedra in each layer is twenty-seven octahedra

Abstraction of two systems of ten locations each

252

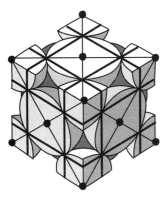

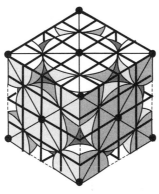

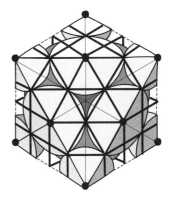

Primary lines of division
of the cube are within
the spheres.

Secondary lines of division.
Four-frequency cube
includes the spaces between
spheres.

Four-frequency tetrahedron
face shows the triangle grid
in the 4-frequency cube to
the order of 14 spheres.

Above. By connecting tangent points and center points of the spheres, all primary lines can be defined within cubic division. From this information each edge of the cube can be divided equally in 4 sections by bands of perpendicular parallel lines, using tangent and center points. Each square face is a division of 16 squares. The total cubic division is 4 x 4 x 4, or 4³, 4 to 3 dimension (7); 64 cubic units. (6+4=10, the first tetrahedron fold in the circle). The diagonal of the cube shows the center/tangent point sphere connections.

When a corner of the cube is truncated, cut away corner-to-corner, the triangle face of one of the tetrahedra which forms the cube is revealed. This is another way to see how the square and the triangle are two different views of looking at the same spatial order of spheres. It is a matter of how one chooses to view and from which perspective we chose to draw information.

Below. Another way to view what is going on is to see a single layer of spheres pictured as circles packed one next to each other. This is the hexagon matrix of endless spheres. Each sphere is the center for 12 tangent spheres; each circle is the center for 6 circles around it. As each of that infinite number of circles/spheres expands like rubber balls at the same rate and time, they will fill the space between and form a cubic "all-space-filling" arrangement. That does not account for the boundary, which would be expanding rapidly in all directions of infinite possibilities without containment. Cube packing is an abstract model for seeing the center space relationship between spheres. Using center points of spheres, tangent points, and the intervals, endless divisions of diverse individualized shapes and forms can be generated within the Whole circle/sphere context.

The spaces between the spheres reflect the same pattern of the spheres themselves. This reciprocal interaction is the in and out breathing of discrete transformation between the sphere and cube. Forms are specific, never fixed, and always connected. Separation is a choice of selective isolation from the greater matrix of spherical relationships.

253

TETRASPIRAL

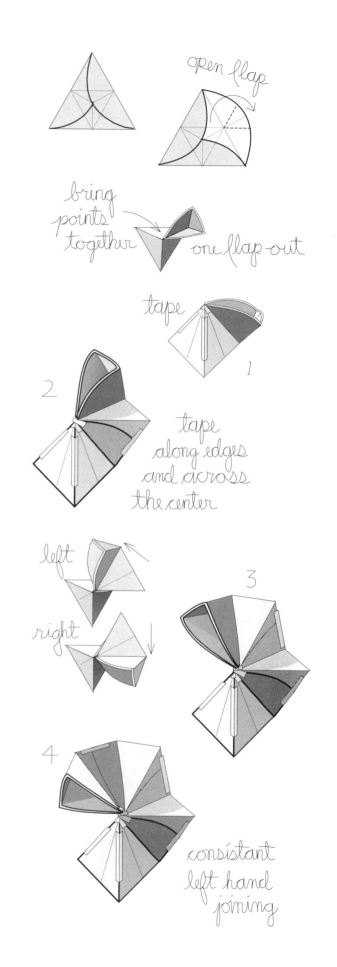

Fold a tetrahedron and open to the triangle.
Open one third of the circumference.

Refold the tetrahedron with the curved flap out
forming a tetrahedra interval (p.230). Tape edges
to closed tetrahedron.

1.) One unit

2) Make another unit the same way. Place one
tetrahedron into the interval of the other
tetrahedron. The top of each successive
tetrahedron can be used as the center location in
keeping track of the direction of the spiral as you
add more tetrahedra. Be aware of the position of
the second tetrahedron when placed in the first
tetrahedron interval.

3) Put one tetrahedron into the open flap of the
previous tetrahedron where the 2 points come
together at the origin of the flaps. The position of
one tetrahedron into the other will determine
whether the directional spin is right-or-left
handed.

4) Add a third tetrahedron to the first 2 and then
add a fourth. They are all the same size. This will
not form a spiral. It is important to do this first to
know how they go together and to see what
happens in a circular form. Make an assumption
about what eight will do moving around a center
location, then make it and compare.

254

Using tetrahedra, as we have just folded, that are progressively smaller in size will generate a spiral formation. The angles of the tetrahedra remain constant, but the interval of movement towards the center depends on how much the diameter of each circle is decreased for each individual tetrahedron.

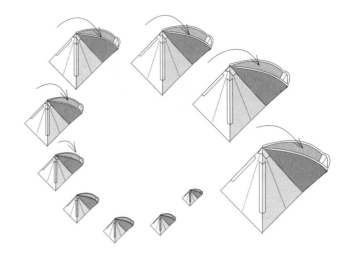

<u>Fold</u> one tetrahedron from a 9" paper plate circle. One third of the circumference will be folded out forming a tetrahedron interval. They will all be folded the same way but from successively smaller diameter circles; 9", 8½", 8", 7½", 7, 6½" and so forth. The tetrahedron of the second circle will fit into the interval of the first and so on in succession. The flaps need to be on the same side of all subsequent tetrahedra. Take it down to as small a circle as you can fold.

Experiment with how much to reduce the diameter each time; 1/8, 1/4, or 1/3 of an inch. Make a spiral to the Fibonacci growth ratio (p.46) where the diameters will diminish sequentially at a constant rate of decrease. There is much to explore in ratio proportions by varying the difference in diameters.

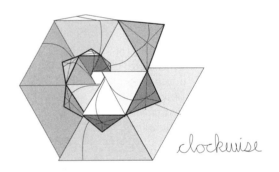

clockwise

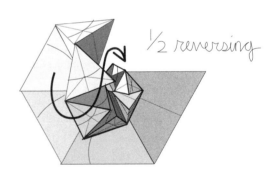

½ reversing

One unusual aspect of this spiral is that it can be reversed from front to back, and at the same time change the direction of the spin from clockwise to counterclockwise. In reversing front to back there are many in-between configurations to explore. This is not a static model of a spiral.

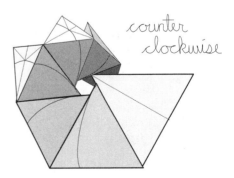

counter clockwise

Spirals move in 4 directions, in and out, right and left. You can make the tetrahedron folding progressively smaller and larger. You might cut large circles, 3-foot diameter, and go down to a 2" circle; or work out the other direction. Play with scale and complexity by combining and branching, making double and triple spirals. This is a fractal process.

Fig.22a 2 models of the same spiral using a diferent number of tetrahedra. *Left*; this has 11 tetrahedra, with ½ inch difference between each diameter of circle. *Right*, the intervals between diameters is about ¼ inch showing 16 tetrahedron units (p.255).

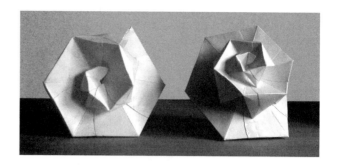

Fig. 22b These are reformed spiral systems. *Left*; is turned so the spiral is twisted to reverse spin putting it to the opposite side. *Right*, this has too many units to reverse itself, the movement is limited.

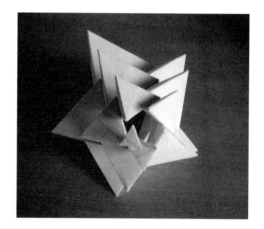

Fig. 23a 15 standard tetrahedral polyhedra with a ¼ inch difference between each diameter are used to make this spiral. Each unit is slid to the half mark of each preceding tetrahedron to form the helix formation By decreasing the size of each tetrahedron in this face to face sliding changes the helix into a spiral pentagon star arrangement with a pentagon center space (p.337).

Fig.23b Here is a different view of the same spiral *Fig.23a*. It shows the half sliding of each attachment that opens up the space.

INSIDE OUT DODECAHEDRON

Here we will form a dodecahedron using the inside/out tetrahedra arranged in pentagons to form an open dodecahedron.

<u>Fold</u> 5 inside out tetrahedra with flaps opened (p.230). Join them flat-end-to-flat-end forming a pentagon with an open space in the middle. Notice how unstable it is as a pentagon without a center point. The opening and closing of the flat end joining will change the angulation of the pentagon shape. (This will get unstable as it develops and you might want to use tape along with the bobby pins to help secure the joining.)

Add 10 more units around the first 5, making 5 more pentagon intervals. Join in the same way.

Make another pentagon set of 5 same as the first, and join to the assembled 15 units to complete 20 tetrahedral units, making 12 open pentagon intervals. Twenty tetrahedra in a pattern of 5 have formed a dodecahedron pattern (p.300, *Fig. 76*).

Adjusting the small flat end-joined flaps to about half open, a small rhomboid space, will help stabilize this system. When everything is equally proportioned, by bringing all joining into alignment, gluing the joints will help to keep it rigid. With the pentagons open there is movement between the tetrahedral units, a point joining arrangement. The bobby pins are good for assembly but they won't hold it very long. The tetrahedra are structural but the system is very compliant because there are no center points to the pentagons and nothing centered to the inside.

The 20 corner points of the dodecahedron are the end points of the 20 tetrahedra. The inside space shows 20 triangle faces and 12 pentagon planes. This is called an icosadodecahedron, one of 13 semi-regular polyhedra (p.176). Truncating both the dodecahedron and icosahedron will generate this same relationship of triangles and pentagons. This shows the reciprocal nature of both the icosahedron and dodecahedron that together form a solid-faced model of spatial organization.

1 tetrahedron

5 tetrahedra

20 tetrahedra

20 triangles

12 pentagons

combined

257

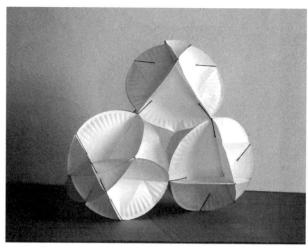

Fig.24a 4 sets of 4-circle irregular units in a tetrahedron arrangement. This end point view shows the 4 points are opened to square intervals.

Fig.24b Triangle plane view shows the irregular and twisted plane.

These are made from 4 of the 8 circles used to form the transformed spherical rhombidodecahedron (p.240). The open part of each half sphere has been opened changing the proportional information allowing it to be combined in multiple ways forming a variety of non symmetrical arrangements.

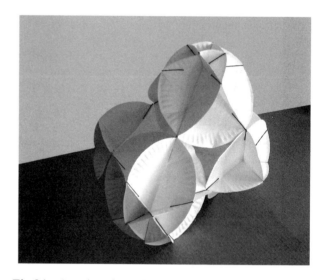

Fig.24c An edge view of the tetrahedron arrangement.

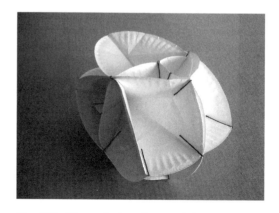

Fig.25a 4 circles used above has been reformed as an oblong sphere.

Fig. 25b The end view (lower right) shows the 4 symmetry.

258

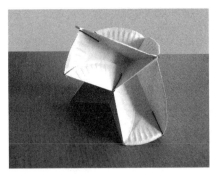

Fig.26a,b,c. 3 views of the same 6 circle unit. *a)* front view, *b)* side view and *c)* view from back. This set is a variation of the folded system on (pp.258, 241). Bobby pins have been used to hold circles together.

Fig.27a,b,c. 3 views of the system above, *Fig.26a,b,c* with a changing placement of a few bobby pins around the outside and the center front. This set is used in the tetrahedron arrangement pictured below in F*i.g28a,b.*

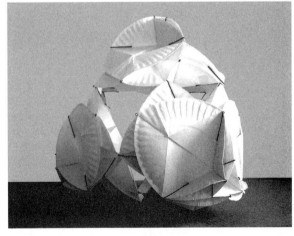

Fig28a. 4 units shown in *Fig.27a,b,c* are assembled into a tetrahedron that has square holes on 2 axis through the center. The units are joined using bobby pins, 2 looped through each other for joining points.

Fig.28b. Another view of the same system pictured in *Fig.28a.* Looking obliquely through the square holes to show the open triangle and the inside octahedral space.

(8) SOFT FOLDING

Soft folding is when the circle is folded but not creased. It is not necessary to crease all folds to show the inherent geometry in the reconfigurations in the circle. There are many variations in soft folding that in combination with creased folds reveal a rich expression that extends modeling towards the more diverse forms observed in nature.

Touching two furthest-most points on the circle exactly will bring together an open tubular form where all longitudinal lines are parallel at right angle to the direction of curving. There is an inside and outside surface, two separate edge lines, one straight diameter and one point of contact. By rolling the contact point on the circumference, the cylinder will form a cone configuration of varying degrees of openness at both ends. When the two points overlap, the inside surface will slide over the outside surface and the tube diameter decreases. Very simple. It is a good a exercise to describe the functions.

<u>Fold</u>   3 internal chambers funneling down centering the circle. The curved spaces slide against each other, proportionally changing relationships. Divide this cone into 3 equal sections. Continue to reform without creasing. Hair pins can be used to hold positions.

Explore some possible variations by soft folding and curving the circle. Use your fingers to curve the surface. Paper is very pliable in that way. Use hair pins to hold reconfigurations. Become familiar with curving the circle, that is what it does most naturally.

<u>Fold</u>   the 3 diameters and triangle (p.102). Using creased lines, fold up the longer parts of the diameters from the center-to-end points. Fold in the short parts of each diameter making an in/out three arm- star configuration (p.216). Use finger to open the space, spreading each arm and rounding up to the center. Place bobby pins to hold the center.

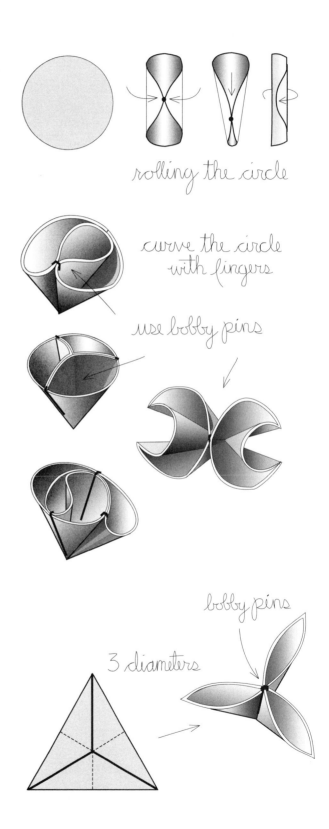

rolling the circle

curve the circle
with fingers

use bobby pins

bobby pins

3 diameters

This reconfiguration of the circle is an irregular tetrahedron. The pinned axis of the unit allows arm rotation changing the angles.

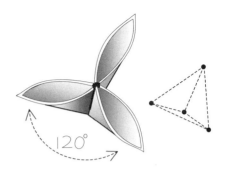

Put 3 units together in a triangle using bobby pins to hold them together edge-to edge.

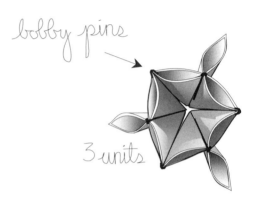

Add the fourth unit edge-to-edge, completing the tetrahedron arrangement. This forms 6 points that form 8 equilateral triangle planes, which is also the octahedron relationship.

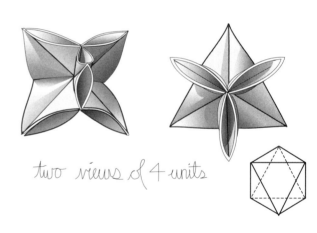

In joining 4 sets of 4 units each into a tetrahedron, it only appears as a VE arrangement because of the image compression. This system suggests other ways to explore the tetrahedron and octahedron combination.

<u>Fold</u>   and join a 4-unit set from previous page.

Add 4 more individual units to the octahedron assembly. They fit curved edge-to-edge into the 4 triangle intervals. Use bobby pins to hold the curves of the arms together. Bobby pins can be placed at the end points or slid down the edges, which will open the ends giving a small but interesting detail variation, as shown to the right.

By opening one arm of each of two sets of 6 arms (eight circles each), they can be joined. Slide bobby pins down on one arm and remove bobby pins from the other. One arm can then slide into the arm of the other. This is a variation of the in/out octahedron forming (p.129). Use bobby pins or tape. Glue will work best in this case. If you glue them together, the individual systems being joined should be glued in process, making sure there is consistency in proportions between units. Each system will have unique joining requirements, which provides practical mechanical problem solving.

Eight sets of 8 units each will join together making a cubic relationship. By continuing to add more cube units to the extended arms the cubic matrix can be formed, which is simply a division into, in an outward movement.

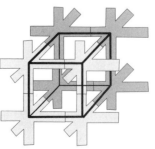

263

There are many interesting directions using the 3-diameter, 3-arm star-curved unit.

<u>Fold</u>   one end point to the center, forming a small closed tetrahedron from the open pedal shape. Do not crease, simply fold over letting the fold round and the edges come together with the tip to the center. Tape and/or glue on the edges.

Do this to all 3 arms. This forms a curved, hexagon-based pyramid. (There are 3 formed triangles and 3 intervals.)

Make 2 more hexagon-folded units in the same way. Join all 3 together point-to-point forming a large and small open triangle. Tape across the points. If you wish, add glue to make joints more permanent.

Make a fourth unit and join it to the 3 already joined. This will complete the open triangles in a tetrahedron arrangement. Secure the corners as above.

This form shows a combination of the tetrahedron and truncated tetrahedron pattern at the same time (pp.176, 248). It also shows hexagon faces stellated rather than the triangle planes. There are 16 convex points, 20 counting the inside 4 points in a tetrahedron arrangement.

264

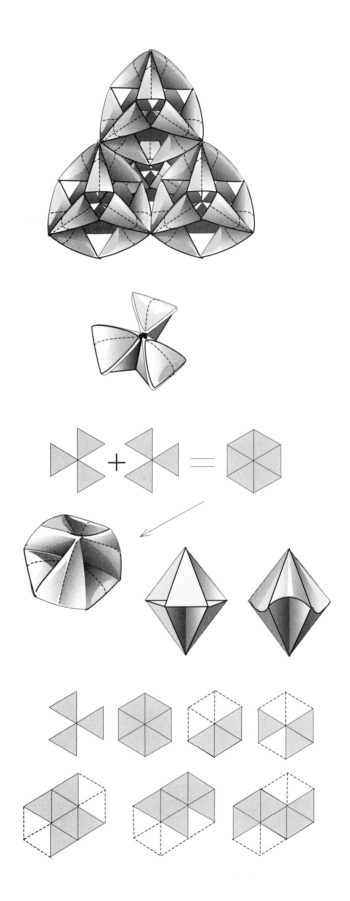

Fold and make 4 sets of these tetrahedra and join into a 2-frequency tetrahedron. This makes a unique formation of that pattern. Tape and/or glue, even looped bobby pins will hold the 4 sets together at the 6 points (p.266, *Fig.32a,b*).

Fold again, making the same single 3-arm unit of folded-over points.

With the hexagon pattern of 3 formed and 3 open triangles, it suggests that we could fill in the open triangle with the formed triangles of another unit.

This makes a bi-hexagon pyramid of 6 alternately formed tetrahedra and 6 open tetrahedra intervals. Tape across where the points come together. There are variations to explore by curving the edges.

There are other combinations for joining two of these hexagon units. They can be joined by opposite intervals, joining on edges and face-to-face using the same system. Each variation has in turn a number of different directions for systems development and can be assembled in a variety of different combinations. Use the axial movement.

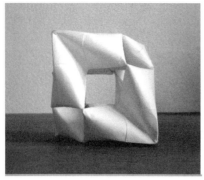

Fig.29 4 circles in a tetrahedron arrangement viewed from edge (p.264--).

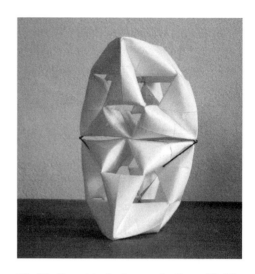

Fig.30 Two tetrahedron units from *Fig29* joined on a common edge

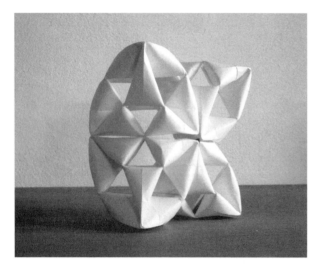

Fig.31 Four units *Fig29.* and 2 sets of *Fig.30* in a square planer arrangement seen from side angle.

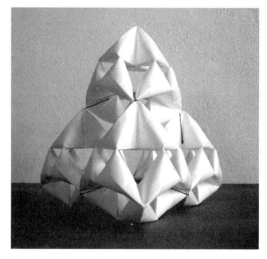

Fig. 32a 16 circles soft folded and joined in to form a tetrahedron. The 6 edges of the tetrahedra have become open rhomboid shapes.

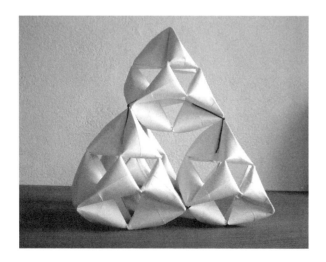

Fig. 32b Another view of *Fig.32b* . The folding of this unit is on (p.264). Because of the curved surfaces it has greater strength than if it were all flat surfaces. The individual units are glued together and the 4 sets are joined point-to-point using hair pins looped through each other.

Fig. 33a An arrangement of 32 tetrahedra form a tetrahedron net. Each tetrahedron is reformed to the unit shown on (p.264) and (p266,*Fig.29*). The connections between units are made by folding back on the 3 arms where the triangle flap of one closes the corresponding triangle opening of the other. The net is interrelated hexagons, triangles, and rhomboids. Tape and hair pins are used to join units together. This is a flexible, transformational system.

Fig. 33b Here the net is curved into a half circle.

The two picture below show two views of completing the curving in and joining on the 4 corners. This forms an unstable tetrahedron arrangement (*Fig 33c*). By closing the open spaces between the edges the tetrahedron becomes stable (*Fig33d*).

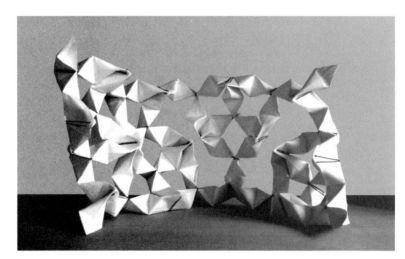

Fig. 33c Viewed from the opened edge, partially closed towards a spherical arrangement.

Fig. 33d Viewed from end point of transformed net fully closed into a complex tetrahedron arrangement.

Here is another direction using the folded over curved petals.

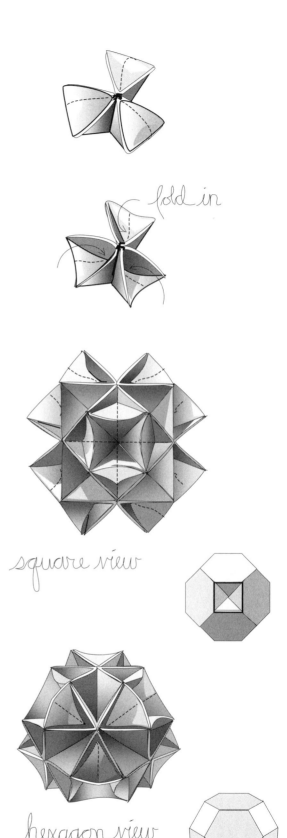

Fold another circle with 3 petal ends being folded over and down into the center of the circle, forming a hexagon-based pyramid (p.264).The curved edges get folded over to the inside.

fold in

Make 8 of these units. Arrange 2 sets of 4 with each set in the shape of a square interval joining on the edges. Put the 2 sets together edge-to-edge. Use bobby pins to join. This is similar to the process of forming the first vector equilibrium sphere (p.89).

square view

This end point view shows the hexagon faces of the truncated octahedron arrangement (p.176).

hexagon view

<u>Fold</u> the full tetrahedron, then flatten to the 2-frequency triangle. Then go back and fold the 3 armed star with the arms curved out like pedals, (p.261)

Make another the same and slide the arm of one into the arm of the first one. Use the creased line to know when to stop. Line up spine on the arms and glue and/or tape. This is related to the slid joining on p.249.

Make another set of 2 in the same way. Put the 2 sets together forming a square, by sliding each arm into the other. Glue and tape as before.

Make another set of 4 and put them together in the same way forming a cube. When all the glue is dry notice how it still wants to collapse, spiraling along the diagonal lines because of the pinned corner axis. This demonstrates how unstable the square/cube is without the diagonals, which usually take the form of diagonal struts or the diagonals that are inherent to a solid plane.

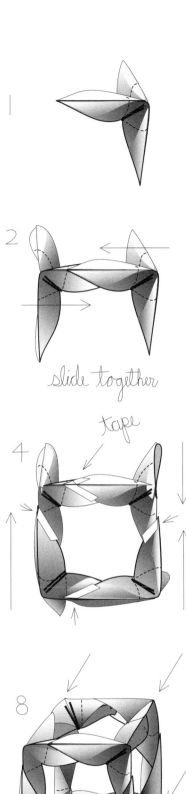

slide together

tape

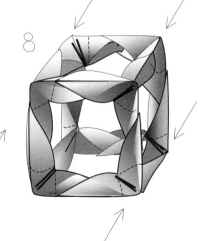

<u>Fold</u> the tetrahedron and refold into the 3-armed star with the circumference on the outside of the diameter folds.

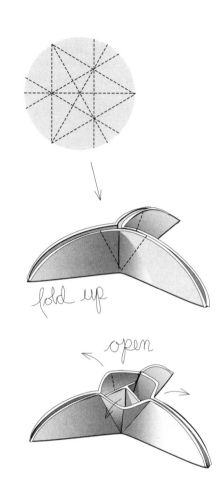

Locate the 6 creases on each side of the center axis, 3 sets of 2 that meet at the folded center of the circle.

Pull the center crease between each set of 2 creases out away from the center axis. They will form a hexagon open space with the folded diameters showing inside.

Turn the unit over so the inside of the arms are facing up and the open hexagon facing down. Put 3 bobby pins in the creases to hold the hexagon tight. Open the arms with your finger.

Place a finger underneath on the diameter spine between the folded-over curved sides and push it up, forcing it in the opposite direction. It will pop up and force open the arm making it very wide with a somewhat flat triangle on the end and then it turns inward to form a concave tetrahedron space in the opposite direction.

Make 3 more units the same way, having 4 altogether. Join 2 of them together on 2 ends by putting the flattened triangle edge of one arm end next to the same on the other unit. Put a bobby pin holding the two creases together.

Do exactly the same thing on the other 2 arms, one of each, and attach with a bobby pin. This will make an arrangement similar to an angled rhomboid that is half a tetrahedron.

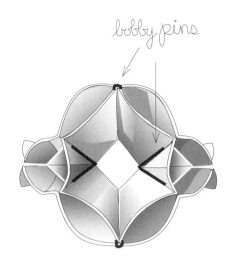

Put 2 sets of 2 together in a tetrahedron arrangement joining the end-curved points of the triangle unit. This shows a relationship of the dual tetrahedron.

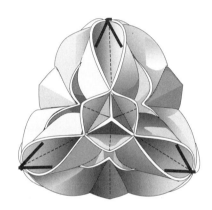

Make 4 of the units above and bring together in a tetrahedron pattern, joining at the end points with bobby pins.

All triangular-formed units can be joined into the tetrahedron, octahedron, icosahedron and through reformation the duals. The circle models the inner and the interrelated nature of forming what is spherically Whole.

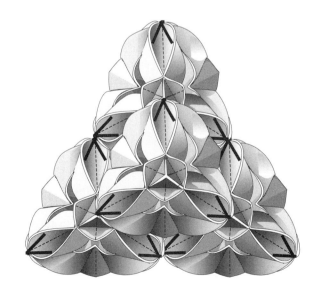

271

<u>Fold</u> the tetrahedron and reform as shown (p.270). There are many directions to exploration using this configuration.

With the 3-arm star configuration pull out the 3 center folds showing the hexagon shape, revealing the center of 3 diameters (p.270).

Leaving the center hexagon open, fold 2 of the arms towards each other using the folds that are there. At this stage there are other options for reforming. Follow and explore the creases as they want to move.

Bring the 2 arms together, closing part of the hexagon opening. Play with opening and closing, moving different parts in relationship to others, always using the creases that are there. Use hair pin or tape, they are easy to remove when you want to make changes or add another unit.

The closed form can be seen, used as parts of a system combining different reconfigurations (pp.279, *Fig.38a,b*, .282, *Fig.48a,b*). The open form joins the lower arms curved in by slipping one end into the other. Tape to hold.

Again start with the 3 diameters and the folded triangle. Sometimes it is unnecessary to fold all 9 lines of the tetrahedron, but it is always important to understand the context of what lines you do use. In the same way it is important to understand the context of anything you are working with.

Fold the equilateral triangle from the 3 diameters. Push up on the bottom and down on the short ends of the diameters, the midpoints of the 3 edges pushing up the circumference, forming 3 arms (p.270).

Curve 2 arms towards each other using your fingers to form them.

There is a double layer for each arm so as they come together one will slip inside of the other. When bringing the ends together into a circle, use tape or glue.

There are a number of variations to this reconfiguration of which only a couple are shown. The advantage of multiple folding over of surfaces is that they can be opened and unfolded in many different ways. As far as I can tell there are about as many possibilities in reforming units as there are numbers of systems that can be developed using multiple units.

The circumference can be spread in an open configuration or closed, bringing the edge together in a closed form. Glue or tape the circumference edge together. This unit is used for the model on page 283, *Fig.51*.

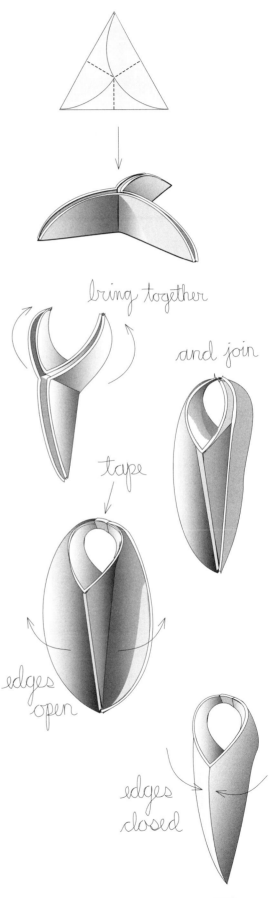

bring together

and join

tape

edges open

edges closed

<u>Fold</u>   the 3-armed star and curve 2 arms down and around on top of the third arm, forming an open circle.

Curve into a circle in front of the straight edge. Overlap the ends. One will fit into the other; tape and or glue ends together.

Curve into a circle behind the straight edge on the circumference edge. Explore other variations. The individual units may not be of personal interest, but the many systems that can be developed from multiple use of any form and combinations are always interesting and varied (pp. 277, *Fig.36*; 282, *Fig.48a,b*).

1) <u>Fold</u> a variation by starting with the 3 diameters and the large triangle. Fold the inscribed hexagon (p. 344). Open the circumference flaps showing the 3 vesicas in the triangle form.

2) Fold under one-third of the folded triangle to the opposite side.

3) Push in on mid-point until the folded crease touches the opposite crease of the same diameter. Push the mid points of the two other triangle edges behind until they meet. Use a hair pin to hold the edges together. It will look like what is pictured on the following page.

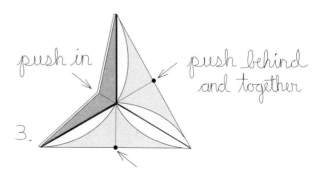

4.) The bobby pin will hold the 2 sides together that have been folded to either side of the third arm. This is what it will look.

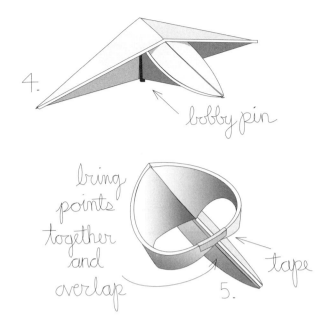

5.) Bring the end point together on top of the vesica shape. Tape or glue when overlapping or slipping one inside of the other. Remove the bobby pin behind, that will allow flexability and further variations in reconfiguring. This is a variation on the circle from the last page. To see examples of systems developed using this unit go to (pp.280, *Fig.45;* 281, *Fig.47a-f;* 303, *Fig.83a,b*).

If the 2 arms can curl out they will curl in, if to the right, then to the left, if in the front then to the back. Explore the obvious symmetries and then from symmetries in combinations, develop assymmetrical systems.

Put the 2 arms together and tape along the circumference edges. Open the vesica-shaped inside folding.

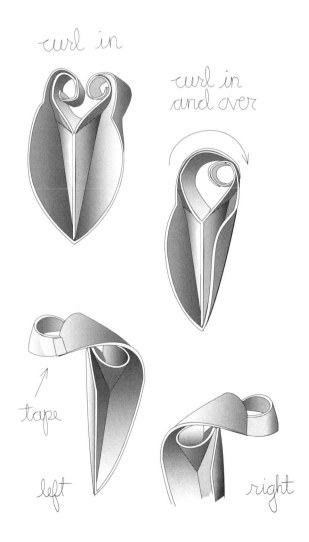

The 2 arms have been curved towards each other and one arm has been twisted 180° before sliding it into the other. This forms a Mobius surface with the 2 arms (p.369). This will take a little doing since one of the arms needs to be curved in the opposite direction from the other, yet still coming together. Feel the paper and work with it.

Here is a variation in forming the 3 arms, which changes the form of the Mobius surface. There is a left-and a right-hand curving.

There are other variations in using the Mobius function. After folding the 3 diameters then fold the inscribed hexagon (p.344). Continue to fold as before, twisting the two arms 180° before attaching the end points together. There are many alterations that can be made along the way. The soft folding allows easy adjustments to be made when joining multiple units together (p. 277, *Fig.35a-b*).

Look for variations on (pp. 278, *Fig.37a-d*, 279, *Fig.40-41*; 300, *Fig.75*).

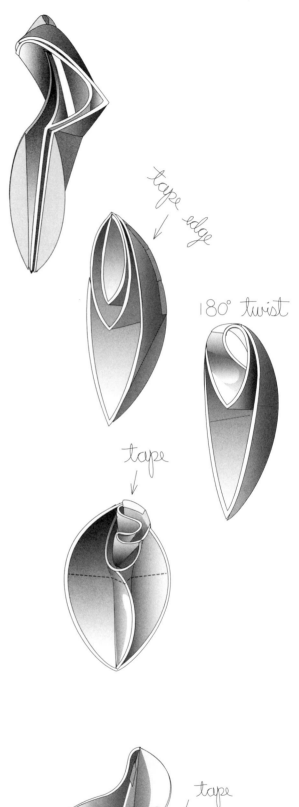

tape edge

180° twist

tape

Here the Mobius, with a doubled surface, is opened to show a few more curves. There are now multiple sides disrupting the flow of the single surface. Using variations of these units explore joining multiples by working with fundamental relationships of polyhedral association and organizing into planer, linear and spherical systems. Because of the complex surfaces they easily assemble into many kinds of helix and spiral systems.

<u>Fold</u> the triangle again and curve the circumference half up and half down on each of the 3 divisions of the folded arms. Tape the up-folded side to each of the 3 arms. The down-folded part has nothing to tape against. Explore various systems using multiples of this unit.

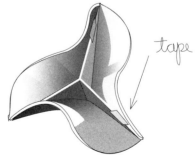

tape

276

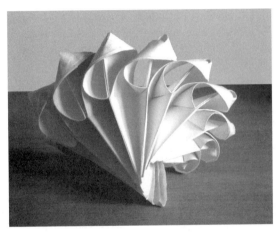

Fig. 34 This is another view of the spiral shown on the front and back cover of this book. There are 32 circles of decreasing diameters from 9" to 3"at ¼" intervals. It is made using the folded units shown on (p.296). Other forming of the same unit can be seen on (pp.304,305). Each unit slides into the next as diameters sequential decrease. There are many ways to fit these units together.

Fig. 35a Frontal view of 14 circles in a spiral form with diameters decreased 3/8" from 9" down to 2" . The circle formation used is shown on (p.275).

Fig. 35b Opposite view of spiral *Fig.35a*.

Fig. 36 Detail of icosahedron sphere. Made using 120 9" paper plates. The straight edges form an arrangement of the Rombicosidodecahedron, that shows triangles, squares, and pentagons. The hair pins can be seen that are used to hold together 60 sets of two. There are variations of this unit and many types of systems that can be developed.

277

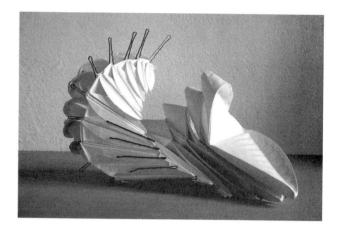

Fig.37a. A spiral formation of 17 circles from 9" down to 3" in diameters. The individual unit is a variation on a mobius folding, (p.276). The hair pins are holding the individually formed circle together before they get glued. In some cases pins and other means of attachment can add interesting elements to the development of the form.

Fig. 37b. Another view of the same spiral formation. Each diameter circle is folded the same.

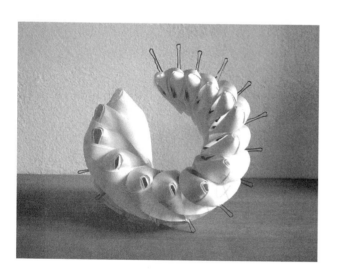

Fig.37c. This is another view of the same spiral (*Fig.37a,b*). Forms can look very different from selected views that completely change how we interpret the information about what we are looking at the image.

Fig. 37d. This is another view of the same spiral with hair pins removed, it has been glued together. Though it is not evident there is a lot of movement and flexibility in opening and closing into a tight spiral.

278

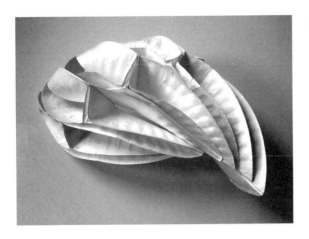

Fig. 38a. <u>Above</u> 6 paper plates circles formed to a mobius curved unit at one end and joined in an open helix pattern (p.276).

Fig. 38b <u>Right</u>. End view *Fig.38a.*

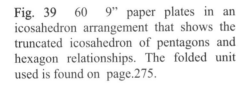

Fig. 39 60 9" paper plates in an icosahedron arrangement that shows the truncated icosahedron of pentagons and hexagon relationships. The folded unit used is found on page.275.

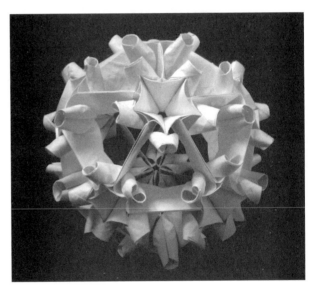

Fig. 40 17 circles from in a mobius curved unite (p.276). The circle diameters spiral from 9" to 3".

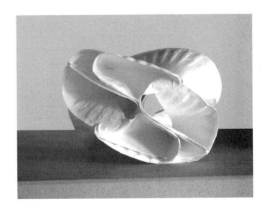

Fig. 41 4 circles formed to a partial mobius surface arranged to a tetrahedron (pp.276; 300, *Fig.75*).

279

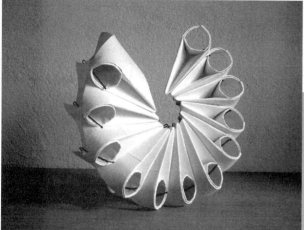

Fig.42 12 circles in a helix form with another variation of soft forming (p.273). Hair pins are used to hold units together. The same unit is used differently in the rearrangement at the bottom of the page, *Fig.46.*

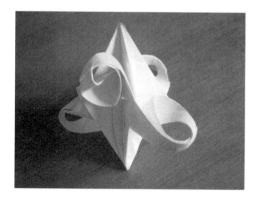

Fig. 43 This model is made from 16 circles folded into the same unit (p.296). Four were then joined in tetrahedral units. Those sets of 4 were again joined into the tetrahedron arrangement making this a reformation of the four-frequency tetrahedron.

Fig.44 Using another variation on the mobius surface (p.285) 4 circles are arranged in a tetrahedron showing both the intersecting tetrahedron and octahedron.

Fig. 45 24 units pictured on next page F*ig. 47a.* in an octahedron arrangement showing a 4-fold symmetry from the end point.

Fig. 46 18 of the same units used in *Fig.42* at top of page. Each unit fits into the opening of the next forming the circle.

Fig. 47a *Left.* A single reconfiguration (p.275). *Right.* A group of 5 joined edge-to-edge in a pentagon. There is good flexibility to the pentagon plane providing changing angles for more diverse ways of assembly, previous page (p.207, *Fig.45*).

Fig. 47b Viewed from underneath the pentagon edge joining is visible. Tape or pins can be used; these were glued.

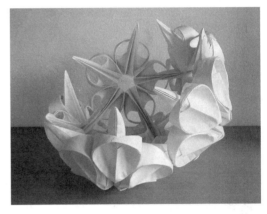

Fig. 47c 3 sets of 5 joined in a triangle. The arms of one slide inside of the arm of the other with both ends under the top flap rim on each others side. That holds them in place without glue or tape.

Fig. 47d 3 more sets of 5 have been added to form a pentagon pattern of 5 around the center making 6 sets. This is ½ of a dodecahedron showing both the inside and outside.

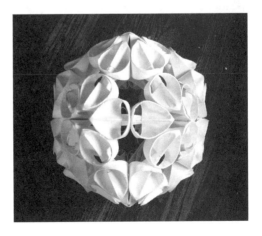

Fig. 47e The completed sphere has 60 folded circles. It shows both the icosahedron of 20 triangles and the 12 pentagons of the dodecahedron.

Fig. 47f View from the icosahedron edge.

Fig. 48a 5 units (p281, *Fig47a*). They have been combined with 5 more differently reconfigured circles (p.272). There are endless combinations of mix and match forms.

Fig. 48b A different view of the same model *Fig.48a*. The form is in a 5-10 axial symmetry. Ten reformed tetrahedra are glued together.

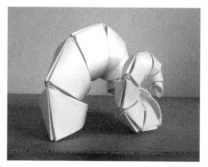
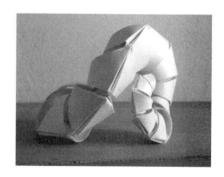

Fig. 49a.

b.

c.

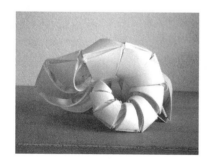

d.

e.

f.

Fig. 49a-f These are different positions of a single spiral movement system where the individual unit diameters diminish from full 9" diameter to 2" diameter. The units that are used to make this are the same as on (p.281, *Fig.47a*) and above (*Fig.48a*). This spiral has 16 circles folded to the same configuration (p.275). Each unit fits under the open circle of the last and is tied front and back of each unit with string making a very flexible movement system. This unit is a good example of how sometimes by changing the method of attachment what is static becomes dynamic.

282

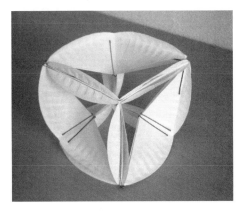

Fig. 50a 4 units joined in a tetrahedron arrangement. The ends are joined by 2 hair pins one connected through other. The hexagon has been infolded to show the small pedals (p.270).

Fig. 50b This tetrahedron collapses flat into a vesica shaped bundle similar to the cube where without diagonal edge definition there is no stability (p.218).

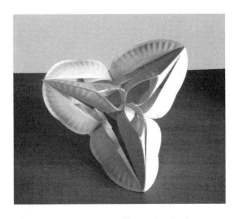

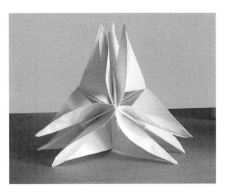

Fig.51 12 reconfigured circles are arranged into a tetrahedron. Each triangular side is a set of 3 circles using the unit on (p.273).

Fig.52 4 reconfigured circles in a tetrahedron arrangement (p.296--).

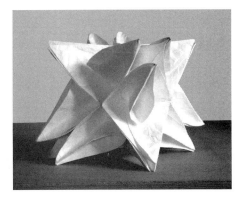

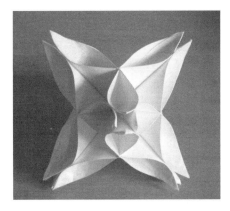

Fig. 53 8 reconfigured circles in an octahedron arrangement. The circles were creased, crumpled up, then flatten and reconfigured to get a crinkled surface.

Fig. 54 Another form of 8 reconfigured circles in an octahedron arrangement (p.261).

283

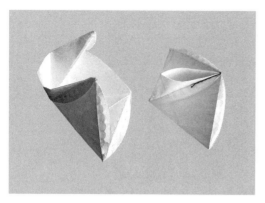

Fig.55 *Left.* Similar to first fold (pp.237,285). Below *Fig.58* shows 2 joined. *Right.* The folding bobby pinned closed. This is the unit for the to the right *Fig.56* and the system immediately below *Fig.57a-b.*

Fig.56 3 pinned circle shown to the right in *Fig.55*, are joined in 3 sets of 3 to make this model.

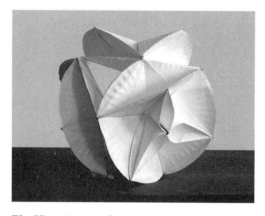

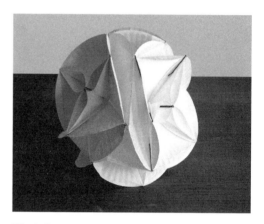

Fig.57a 4 sets of 3 in previous *Fig.56* are joined in a tetrahedron relationship that shows the truncated tetrahedron where the 4 triangles are open to the inside.

Fig.57b Another view of *Fig.57a.*

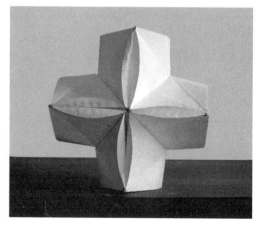

Fig.58 2 sets of 2 are joined where the 4 triangles form a stellated tetrahedron *Fig.55, left.*

Fig.59 8 are joined to form an octahedron using 4 sets of 2 units made from the circle pictured in *Fig.55, left.* This is a another forming of 3 intersecting rhomboid prisms through an octahedron center (p.347).

Another variation of a soft-folded unit from the three-arm star.

Fold the tetrahedron with an inscribed hexagon (p. 344). Reform the large triangle and fold in half on all 3 diameters.

Fold the small flaps out from the triangle. The small circumference flaps become a detail having their own variations.

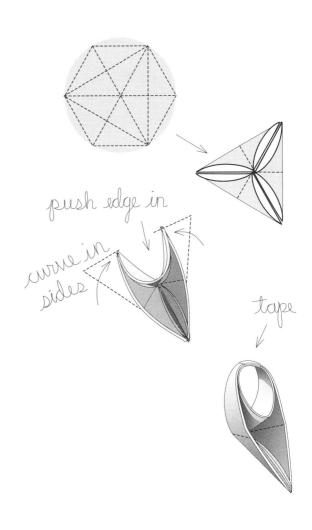

Fold the short part of one diameter in at the same time curving 2 arms toward each other.

Complete curving by bringing the 2 arms together and overlap them a little bit making an oval kind of space. Tape and/or glue them together.

Using your fingers to round out the inside edge of the open space giving an even-curved shape.

Fold 5 more circles the same way, 6 all together. When tapping each circle make overlaps all the same distance. Consistency is important.

Make 2 sets of 3 folded circles as shown. Join 3 straight edges together as a central axis and use 3 bobby pins to join one to the other at the pointed end. The end opposite from the bobby pins remains open. These 3 units form a tetrahedron relationship with a point symmetry axis.

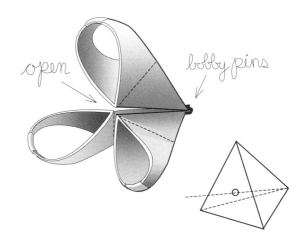

With the open axial ends (opposite ends from the bobby pins) facing each other slide the 2 sets together alternating between each unit.

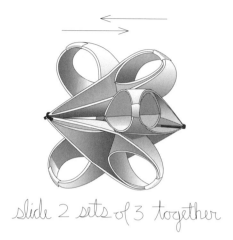

slide 2 sets of 3 together

Slide them all the way up until stopped by the bobby pins, preventing them from passing through each other. The units will open as they slide together, forming an octahedron relationship of varying proportions.

slide ends to center

Explore other reforming variations. Pushing up on the end point inside each hole will form a tetrahedron coming out the top. This will slightly change the look of the open heads and further stabilize the system; all movement affects other parts. Where there is overlapping there is potential for opening and closing.

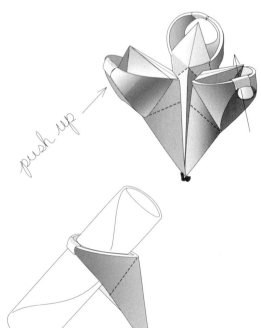

push up

When holes are formed by curving arms together they can be used to hold rolled circles (pp.287, *Fig.60a-f*, .288, *Fig.63-4*). Circle tubes can serve as connectors and are functional as rotational axial for movement systems.

286

FIG.60a.

b.

c.

Fig. 60a-f This movement system is made from the same unit used on (p.280, *Fig.42,46*). Circles have been rolled into tubes and used through the open holes for rotation and extension. Every other movement joining is an edge-to-edge hinged joint that has been tied together. There are a variety of forms that can be used in this kind of system.

d.

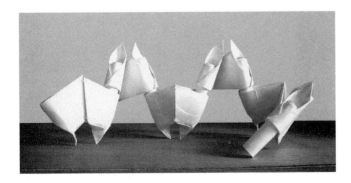

e.

f.

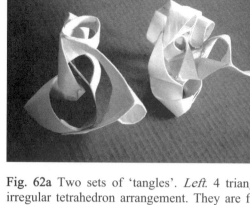

Fig. 62a Two sets of 'tangles'. *Left.* 4 triangles in an irregular tetrahedron arrangement. They are folded into the 3-armed star and curved and glued into a closed system. *Right.* Another view of *Fig.61.*

Fig. 61. A 'tangle' of 6 circles patterned to an irregular octahedron. It has one unattached arm seen in *Fig.62a, right.* For other examples of this curvilinear form see (p.379-381).

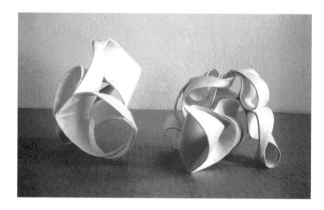

Fig. 62b A another view of pieces above *Fig.62a.*

Fig. 64. 5 rolled circles are inserted through some of the round openings clearly showing the tetrahedron nature of the tangle. *Fig.62a-b, left.*

Fig. 63 In *Fig.62a-b, right* further developed by adding on a rolled circle (p.261) and a spiral piece (p374).

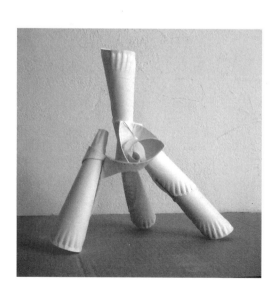

Here is another variation on folding three arms.

<u>Fold</u> the tetrahedron and open the circle out flat.

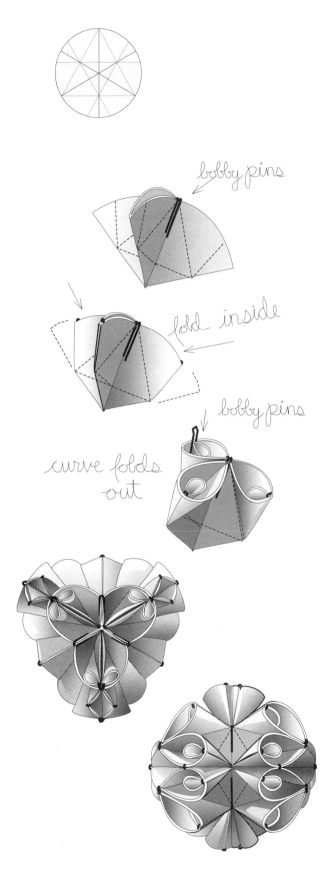

Refold the star configuration, pushing out the long diameter segments with the short segments pushed in (p.75, #9). Hold it together with 3 bobby pins from the circumference end.

Fold in on the outer 30° creases of each arm. Use a bobby pin on the inside to hold them closed.

Placing your finger into the space where the folds come together opens a large circle and a smaller circle inside of the large one. Do this with each arm of the star.

Net make a number of multiple units and explore various ways to join them; in sets of 2, 3, 4, 5, or 6 or what every looks interesting to pursue, in light of what you have folded so far.

<u>Right</u>. Two different systems; the upper image is 12 circles in a tetrahedron arrangement (4 sets of 3). The lower image represents 16 units in an octahedron arrangement (4 sets of 4). Use bobby pins and/or tape as appropriate for attachment.

Fold a tetrahedron and open the two-frequency triangle. Curve over one of the corners and push the end down towards the curving the surface. Curve all 3 corners the same way. Use your finger to get an even curve.

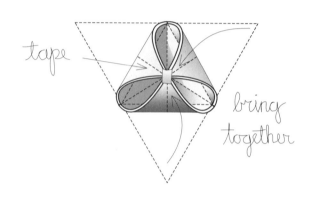

Point-to-point. Curve each point to the center forming 3 open vesicas to the end points of the base triangle. Put a piece of tape under one point as it curves over and bring the other two points together using the same tape. Put another piece of tape on the top. That secures the joining well. Push in the web of tape that spans the openings. Another good way to attach points is to sew them with needle and thread (p.298, *Fig.68*).

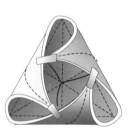

Point-to-edge. Slide the end points in one direction down one-half of the length of the adjacent edge making three smaller openings. It forms an open triangle in the center. Tape or sew together as before. For strong joining, use tape on both sides or twisty ties (p.299, *Fig.69*).

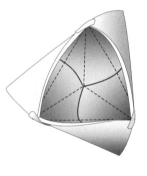

Point-to-opposite-point. This shows points slid all the way down to the opposite point of the same edge forming 3 tighter circles and a larger open center triangle interval (p.298, *Fig.66*). It does not matter where, along the edge, the joining happens. In the beginning be consistent for all 3 and for any multiples in the same system. Later you can begin to mix things up.

Edge-to-edge. Roll the above point-to-point so the length of the edge lies on the adjacent edge of the two-frequency triangle forming 3 small cones. The best way to hold this is glue, edge-to-edge (p.298, *Fig.66*).

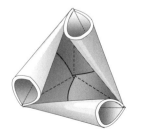

To take this edge-to-edge curling a little further continue to roll towards the center until the opening becomes tighter and flatter.

1 circle

Fold and roll 4 circles into the above configuration and put them together in a tetrahedron. The flattened curls come together forming 6 rectangular planes, corresponding to the 6 edges of the tetrahedron but angled differently.

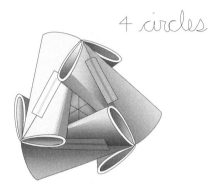

4 circles

Disregard the 4 points of the tetrahedron and with an imaginary line or using string, connect the outer points of the rounded ends of this form to show an irregular icosahedron relationship. The model shows 4 tetrahedron-folded circles with 3 sides curled and arranged in a tetrahedron which will form the 12 points of the icosahedron. Other reformations of the same tetrahedra will inherently hold the icosahedron that becomes visible through differently formed relationships.

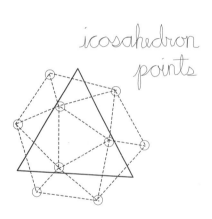

icosahedron points

O *verlap spin.* Here is another variation, which starts with the point-to-point touching, and the triangle continues sliding over/under each other, keeping the center closed to a common point of crossing. This creates a 3-arm center with open ends. As with most soft folding you will need to curve the edges with your fingers to get fluid curves, which will also help hold the parts in place when gluing.

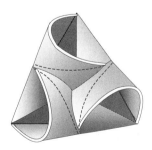

291

Overlap joining. Overlap the center points slightly and glue them together. This is a much stronger joint than taping point-to-point. It is not as clean as needle and thread.

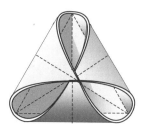

Increase the overlapping, making the center triangle larger as the 3 vesicas become smaller.

The points can be pushed to the center of the circle creating a tetrahedral indent. This will take glue, using tape to hold it until the glue dries.

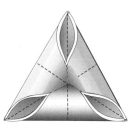

Finish the movement by pushing the three open vesicas closed. Tape or glue edges together.

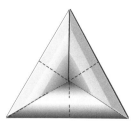

The overlap can go all the way until the vesicas close up. This will form a round-edged layer of a partial truncated tetrahedron (pp.151, 331).

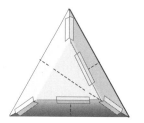

292

The more you play with the soft folding the more variations appear. This happens particularly when you combine them into sets and in multiples to larger assemblies. Some variations will form easily and others not so easily. Each small difference will always change what the forms will do within the symmetry restraints of specific systems. Reforming always pushes the boundaries of a system, which is about accommodations towards growth and change.

Start with some of the folding you see here, staying with basic patterns and systems. These will always yield variations, some more interesting than others. Go towards what looks interesting to you. Each different form is important in what it has to reveal about pattern and the forming process.

<u>Fold</u> 8 triangles *point-to-point* and arrange to an octahedron; units joining on the edges (p.165). It shows the stellated octahedron, dual tetrahedron, cube, all in rounded arabesque forms. For a variation see (p.298 ,F*ig.67*).

Explore forming the tetrahedron and icosahedron arrangements using this unit.

<u>Fold</u> 4 circle tetrahedra, open to the triangle and form each to this *point-to-edge* configuration.

Put them together with the triangle bases to the inside forms a tetrahedron. This is another variation of an inside out tetrahedron that is soft-folded and pictured on page 298, *Fig.66*.

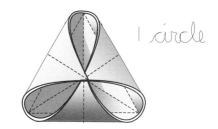

1 circle

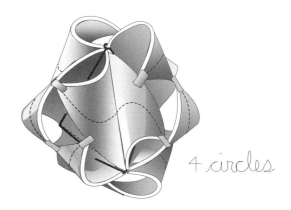

8 circles

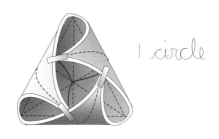

1 circle

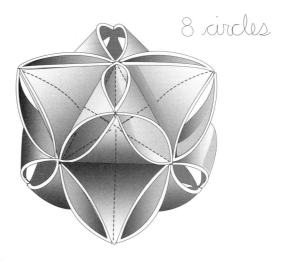

4 circles

<u>Fold</u>   4 tetrahedra and overlap the 3 ends forming a triangle connection.

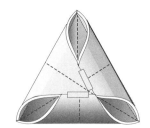

Put the top triangle towards the center, forming an internal tetrahedron with the large triangle base to the outside. Use bobby pins to hold them together.

This is an example of how, when 4 triangles of the tetrahedron open out, the 6 edge lines become 6 rectangular intervals. The 6 quadrilaterals plus 4 triangles, 4 triangle intervals, and 8 triangle planes reveal an irregular VE (pp.99, 134).

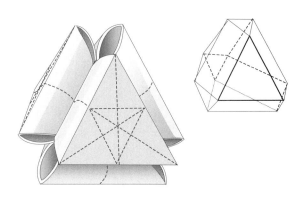

There are many ways things can go together. Congruent shapes, edges, openings and proportional relationships are key to determining potential options of joining. Consistency to process and purpose is an important guide.

<u>Fold</u>   4 tetrahedra reforming to this *point-to-opposite-point* unit.

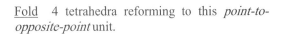

Arrange these 4 triangular units in a tetrahedron where the center triangles are to the inside joined on edges. The VE relationship is similar to what is above but may not be quite so obvious.

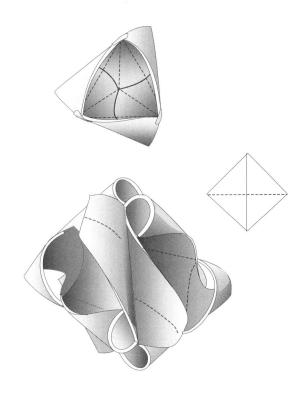

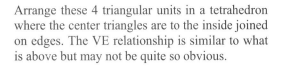

<u>Fold</u> the tetrahedron into a triangle and tuck the flaps one under the other to keep them out of the way. Explore using only 2 openings.

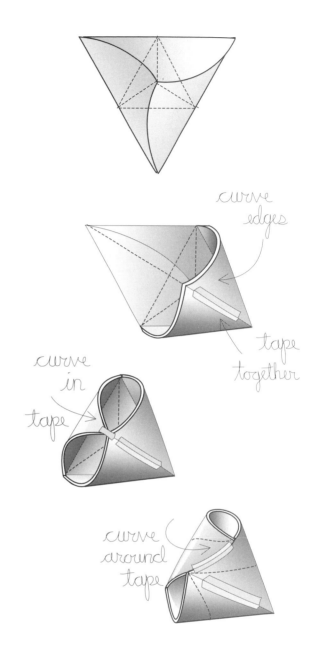

Tape 2 edges together along the length as in taping a tetrahedron. Curve in on the taped edges of one side of the triangle using your finger to round out the edges. Tape on other side for greater strength.

By folding the third point to the point where the 2 come together 2 openings are formed. Tape across the top and underneath for maximum strength. This is like a tetrahedron squished open on 2 edges.

Before taping slide the point to one side or the other about halfway down the edge and curl it until the edges touch. Tape edges together. This makes 2 different openings on each side. Sliding to the right or left determines the right-or-left-handedness of the unit.

Another movement that achieves a dual opening is to fold the full length of two adjacent sides curling from a common end point until the circling meets the mid point of the third edge. This divides the opening into two equal unconnected cones. Spreading the curled ends where they come together will further divide the space into three openings.

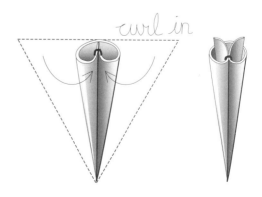

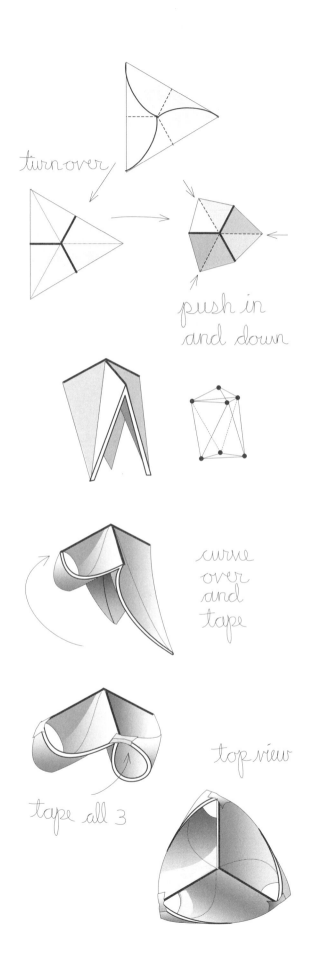

<u>Fold</u>   full tetrahedron is unnecessary, only the large triangle. Tuck in the curved edges so they all face in the same direction locking into the center crossing.

Turn the triangle over so the circumference is to the bottom. Push all 3 kite shapes in, folding on the radial bisectors. The short segments of the diameters become the 3 spines.

This will fold an in/out edge forming an octahedron/tetrahedron combination. It features the 7circle points.

Curve each bottom point to the end point on the top spines. Either the right or left side on top will work, but do all 3 in the same direction. (Later you can explore mixing up directions.) Tape works fine but is temporary. This is also good to sew with thread or kite string.

All 3 tapped point-to-point in the same direction forms 3 circle openings, much like a pinwheel. Two examples of different systems using these units in multiples   (pp.304, *Fig.87a-c*; 305, *Fig.89a-d*).

296

The paper will retain a memory of the curve when you remove the tape after a time. Or just use your fingers and impress the curves into the paper.

Tape the points half way to the center, or at any point on the short part of the diameter.

Play with the reconfigurations, discovering other variations and the systems that can be formed with them (p.301,*Fig.77a-c78a-c*, P.302, *Fig.79-82*).

Play is a crucial part of exploration, it not only is a means for revealing the unknown, but also provides balance and lightness to the "importance" of discovering something we just did not notice before.

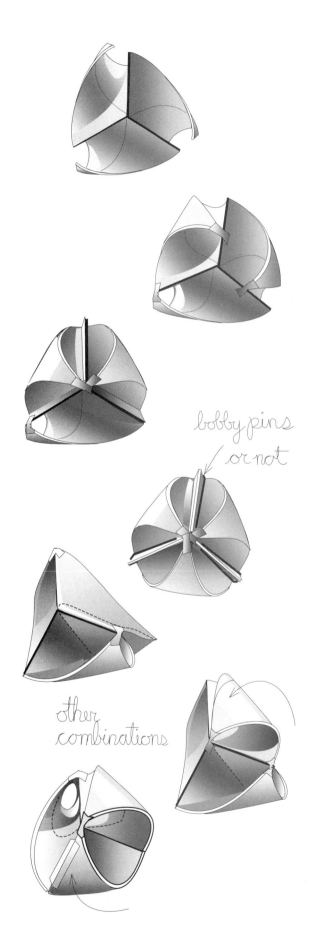

bobby pins
or not

other
combinations

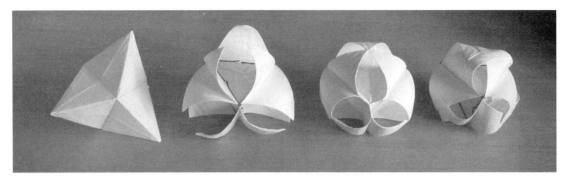

Fig. 65   A progression from the stellated tetrahedron to opening the stellated points, closing the points leaving the edges open. The model on the *right* shows the points overlapped (p. 292).

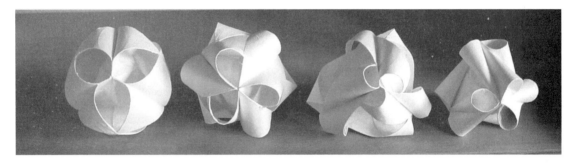

Fig. 66   Here a progression starting from the third model above *Fig. 65* and the points of attachment slides along one edge until curling to the opposite end of the edge from where it started (p. 291).

Fig. 67 Stellated octahedron with a variation to the opening of the stellated points where each curved surface is attached to the adjacent surface. There are 8 open tetrahedra (p. 125).

Fig. 68 The 3rd reformation in *Fig.65* are in an open stellated icosahedron arrangement. There are 20 tetrahedra in this model (p. 292). The points are taped and units joined with hair pins.

Fig. 69 Here is a single tetrahedron opened with end points of one triangle connected to the half way point on the edge of the adjacent triangle. Holes have been punched at the ends of the 3 points with corresponding holes punched midway on one edge. All 3 halfway holes need to be on the same side of each triangle. Twisty ties are tied through the holes joining the triangles leaving an open triangle space between them. Reforming the tetrahedron in this way changes the pattern of the tetrahedron form to a twisted truncated tetrahedron.

Fig. 70 A stellated octahedron formed using 8 of the twisted truncated tetrahedra units above *Fig.69*. They have been joined using hair pins.

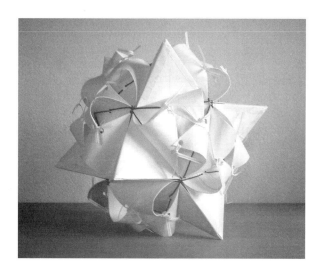

Fig. 71 An icosahedron using 20 tetrahedral units *Fig.69*. Start by putting 5 together, continue to add until the sphere is complete.

Fig. 72 8 units have been removed equally around the icosahedron having been replaced with regular tetrahedra. They show the 8 stellations of the octahedron that reveal the cube relationship with an icosahedron center (p.166).

Fig. 73 This is a form variation using the typical soccer ball arrangement of pentagons and hexagons, a truncated icosahedron. There are 120 tetrahedra that have been assembled into octahedra. The octahedra are formed by and open and closed tetrahedra (p.148). Five octahedra have been arranged into a pentagon, the rest are added until the sphere is completed.

Fig. 74 20 reconfigurations are used to form this dodecahedron designed sphere of pentagons and hexagons. The units are held together by hair pins. (p.270).

Fig. 75 24 paper plates reformed to a soft folded mobius (p.276). Three units have been joined to form a triangular set. Eight sets have been arranged into an octahedron pattern. Notice the small square depression where 4 triangular sets come together. The give placement to the 3 major axis.

Fig. 76 20 tetrahedra with the insides out have been joined to form an icosadodecahedron pattern (p.257).

300

Below. 3 views of a tetrahedron made from 4 triangle folded circles (p.297) In *Fig.77a-c* the 3 points of the triangle are curved over and attached to the center point, by sewing in this case. Bobby pins are used to hold the units together. In *Fig.77a* the vertex is reformed to an open hexagon star shape where the 3 planes come together. *Fig.77b* shows the triangle plane where the inside and outside of the surface comes together at a center point. *Fig.77b* is a view from the edge showing the attachment along the length of the transition surface of inside to outside, The 3 diameters that have become raised ribs which begin to reveal the dual tetrahedra as it moves towards the cube form. The plane becomes a point and the diagonal immerges as an edge.

Fig. 77a   b.   c.

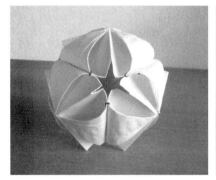  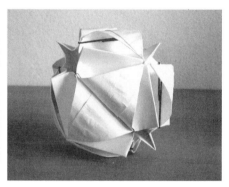

 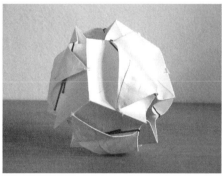 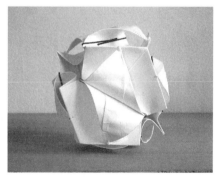

Fig. 78a   b.   c.

Above. *Fig.78a-c* is the same system as *Fig.77a-c* but the formed unit has been altered so that the 3 end points are attached to the half way point on each of the 3 ribs rather than the center of the circle. The ribs are the short segment of the diameter bisector of the triangle. This shifts everything into irregular symmetry, being somewhere between a tetrahedron and a cube. *Fig.78b* shows the half joining and making it easier to see the forming of the cube. Even though the angles are not correct for the regular form of the cube the pattern is clearly discernable. These are only 2 of many variations in this line of development. The vertex of 3 planes can be opened to 4 and 5 planes going into the octahedron and icosahedron symmetries. When opening further to 6 planes it becomes a flat formed system.

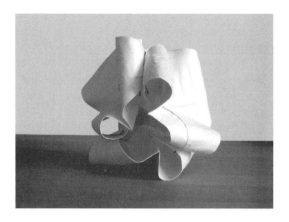

Fig. 79 This picture shows the model (p. 301, *Fig. 77a*) reassembled with the inside on the outside.

Fig. 80 Here the 4 components of the tetrahedra model are reassembled with the inside on the outside similar to (p.301, *Fig. 78a*).

Both of these reassembles are tetrahedral in the relationship of parts. As with regular tetrahedra, there are a number of ways these odd variations can go together combining in different ways. The uniqueness of the forms limits how many ways there are to join them, but there is flexibility in the soft folding to accommodate a lot of adjustments.

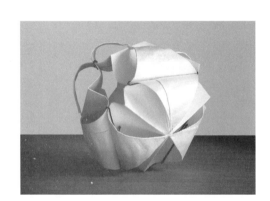

Fig. 81 Here the tetrahedron of 4 units has 2 of each of the units above, *Fig. 79-80*. Two different proportional systems are working together in an interesting combination. There are no set congruent angles. The open flexibility of the curved forms allows joining between the two configurations that could not happen were they rigid in formation.

Fig. 82 Here is one of many possible combination of how multiples of differently formed aspects of the same system can be reassembled. There are 4 tetrahedral units in a tetrahelix arrangement joined edge-to-edge (p.202, *Fig.15*). Each of the two differently formed proportional reconfigurations (one outside and one inside) are combined in this system (*Fig. 79, 80*).

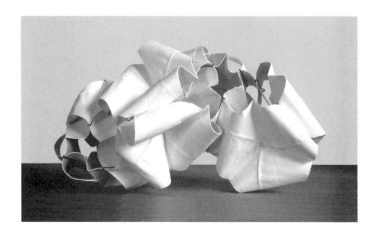

Fig. 83a Edge view of a tetrahedron arrangement of 12 circles in 4 groups of 3 each. The folded unit used is found on page 275.

Fig. 83b Arrangement from the triangle plane showing 3 circles coming together at the center of the plane.

This model starts with *Fig.83a,b* as a arrangement that is not discernable as a tetrahedron form and goes through an opening and reforming transformation that clearly shows both the interpenetrating tetrahedron and octahedron forms *Fig.85*.

Nothing is added or taken away from this system. Some parts have been disconnected and repositioned to other parts in the system changing the form. This reflects similar movement transformations observed in nature.

Fig. 84 The circles (*Fig.83a,b*) above, have been opened changing the form.

Fig. 86 Here the circle arms have been joined to the arms of the adjacent units *Fig.85.* showing the octahedron.

Fig. 85 The open circles above, *Fig.84*, have been opened more.

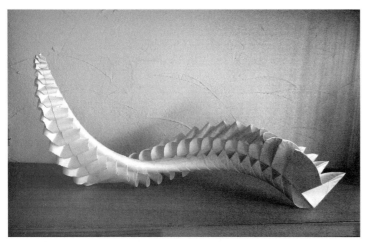

Fig. 87a

Fig.87b　32 circles all folded in the same way with variations in configurations towards the larger end. 15 units diminish in size from 9 inch diameters down to 1 ½ inches (p. 296). The helix part is individual units the same sizes and then the diameters of the circle change and it spirals. This helix-spiral from different views shows a very different look to the same system. There is an opening of the form as it progresses from the helix to the spiral. Other systems from this unit (pp.280, *Fig43*; 305, *Fig.89a-d*; 306-7).

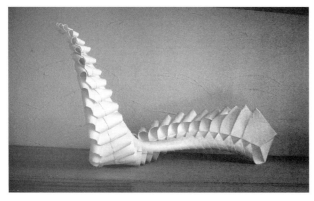

Fig.87c　While this does not look like *Fig. 87a* and *b* it is the same model seen from a different angle.

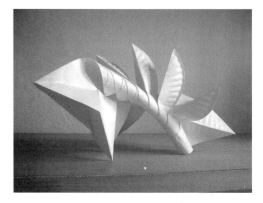

Fig.88 These 8 units are a segment of a continued sequential opening of the folding used in the helix to the left, *Fig. 87a-c.*

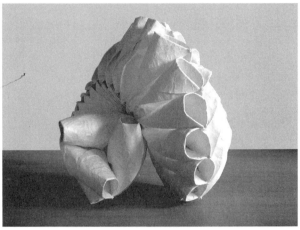

Fig. 89a While this looks like a spiral because of the angle of viewing it is in fact a helix form. Twenty paper plates the same size, have been folded to the tetrahedron, then crumpled, flattened to the tetrahedron and reformed. The crumbling give a nice textural look to the model. This view is looking down the helix from one end.

Fig. 89b This side view clearly shows the folded unit. It is folded into a tri-fold tubular form in multiples and joined one opening into the next (p.296).

Fig. 89c This other side shows 1½ turns of the helix. All units are 9" diameter circles. The spiral is often thought of a flat images. The spiral image can also be the distortion of the compressed helix, as this model indicates in *Fig 98a*. The separation of 2 and 3-D often causes confusion when there is little understanding of what gets distorted during the compressing process.

Fig. 89d A view from the opposite end of the helix shown in *Fig.89a*. It shows the other side of the unit.

Fig. 90 This dodecahedron is a combination of the units shown on (p296) and the soft folding used for the spiral (p.304-5). It is the same procedure used on page 226, only with differently formed tetrahedra. This has been taped together.

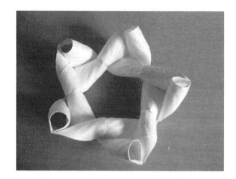

Fig. 91   5 reformed circles are put together in a pentagon showing the curved over side up. The unit is curved a little more than in *Fig.90* the same as on (p.304).

Fig. 92   3 pentagon units from above *Fig.91* are joined together in a triangle relationship. The open tubes connect the outside to the inside.

Fig.93   3 pentagon units placed on the dodecahedron in *Fig.90*. To continue this covering would give the system an outside form that is the same as what is facing in.

Fig.94a 60 folded paper plates refolded and arranged to a dodecahedron. This is the inside out of the dodecahedron in (p.306, *Fig.90*). Each circle is folded in the same way. Five units were assembled into a pentagon and then 12 pentagons were put together forming the dodecahedron.

Looking through the pentagon opening the opposite pentagon is not rotated half way as with the regular dodecahedron (p.167) but turned about a third of the way. There is a partial twisting of the axis due to the over under folded tubes.

Fig. 94b A vertex view from the point of joining 3 pentagons (*p.306, Fig.91*).

With the assembly of 2 units a tube space is formed that is open between the inside and the outside surface making it a single inside/outside surface. Joining 5 together form 5 open tubes that weave over and under each other.

Fig. 94c The dodecahedron from the edge view showing 2 adjacent pentagon openings.

Fig.95a Branching system.

Fig.95b Detail of spiral ends.

Fig.95c Detail of connecting spiral ends.

Fig. 95a-d This is fractal demonstration of growth using only one folded unit in multiples with some diminishing scales. The main branches are helix in form, ending in spirals of different proportions. The diameter of the circles decrease as the spiral tendrils from 9" diameters down to 1 ½" diameter circles There are 97 paper plates used. They were first hair pined in sections, then the sections glued and assembled. The unit used is a mobius folding that allows one to nestle into another (p.374).

Fig.95d Another view of branching.

9 POCKET POLYFORMS

We have looked at polyhedra with the circumference folded in, folded out, with curved surfaces and soft folding. Now we will explore curved surfaces by joining sections of the circumference forming *pockets*. All folds remain consistent to the 9 tetrahedron creases.

<u>Fold</u> the tetrahedron. Open to the circle. Fold over one diameter in half. Tape the circumference together using short pieces of tape all along the edge; one half of the tape on one side folded over to the other side.

With your fingers push the 2 points (as shown) towards each other finding the crease that is one side of the center triangle. One half of the dark line is on one side; the other half is on the other side of the crease. Straighten the folded line by pushing in on the crease line and the rest will pop out into curved surfaces on both sides, forming a tetrahedron interval. By pushing on the circumference and the center of the line opposite, the pocket will pop out perpendicular to the direction of pushing. This is a different kind of forming than what is used in soft folding. This is using the tension/compression nature of the creased lines in relationship to the circumference.

Close the tetrahedron space by pushing the 2 triangles together face to face. Open it back up pulling the points in the opposite direction. The axis of open and closing will change lines and the radius of the circle becomes the predominate fold. This pushes the triangle plane into the center causing a concave space.

Make 2 circle units of the pocket tetrahedron. Each unit has 2 triangles forming the tetrahedron pattern, half the tetrahedron form. Join the 2 open units together forming an interior tetrahedron. The inside triangles of one fit into the open triangle space of the other, (pp.108, 312, *Fig.96b*).

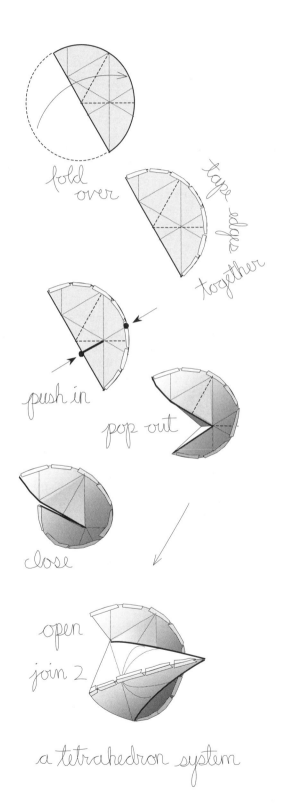

fold over

tape edges together

push in

pop out

close

open

join 2

a tetrahedron system

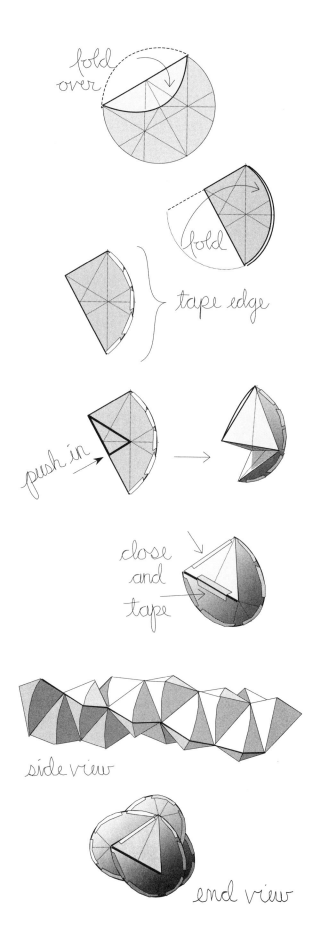

<u>Fold</u>   a tetrahedron and open flat. Refold one third of the circumference to the center point on one of the creased sides of the large triangle.

Next fold in half on the diameter perpendicular to the folded triangle edge.

Tape the 2 sides of the circumference together using short pieces of tape along the edge, half on one side, fold over and stick other half of tape to opposite side. Leave folded-over end untapped.

There is now a pocket-like shape. Put your fingers inside and open it up using the crease already there: this movement will fold one side of the middle equilateral triangle. Look for the right-angle triangle, half of the equilateral triangle, shown to the right. Push the right angle in until it pops into a tetrahedron with a tetrahedron interval defined by 2 sides, similar to the previous page but differently proportioned.

This shows one formed tetrahedron and one tetrahedron interval, 2 tetrahedra with a circumference spine. As the opening closes there will be a change in the shape, a lung-like action of expansion and contraction. Close the interval tapping the edges.

<u>Right</u>. Explore ways to combine this unit in multiples. One direction is the helix, seen from the side and end views. They are joined on the triangle surfaces, each unit rotating one third to the other forming an irregular tetrahelix.

310

Before taping the tetrahedron edges closed push in on 2 corner points to open and curve the triangle faces. (Sometimes the curved surfaces get partial pushed in. You may have to put your finger or a pencil inside to pop the surface out.)

Make the opening more circular by using your fingers to curve the surface and edges. Opening one side of the tetrahedron changes the angles as the surfaces curve.

Make 2 the same way, with oval-shaped openings.

Join these 2 units together with open tetrahedral intervals fitting together, one onto the opening of the other, joining edge-to-edge. In forming a round opening you have two 360° rotation attachment positions.

In closing the opening a tetrahedral interval is opened. One edge and 2 points come together forming an open tetrahedron space. Tape the 2 units together where straight edges join. A hinge joint will allow it to reform. The 2 pockets will collapse to a flat position and can be popped back out to the curved surfaces (p.312, *Fig.69a-69c*).

In closing the openings to form straight edges, there will be 2 adjacent triangular planes that reflect an outside angle close to the icosahedron. The many congruent surfaces, openings and angles that provide a number of in between variations and combinations in multiple joining of the curved surfaces.

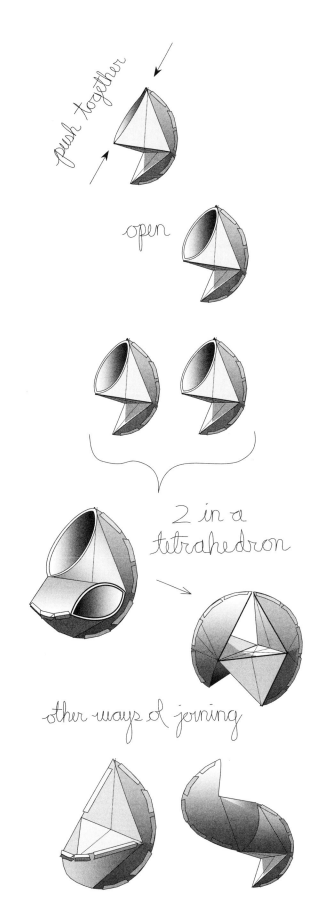

push together

open

2 in a tetrahedron

other ways of joining

Fig. 96a 2 circles pocket folded and joined together by tapping on a folded edge. They are flat (p.309).

Fig. 96b The 2 circle pockets are pushed out and moved into an internal interlocking tetrahedron arrangement.

Fig. 96c Tetrahedron is opened half way and the interval forms 2 open tetrahedra joined by a common edge.

Fig. 97a 2 sets of 2 tetrahedra *Fig. 96b* are joined on a straight edge and taped together with a hinge joining.

Fig.97Bb The system of 4 circles is partially opened as in *Fig. 96c* to reveal the enclosed tetrahedra spaces.

Fig. 97c Opened more.

Fig. 97d Here the system is totally open showing the 4 individually formed circles. It still can be collapsed flat *Fig.96*.

312

Fold the tetrahedron, open to the circle and bush one third of the circle in on one of the radii from triangle point to center point.

This will fold 2/6 of the circle to the inside, leaving a quadrilateral opening. Flatten the folded-in flap against the inside surface (p.74 #4).

Fold flat with the folded parts to the inside and the circumference edge together.

Tape the circumference together with short pieces of tape folded over on both sides.

As before, put your finger in the open side and push out, forming the tetrahedron and rounding out the curved surfaces by squeezing gently on the triangle base of the tetrahedron and the opposite curved edge.

Make another one the same way. Join the open spaces together and tape. Explore other combinations of joining.

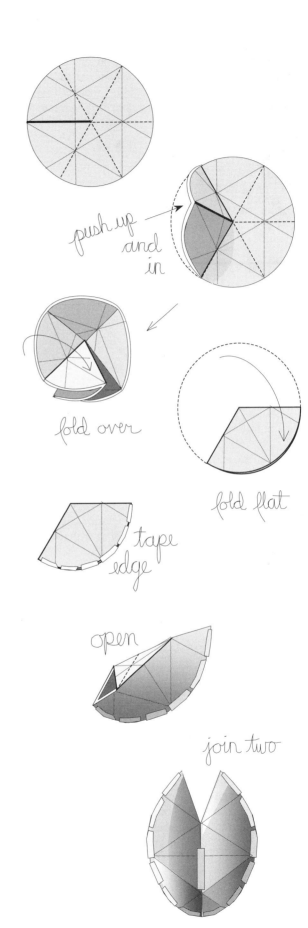

313

Join 2 single units together with one open end slipped over the other so the opposite end points are open to the length of the regular triangle edge. This forms an elongated tetrahedron interval.

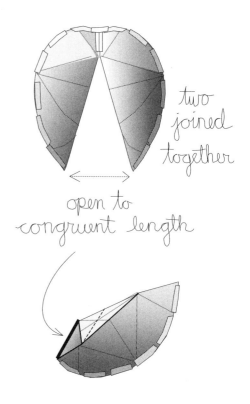

two joined together

open to congruent length

Make another set of 2 taped in the same way. Put them together so that the triangles of one fit into the triangle spaces of the other in a tetrahedron arrangement. Adjust the open end of each to the width of the other so they are closed, and tape all four joining points (p.129).

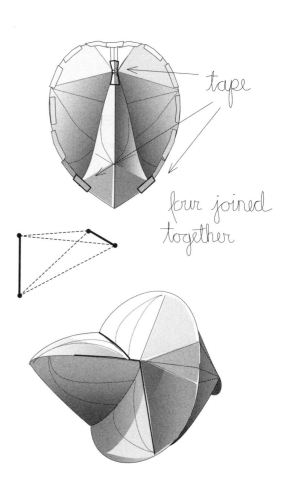

tape

four joined together

This is an irregular tetrahedron with all curved surfaces folded out, similar to what we find in the tennis ball covering (p.117) but with a very different form. Two curved edges perpendicular to each other divide the form into four folded-in areas, making it a complex discontinuous surface folding into itself rather than out like a ball with a continuous convex surface.

Further explore this folded unit by locating the centerline that runs in the long direction of the triangle plane. Fold in on this line causing the plane to become inverted, decreasing the width of the unit until it is flat. Flatten it out so the inside folds in.

Fold another one and flatten the same way. Tape the 2 together joining on the open end. Turn over and tape on other side the same way.

Make another set of 2 taped the same way. Join the sets, one intersecting with the other as shown to the right. Tape folded edge of 2 equilateral triangles, shown in creases, forming a rhomboid. Tape on the other side of the same edge making a hinge joining. This makes a flat rotating arrangement.

Move the 2 parts on the hinged edge so they are perpendicular to each other. One at a time, push in on each circumference edge simultaneously. This will spread the 2 open spaces until they close on each other, forming a tetrahedron pattern of two interlocking inflated forms.

Release the pressure on both circumferences and continue to move in the direction you started on the hinge joining. When it begins to look flat, bring the two end points on the circumference together. This will complete the circle with an open triangle in the center. Because of the double circumference edges at the bottom it will stand up and even rock a bit; a circle with a triangle hole in the middle.

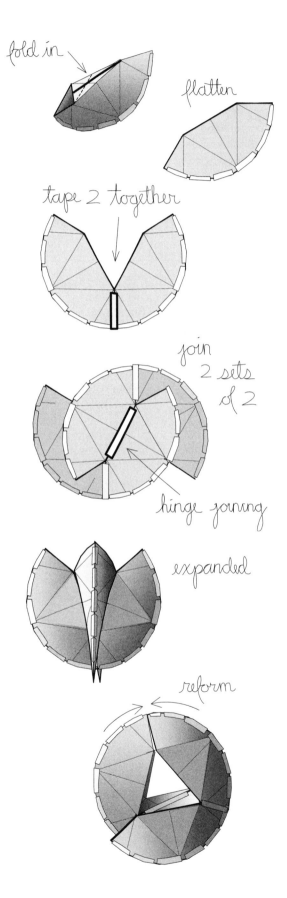

315

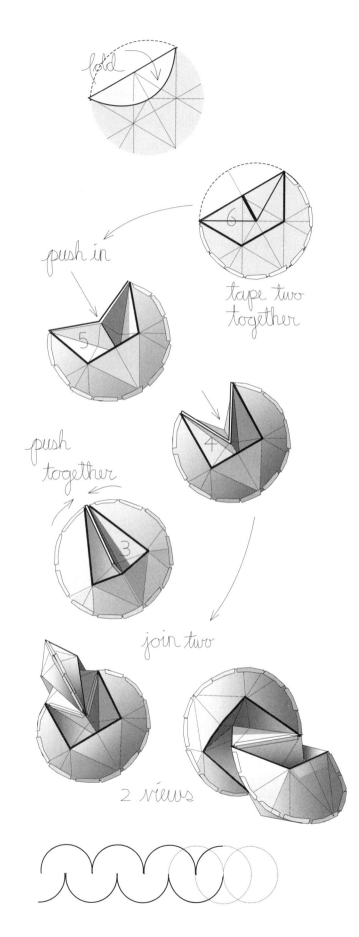

Fold   2 circles where the diameter end point touches the center point, One third of each circle is folded over.

Put the 2 folded circles together so that the folded-over sections are together on the inside. Line up the straight edges and the two thirds of the circumferences. Tape the 2 circles together all the way around on the circumference leaving the straight edges open. Look for the trapezoid with the long edge on the folded edge.

This next reconfiguring might be a little difficult at first. Spread the open sides of the trapezoid, pushing out on the center line, in on the edges of the large isosceles triangle where the straight edge is the long side. This forces the 3 lines out, shown at right as darker lines. The three dark lines are 3 sides of the hexagon; when pushed out the angle changes and they become 3 sides to a pentagon. Continuing to push the end points of the circumference together the lines become 3 sides of an almost square. Continue to push the ends of the circumference together and the 3 lines form a triangle. Half a hexagon is reduced through a pentagon, a square, to a triangle.

Make another set of 2 in the same way. Join them both with the opening of each configured to half a tetrahedron. Put the 2 openings together edge-to-edge forming a tetrahedron. This puts the 2 circles at right angle to each other as seen to the right. This is not unlike the models on the previous pages, but with a different form.

Track the rolling movement of this system on a flat surface. One circumference rolls around a stable point of the other, then changes rotational points to roll on the opposite circumference creating a wobbly movement in a straight line. This is the movement of the tetrahedron in a straight line

316

Each pictured to the right shows changes as this system is sequentially reconfigured. Every movement offers possibilities not initially obvious.

Push the creased lines in and out, changing the four-circle formation. Using congruency and bisection, explore the shapes, angles and intervals that get tucked in and pushed out. Move the system symmetrically into and away from symmetry. Tape the forms that are of interest, make 2 or more and explore possibilities of greater complexity.

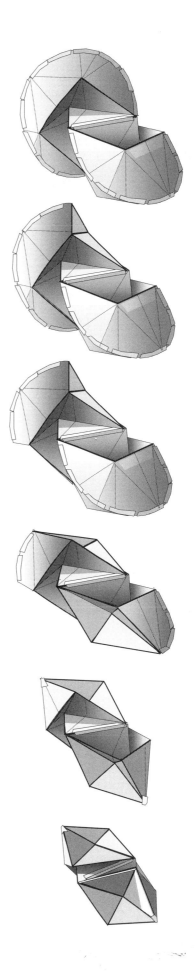

It is sometimes important to look for where movement stops, were it will no longer change form, where all parts lock up.

317

<u>Fold</u>   the tetrahedron-folded circle in half.

As you bring the circumference together tuck in the triangle corner fold.

Tape the circumference together, leave the tucked in edge untapped.

When you push in on the center point of the folded circle, the radius from the circumference to the center point pops out and the short part of the line, from center to folded-in part, goes in. Feel what the folds want to do, because that will be your directive.

When you pull the two ends of the circumference in opposite directions it will flatten back out.

Make 2 pushed-out units the same as above. Put them together side-by-side so they are vertically in opposite directions, one going up and the other down. Look for the pushed-in long and short edges to find orientation.

Tape the 2 lower, straight edges together as shown. Tape on both sides making a hinge joint that moves. Bring the 2 edges together as shown to the left. It doesn't move very far.

Make another of this system of 2-circle units.

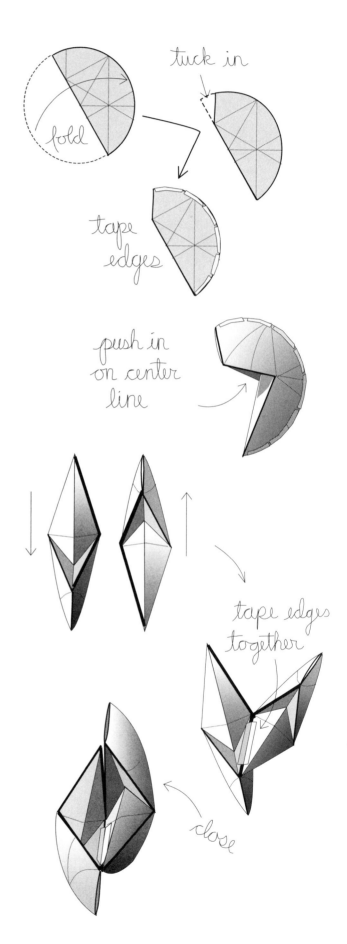

318

Put 2 sets in the same orientation.

Slide them together so the lower inside edge of one joins the upper inside edge of the other keeping the same orientation.

Tape the straight edges on both sides in a hinge joining. This set of 4 has a lot of interesting reconfigurations as each individual form is moved in relationship to the others.

When all the straight edges are moved to the inside with circumferences to the outside an internal octahedron pattern is formed indicated by the 6 points of coming together. By making another set of 4 and putting them together all in the same way a helix begins to develop. This helix is particularly interesting because of the inside and outside form difference (p.320, *Fig.98-100*).

Using hinge joints will allow this helix system to collapse flat by pushing in on the sides and the long ends moving out on each individual unit. It can be reformed in the same way that each individual unit was initially formed.

Joining 2 sets of 2 hinged together will fold into an octahedron arrangement in a very un-octahedron form.

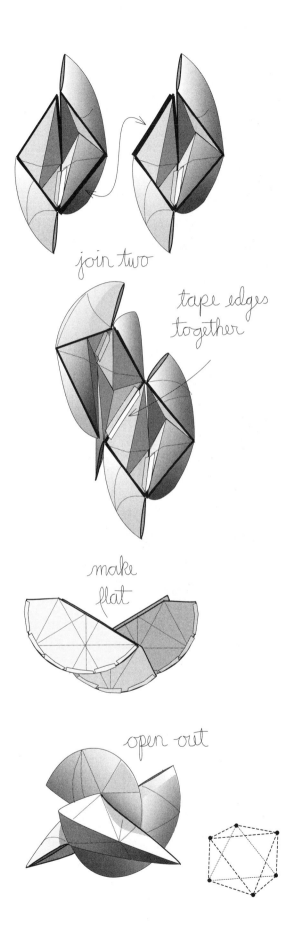

join two

tape edges together

make flat

open out

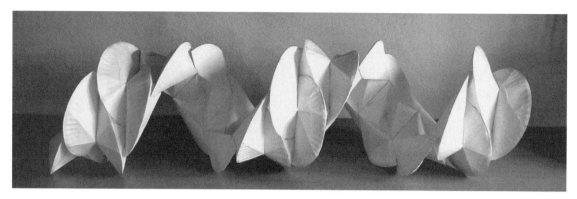

Fig. 98 The helix model above is made from the units joined together (p.319). This makes a strong open helix curved on the outside and the straight folds on the inside.

Fig. 99 8 circle units form a small segment of the helix above *Fig. 98.* It is not quite a full turn. It could easily be extended as the one above and the two could then be fitted one into the other *Fig.100.*

Fig. 100 The small unit *Fig.99* is positioned into the middle area of the model in *Fig.98* at the top of the page. Each is independent and one fits into the other. They are the same helix. One fits into the other forming a double edge filling in the outside while the central core space remains open.

Here is another curved surface variation to the tetrahedron.

Fold   2/3 of the circumference to the center on the already-creased lines forming 2 sides of the large triangle with 1/3 circumference out.

Fold this shape in half and tape only the 1/3 curved edge together.

Remove the tape from the pocket so that 2 surfaces of the tetrahedron are curved. Pushing on the curved edge and the triangle bottom will help pop out the natural curve; you use your finger on the inside to give a nice rounding on each side of the triangle.

Pull the curved edges in and tape together to close the tetrahedron, or leave it open. Make 4 of them.

Join 4 curved-surface pocket tetrahedra in a two-frequency tetrahedron arrangement. Orientation is important. Start with keeping direction consistent.

Rotate the orientation of the first 3 tetrahedra around the triangle. There is a twisting of symmetry where no tetrahedron is in the same orientation as any other, even as the fourth tetrahedron is placed on top.

There are other developments to be explored by using this pocket tetrahedron unit as well as the bi-tetrahedron variation to the right.

321

DISCUSSION: SOFT FOLDING AND POCKET POLYFORMS

Soft folding and pocket forming of circles works well because of the inclusive nature of the circumference. Both curved surfaces and creased straight lines work from the same pre-formation of pattern. Straight and curve compliment each other in ways difficult to calculate using traditional molding techniques and construction methods. To fully advantage the nature of the circle we must spend time moving it, folding it, feeling what it does, observing and thinking about it.

Soft folding is curving and reforming the circle surface with minimum or no creasing. There need not always be well-defined creases and points of intersections. Pocket-like spaces can be curved and reformed using the creased folds or not. With both soft folding and pocket polyforms, surfaces are formed into arcs and curves regulated by right angle movements and the straight-line dynamics of the circle. The soft folding corresponds to the same triangulated relationships observed in traditional polyhedra but the forms generated are quite different. The richness of the circle is in the combinations of the straight line grid folding and the soft curving of the surface.

There are many forms where the doubling of surfaces can be opened and closed, expanded and collapsed, by soft curving of edges and planes. This is not unlike what happens in nature with the wrapping and unwrapping of curving surfaces. It is much like what we see with leaves, the unfolding of flowers, shell growth, and in folding and unfolding of cellular functions and flow-forms of many kinds. In nature when two surfaces are together there is a temporary pocket of protection, a place from which growth can emerge. The pocket has similarities to the flat shape of the vesica; it is a divisional opening. There is waste to double surfaces where there is no functional reason. It is not efficient to use two where one will do, nor is it effective to use one when two are needed.

Traditional polyhedra are abstract representations of generalized patterns rendered in points, straight edges, and flat planes. Through soft folding and forming pockets we can begin to easily model many of the organic forms we observe in nature where curving and wrapping of surfaces show a more fluid and naturalistic form of modeling. Each straight-line edge that is curved references back to the continuous boundary of the circle echoing spherical origin.

Folding the circle expands our form vocabulary, bridging traditional polyhedral folding with something more natural and comprehensive. The circle removes the limitations of the square in traditional origami. It is important to be able to explore all the possibilities without limiting the means of how to attach, joining the circle to itself, either single or in multiples. How to touch and attach is always a consideration. That is part of the problem solving in making any kind of model. There are no constrains in nature as to how things get attached. It makes no sense to place limits on the open exploration of the comprehensive nature of the circle. There is only what is intentionally appropriate to principle. The integration of creasing and curving helps us to actually see the natural forms of our physical world as they intersect with the hard edges and flat planes of the geometric and the mathematical world. Being able to model all kinds of connections make it easier for the mind to bring into balance the experiential beauty of the physical world with the beauty of the mind. With a more extensive exploration of Wholemovement through folding the circle, our perceptions of a fluid, moving, continuous life expression and the more abstract ideas of discrete, discontinuous, individual, and separated parts can more closely be brought into alignment.

 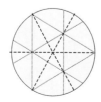   

COLLAPSIBLE TETRAHEDRA

This multi-frequency tetrahedral system will open out and collapse to a flat bundle in a way not possible with solid tetrahedra (p.122).

<u>Fold</u>   the two-frequency triangle. Tuck one half of each flap under the other locking the flaps.

Fold it in half with bisecting diameter re-creasing the right triangle.

When opened an irregular tetrahedron is formed using only 2 of the 4 triangle planes.

Fold 3 more the same, arranging all 4 into a 2-frequency tetrahedron. Tape them at the end points, or punch holes at the ends and use string or twisty ties. These are the first 2 layers. Make 3 more sets of 4 and add another 2 layers following the same process.

When a multi-unit tetrahedron system is made with only 2 faces the system will collapse in the same way the single circle triangle folds flat. As a multi frequency system it has great flexibility beyond simply opening and folding flat.

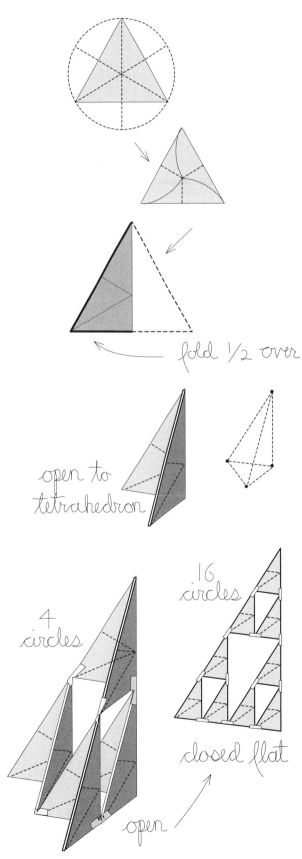

fold ½ over

open to tetrahedron

4 circles

16 circles

closed flat

open

323

<u>Fold</u>   tetrahedron forming the triangle.

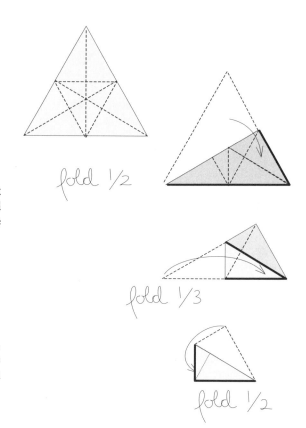

Fold it in half again on one diameter.

Notice the division of the 3 large congruent right angle triangles. Find the isosceles triangle and fold one half onto the other half forming a kite shape with the right triangle left over.

Re-crease the quadrilateral into the right triangle. There will be a lot of layers to fold over. Crease well. (Either of these folding is a tetrahedron relationship at different scale.)

Open the in/out folding that is doubled over, a zigzag in and out form.

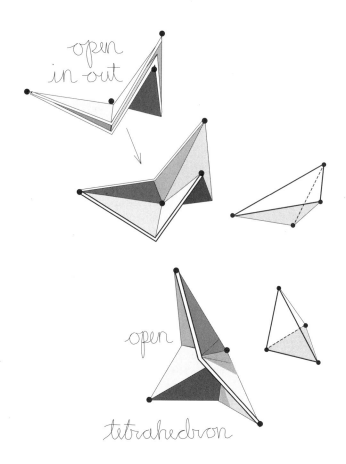

Unfold the doubled-over part a little. It doesn't matter how far, as long as it is not opened a lot. The 3 end points and the circle center point are the 4 points of the tetrahedron.

In this configuration the tetrahedron is a twisted arrangement that will allow sideways twisting in the larger system. Of the seven points the 4 outer-most points will define the tetrahedron. This tetrahedron moves and collapses. The only rigid edge connections are the long segments of the diameters. The connections between the points of the large triangle plane are now jointed.

Start with one circle folded to this tetrahedron configuration. Keep the same orientation when putting the multiples units together. Punch a hole in the 4 corners where the dots are shown.

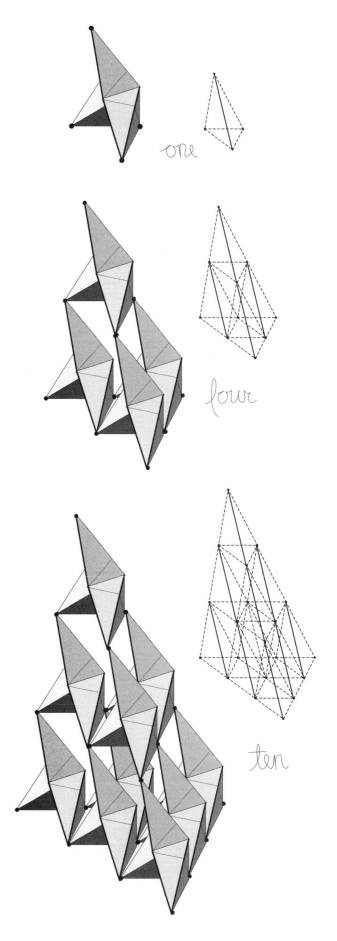

one

Make 3 more of these same units and put them together in a two-frequency tetrahedron. It is important they all be opened about the same amount so you don't lose track when putting them together. Use twisty ties or string to hold them together. Too tight restricts movement, too loose and it will be floppy.

To count the number of tetrahedra you will need, count each layer adding the previous layer:
the first layer is one, 1+0=1.
the second layer is 2+1=3.
the third layer is 3+3=6
the forth layer is 4+6=10
the fifth layer is 5+10=15
and so forth ...

Or it can be:
1st layer is 1+0=1
2nd layer is (1+1)+1=3
3rd layer is (2+1)+3=6
4th layer is (3+1)+6=10
5th layer is (4+1)+10=15
and so on ...

four

Adding each layer makes it easy to know how many tetrahedra you will need to fold as it gets larger (p.124). The larger it gets the more compliant it will become. Movement in the single unit is limited but is amplified as the frequencies increase. In the same way it started out flat it can be collapsed flat regardless of how many layers of tetrahedra there are.

ten

325

FIVE HINGED TETRAHEDRA

<u>Fold</u>   5 tetrahedra and hinge-join them together in a line. This system can be formed into a stellated tetrahedron, a tetrahelix and other in-between arrangements.

Five tetrahedra are joined in the linear net. Here the design is colored to the flat circle matrix making it instructive to circle reformations as the flat planes reconfigure. The surface designing can be informational to both spatial and flat relationships at the same time adding surface beauty.

Wrap 4 tetrahedra around one tetrahedron to form the stellated tetrahedron, generating a larger tetrahedron pattern. Start from either end covering all 4 sides.

From the stellated tetrahedron moving 2 will reform the system into the tetrahelix. First move the last-moved tetrahedron back.

Move the opposite end 2 tetrahedra as a single unit back again. When folding with hinge joints do not force the movement, let the hinges direct the movement.

The form now shows 5 tetrahedra in a helix formation with a specific directional twist. This has been colored as shown above, to show spherical sections as the circle matrix changes.

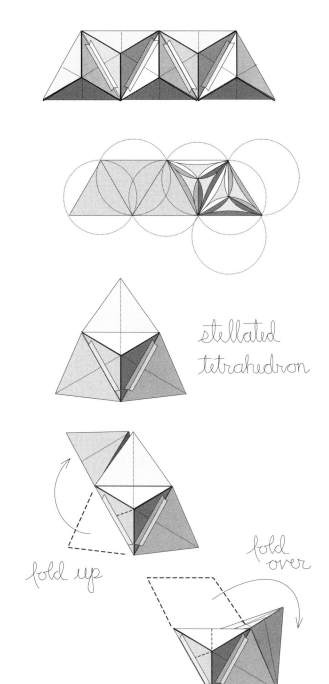

stellated tetrahedron

fold up

fold over

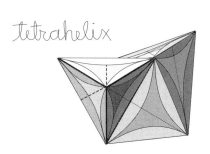

tetrahelix

326

Look for 3 individual spines of the helix where the edges fall into alignment. They will show the direction of spin. This may be difficult to identify with a short segment if the helix is an unfamiliar object.

This is similar to the helix arrangement in a different kind of movement system (p.201).

Move the 2 end tetrahedra as a single unit that is at the opposite end from the last movement in forming the helix. This hinging over will reform the tetrahelix so it is twisting in the opposite direction. Look again for the direction of the edges that line up with each other. Notice the in/out movement of every 2 triangles on each twisting surface.

Another way to reverse the spin is to move the tetrahedron on each end of the helix to the other hinge position. This only happens in the five-unit tetrahelix because there is no orientation to the middle 3 tetrahedra. So 1 and 5 become active to establish direction to the middle 3. This is the same function as when adding the 4th to 3 tetrahedra in a helix segment that has no established direction.

The properties of the form stay the same except the directional spin is twisting to the opposite direction.

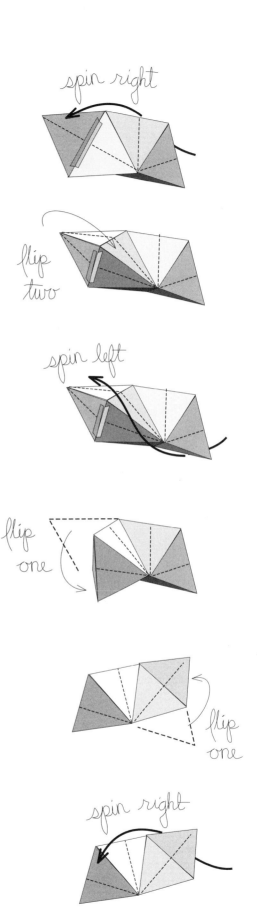

TRANSFORMING TETRAHEDRON

<u>Fold</u> a two-frequency triangle. Fold curved flaps under each other and hold with tape.

Re-crease along one of the diameters.

Fold 4 the same way. Line up the triangles in a net edge-to-edge with the re-creased diameters parallel and perpendicular to the line of triangles.

Tape the edge joining full length. Tape both sides for greater strength.

Turn the net of triangles over so the taped curved sides are facing up. They will be on the inside when folding on the re-creased diameters bringing the 2 opposite edges together. Tape them edge-to-edge. Tape full length as with the other joining.

This will form a flattened rectangle of 4 equilateral triangle circles divided into 8 right angle triangles. There is a lot of information about transformation in this reformation.

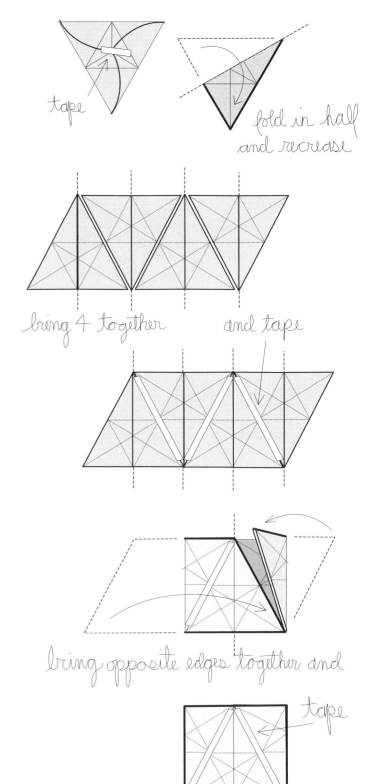

Open the rectangle to a hollow square prism. With both ends open the rectangular tube is unstable and collapsible. This again gives demonstration to the nonstructural nature of the square without the diagonal.

Bring 2 opposite points of one open end together. This forms 2 equilateral triangle sides with 2 right angle triangles and an open square end. Squeeze the opposite end points together perpendicular to the first points closed on the other end. The instability of the rectangular prism becomes stable in the tetrahedron form.

The open-end square can be closed in 2 directions, each perpendicular to the other. Reverse the movement back and forth. It appears to be the same tetrahedron either way. The tape edges change position from one movement to the other. They are in fact 2 different arrangements of repositioning 8 right triangles. Edge length AB moves perpendicular to opposite edge CD, each compressing back and forth into midpoint AB and midpoint CD. The surface transforms and the tetrahedron appears to remain unchanged.

Color any 2 adjacent triangle sides of the tetrahedra. Then in changing the perpendicular polarity the surface changes are apparent. With the reverse change each triangle becomes part of three different triangle sides of the tetrahedron.

Color in 4 center triangles of each side of the tetrahedra. When changing through moving perpendicular from one direction to the other, the design gets redistributed over the 4 triangle faces.

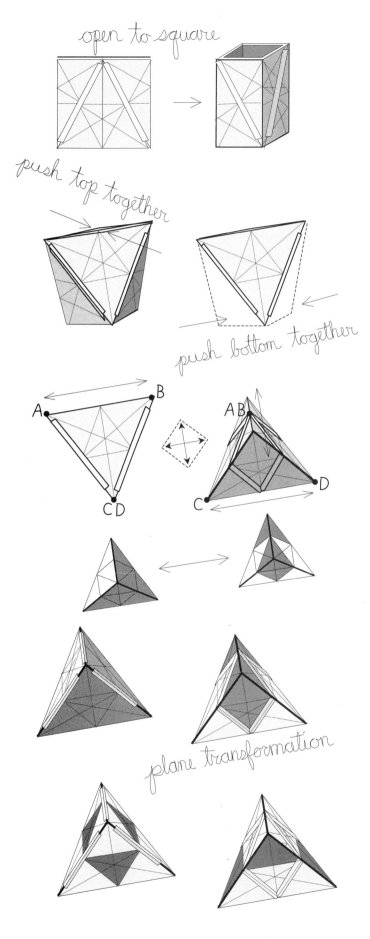

329

Open one edge of the tetrahedron and push in on opposite sides at the crease, which run around the center of the tube forming a square plane. As you do that, close and tape the open end so both end edges are now parallel.

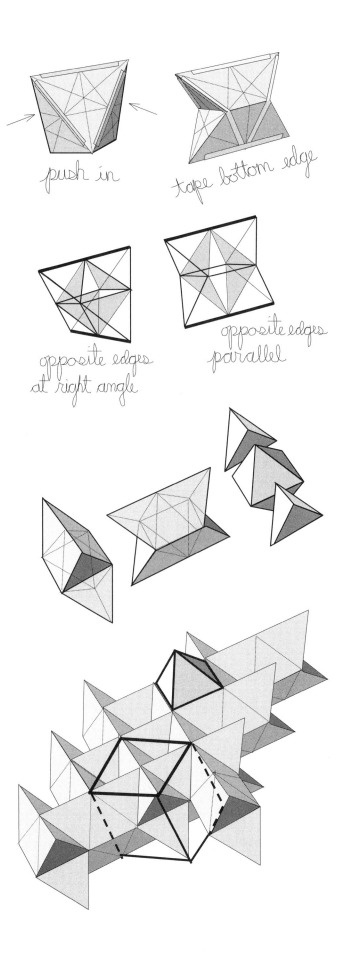

The opposite edges have changed from perpendicular to parallel without changing the central octahedron. The tetrahedron has been divided and turned 180˚. Both forms are arrangements of 4 tetrahedra stellating 4 of 8 sides of the octahedron center; one of alternating faces, the other in pairs of 2, joined on an edge.

There are other ways that also show the right angle movement of the tetrahedron as it is divided through the square center of the octahedron (see p.130). In the separation of individualized parts they will reassemble, each fitting into the others. This happens because of the perpendicularity within the unity of 4 spheres.

Multiples of this form key into themselves with the end fitting on to the side of another, perpendicular to each other. As this is multiplied out with many units it forms square intervals that are half an octahedron space. When these grid units are joined in layers, edge-to-edge one onto the other, the full tetrahedron/octahedron matrix is formed. This can also be formed using only tetrahedra joined edge to edge (p.198, middle of page).

The dark square shows the top plane of the cubic relationship of the tetrahedron star or fully stellated octahedron. This division is inherent in the "all-space-filling" cubic system of expanded spheres (p.253).

To continue with this tetrahedral transformation, remove the tape from the bottom edges and open into a square.

Push the bottom point of the large top triangle in, forming 3 sides to an inverted square (dark lines). This becomes a partially formed octahedron space.

Bring the 2 points of the unformed side of the square together, forming an inverted tetrahedron, bringing everything into triangulation again. Tape the side edges together. This forms the bottom layer of a two-frequency, partial truncated tetrahedron (pp.133, 199, 292).

There are other variations to this folding sequence. Each variation of in/out folds will change the form. Each form has its own unique self-referencing that will either continue to generate or cease. Look for congruencies to other systems that allow continued explorations into developing symmetries.

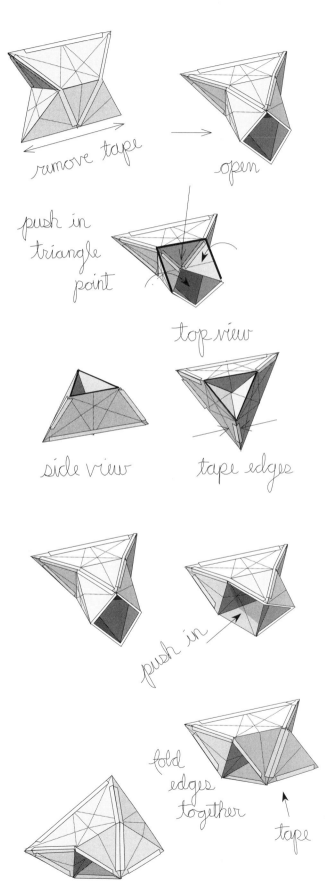

remove tape

open

push in triangle point

top view

side view

tape edges

push in

fold edges together

tape

331

DISCUSSION: SQUARE PRISM / TETRAHEDRON

The folded rectangle is inherent to the equilateral triangle grid pattern in the relationship of the three diameters in the circle (p.112). A grid drawing is usually reduced to only triangles or squares where divisions and context are eliminated. Using construction methods we learned to reconstruct the lines of division that are there in the first place; otherwise we could not construct them. The circle makes everything accessible by having it all in one place and in relationship with everything else. There is too much to show everything at one time so we discover things over time. It just makes it easier when everything is in the same place.

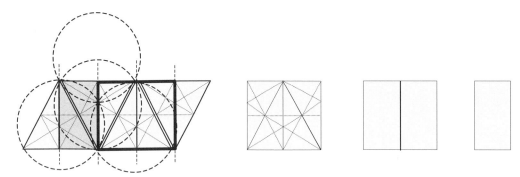

Above. Erasing the origin and contextual information from a quadrilateral relationship and reducing it to a rectangle with four sides is not instructional towards understanding the nature of the rectangle, or much else. We erase instructional information. Measuring is left without proportional understanding. Experience gives way to the image that becomes symbolic to an abstract concept. Without the experience of personal discovery, of touching and making connections, there is no meaningful understanding. Drawings at best are short-cut reflections to understanding that comes from experience. Making our own discoveries from the greatest place possible allows us to understand the information that is left out from what we are taught and what we teach. While individual experience is the beginning of understanding it is not the bigger context.

Folding the circle into the triangle shows the rectangle as part of the right angle movement of the circle. We have seen the square prism and the tetrahedron are not separate. If they are not separate just one time then maybe they are not separate. When we draw and construct information to model generalizations there are always unexpected connections by measuring and comparing lines, shapes, angles and areas that expand our understanding. That should tell us that there is far more than the symbols of generalization we use in describing these things. There is no right or wrong in all of this, for it expresses our experience as we observe what works and what does not work expanding a truth that remains consistent while evolving through the changing and transforming process of constant development.

The fundamental relationship in any rectangle is the diagonal, the right angle division of the first fold in the circle (p.36). In the drawing above the diagonal is the side of the equilateral triangle and the long edge is the bisector of the equilateral triangle, giving importance to the 30°/ 60°/ 90° triangle (pp.110, 346).

Below. There is a closeness of square to this rectangle and the dynamics is in the difference. This rectangle and the equilateral triangle are a function of two intersecting circles, which is an image of two tangent spheres seen from a 45° angle, half way between one in front of the other. Moving to a straight-on tangent view, the rectangular shifts to a square of the longest edge length which is the diameter of the spheres.

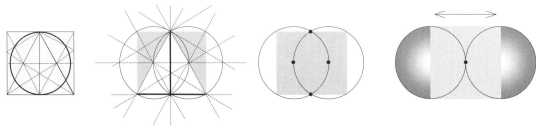

332

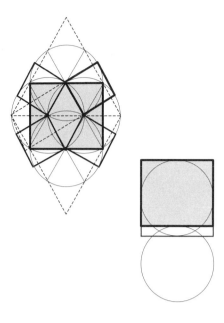

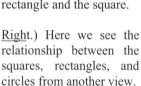

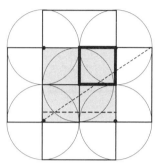

<u>Left</u>.) The rectangle and the square share a radial measure. In one direction is the edge length; in the other the radius is the measure of relationship between two intersecting circles. The difference is with the length of the small vesica found in a tilted square and the large vesica of two circles sharing one radius. The small vesica is the difference between the rectangle and the square.

<u>Right</u>.) Here we see the relationship between the squares, rectangles, and circles from another view.

Below. The rectangle of the four-circles-folded-creases is overlaid on a layer of four tangent circles/spheres to show the same pattern. The ordering of relationship between the center of the spheres and the centers of the spaces created by the spheres are the same. One is defined by the forming of the other. This is a spatial relationship and, when compressed, the flat images look different showing the proportional distortion through compression. The triangle reflects the hexagon matrix of circles with 3-6 symmetry and the square reflects the 4-8 symmetry (pp.90, 194). The further into division and away from spherical origin we get the more distorted the information with greater separation and fragmentation. The image retains a sense of unity to the degree that it reflects contextual origin in the proportional arrangement of parts.

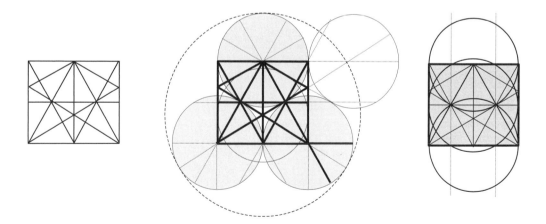

Within the divisions of the tetrahedron rectangle, count the divisions of the three basic shapes; the equilateral triangle, the right-angles triangle and the pentagon. How many combinations are there? How many of different sizes and proportions of each are there? What other combinations of shapes are there? There are many interesting relationships that give primacy to the tetrahedron/circle/sphere pattern.

OCTAHEDRA JOINING

Two congruent, half-opened tetrahedra joined edge-to-edge will always make an octahedron (p.129). This is the means to joining 2 tetrahedra structurally in a point-to-point relationship.

We have seen that the 2-frequency tetrahedron (4 tetrahedra) can be turned inside out forming a solid tetrahedron (p.207). By opening the corner points octahedra are generated. Here is a slightly different approach to making the benzene matrix (p.211).

4 open tetrahedra/ 5 octahedra in VE pattern

<u>Fold</u> 4 tetrahedra. Open them to the large triangle. Fold the circumference flaps under each other, keeping them out of the way. Turn over so flaps are underneath. Tape all 4 triangles edge-to-edge in a 3-around-one tetrahedron net.

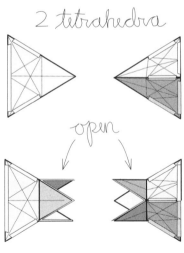

turn over
tape 3 together

Bring edges together into a tetrahedron and tape the 3 edges halfway from bottom triangle, leaving top half of each edge untaped. Make another tetrahedra in the same way with 4 more circles.

2 tetrahedra

open

Open the top corner points of both tetrahedra, each forming an octahedron arrangement with the triangles' flaps. Put the open flaps of each into the other, joining edge-to-edge and tape.

slide together edge to edge

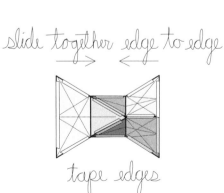

This joining of octahedra is a point-to-point joining of 2 open tetrahedra, showing another way of joining 2 tetrahedra to form 2 octahedra of 2 different sizes. This octahedron connection makes possible a rigid joining of 6 points that otherwise would be unstable.

tape edges

334

Make 2 sets of 2 joined tetrahedra and bring the triangle ends of each set together edge-to-edge. They will only fit one to the other, approximate 2/3 of the edge length. Pictured to the right are 4 sets of two that are joined, offset edge to edge. This sliding edge-to-edge joining is covered a little more extensively in the next few pages.

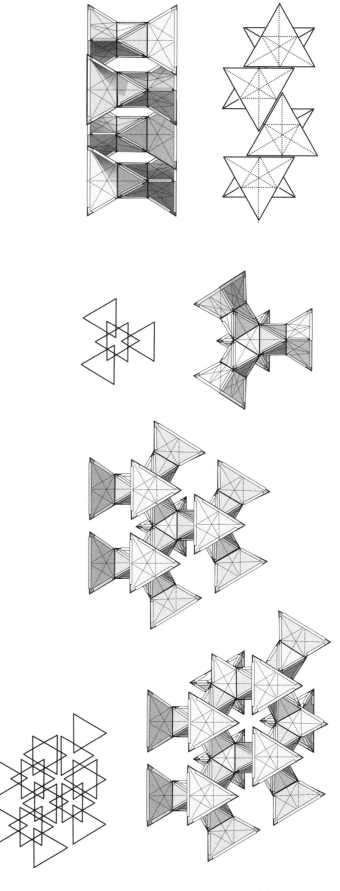

Go back to the 2 joined tetrahedra. Remove the tape from 3 corners of one tetrahedron. Add 3 more open tetrahedra in the octahedra, joining, completing a tetrahedron relationship of 4 around one. There are 9 octahedra, 5 formed and 4 internal.

The 12 end points of the 4 tetrahedra can each be opened to add another layer of tetrahedra in the same manner.

This is an octahedron-centered association of tetrahedra joined in a point-to-point/octahedron relationship where each tetrahedron will eventually be transformed into 5 octahedra where everything is interconnected through an opening and integrating process. This could easily become a fractal form by further dividing the edge length with each level of expansion.

As more tetrahedra are added, more connections are possible, forming an open-centered system of octahedra. This is another approach to forming the Benzene matrix (p.211).

Fig. 101a 103 folded circles glued together. Benzene formation of 3 layers in 4 directions corresponding to the 4 planes of the tetrahedron, the same arrangement for the vector equilibrium (p.99). This shows the hexagon openness and space within the matrix. The spherical cavity is the truncated tetrahedron relationship. Four hexagons are the open spaces and the 4 triangles are the surfaces of the connected octahedra.

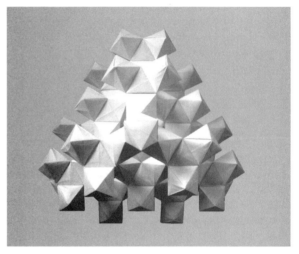

Fig. 101b Looking from the end point of the tetrahedron (center octahedron) showing 3 layers, which are easier seen in *Fig.101a*. From this angle we see very little of the spatial nature of what appears to be solid and closed.

Fig. 101c. From this edge view of the tetrahedron the square nature is revealed in some of the openness appearing as square holes. The squares do not exist in this matrix and are only a function of this flattened one point perspective of triangulation. The 3 layers can be seen by looking at the levels of what appear to be vertical elements in the picture.

336

SLIDING EDGES

There are many points of proportional division along the edges of the tetrahedron. We have explored primarily a 1:1 joining, and 1:2 joining with the octahedron and icosahedron nets using triangles (pp.141, 149). Expanding that to polyhedra the proportional correspondence works the same with different results.

<u>Fold</u> 4 tetrahedra. Join them in a square, edge-to-edge, forming one half an octahedron interval.

Slide one tetrahedron half way out on the edge from the center. The bisecting creases will indicate the halfway point. Tape a hinge joining of edges (p.66).

Tape the other 2 tetrahedra in the same way going in a circle, sliding each one halfway on the adjoining edge of the previous slid tetrahedron. In the drawing it looks like a square, but the first and last tetrahedra are no longer touching edges. The edges of the polyhedral forms do not lie on one plane as they appear in the drawings.

Tetrahedra of the half-octahedron arrangement have become an open helix formation that can move and be reconfigured. The advantage in using hinging joints is the ability to reform the system exploring various transformations.

The 1:2 joining of tetrahedra reflects the same patterned hexagon matrix in the triangle grid.

Make another set of 4 joined in the same way and join to the first 4 extending the helix to 8 tetrahedra. Explore the reformations of this hinged system.

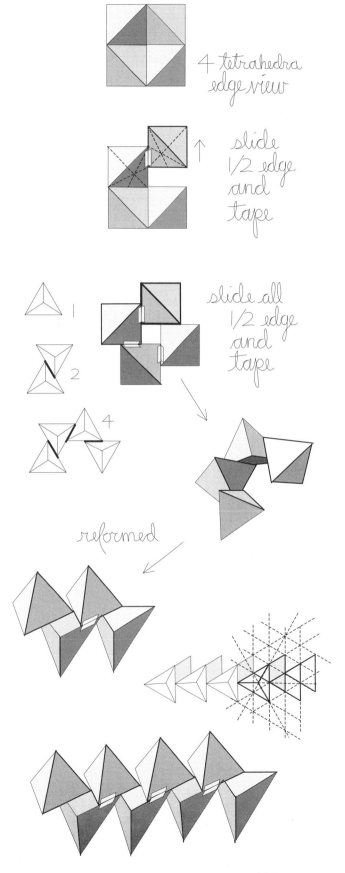

4 tetrahedra edge view

slide 1/2 edge and tape

slide all 1/2 edge and tape

reformed

337

Fold   5 octahedra and put them into a pentagon joined on the triangular surfaces. Sliding a surface is similar to sliding edges. (This works using tetrahedra in the same way.) The area of contact is diminished while keeping the same shape of contact. In the picture to the left the edges are moved in a 1:2 edge ratio, the area is in a 1:4 ratio. The faces keep triangle contact, leaving an exposed trapezoid.

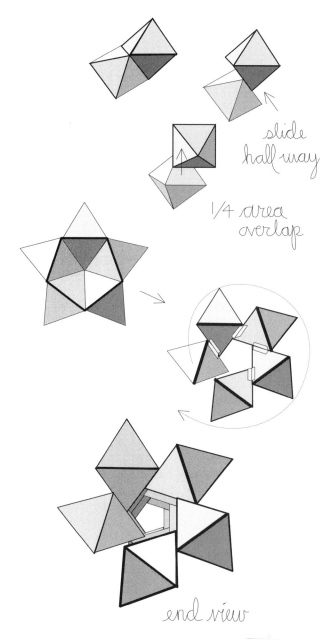

slide
half way

1/4 area
overlap

Arrange 5 octahedra forming a pentagon. Slide each connecting face in sequence as shown to the right and above. Tape or glue surfaces together. (glue works better) When the 5 octahedra are joined halfway, the octahedra will form an open pentagon center. (This can also be hinged joined on one of the edges.)

The amount of slide will determine the openness of the helix. By changing the rate of slide with each successive joining, a spiral center space can be formed (p.342). To spiral the form the diameter of each successively used circle must be diminished (p.256, *Fig.23a,b*).

end view

Make a couple more pentagon helix forms and put them together end-to-end. To the right is a side view of the helix made from 15 octahedra.

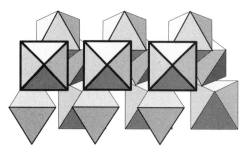

side view

338

Let's slide some torus edges and look at the difference in movement.

Fold 8 tetrahedra, find the 2 opposite edges and mark one-half division on each. Keep the same orientation of tetrahedra as you mark correspondingly the same place on each. There is a right-hand and left-hand way of marking the half edges on two opposite edges; consistency of orientation in development helps eliminate confusion. Hinge tape together only on marked half edges.

Compare full edge taping of tetrahedra helix (p.201) with the now moved attachments 1:2 edge length. This opposite-edge joining is different than the square joining, —adjacent edges— on the pervious page. Each has different twisting properties. The 1/2 joining is the same.

(To make a fully functional sliding system; cut off the tops of 8 ziplock bags, leaving enough plastic on each side to attach to the opposite edges of each tetrahedron. Attach by gluing and taping the corresponding strips to the full edge length of tetrahedra. Or the strips can be attached to the inside as the tetrahedra are folded together. Sewing the plastic to the folded edge is a more permanent solution. In this way you have functional sliding edges that adds another level of dynamic information to the system.)

Bring the 2 ends around in a circle and tape them together at the 1:2 marks so it completes the torus ring (p.203).

Here are 2 tracings reflecting the symmetry in continuous edge connections showing the 1/2 edge joining. The drawings show only 2 different positions. All schematic images can be used for developing many directions in 2-D designing. Movement can be tracked by color-coding faces as they correspond to the full-edge joining of the tetratorus (p.203).

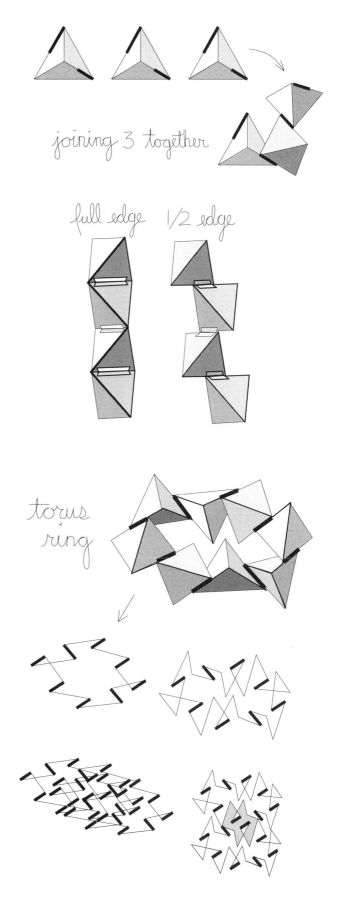

joining 3 together

full edge   1/2 edge

torus ring

The folded two-frequency triangle shows primary divisions both into and out from itself. Using 4 folded triangles and putting them together in the same net, without taping, allows rearranging in many ways using the folded lines. With multiple triangles it is helpful to color them to better observe arranging and designing. This is a good floor activity for young children (p70). There is far more to explore in making your own "pattern blocks" using creased circle than anything that can be purchased. Using the creases, fold the circle/triangles to form other shapes and explore the same grid. There are no wrong ways.

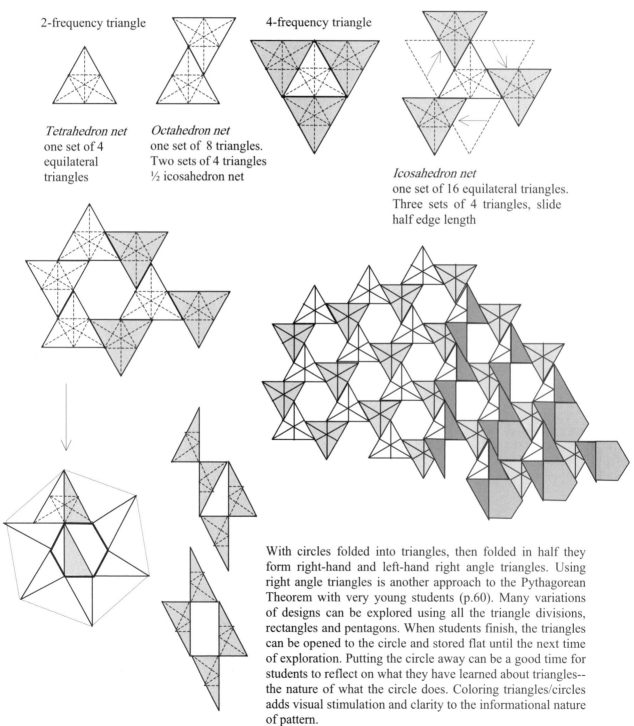

2-frequency triangle

4-frequency triangle

Tetrahedron net
one set of 4
equilateral
triangles

Octahedron net
one set of 8 triangles.
Two sets of 4 triangles
½ icosahedron net

Icosahedron net
one set of 16 equilateral triangles.
Three sets of 4 triangles, slide
half edge length

With circles folded into triangles, then folded in half they form right-hand and left-hand right angle triangles. Using right angle triangles is another approach to the Pythagorean Theorem with very young students (p.60). Many variations of designs can be explored using all the triangle divisions, rectangles and pentagons. When students finish, the triangles can be opened to the circle and stored flat until the next time of exploration. Putting the circle away can be a good time for students to reflect on what they have learned about triangles-- the nature of what the circle does. Coloring triangles/circles adds visual stimulation and clarity to the informational nature of pattern.

340

Using the regularity of the folded divisions in the tetrahedron and with consistent development of that information, there are many possible arrangements of design and scale between individual parts. The flat nature of the triangle makes it is easy to show flat arrangements of spatial formation. As triangles are folded and arranged we are aware of the 3-dimensional origin of the flat arrangements. The math of 2-D is simply 3-D minus 1-D, where 3 is actually compressed into 2; one has not been taken away. There is no 1-dimension or 2-diemsion without the third. This goes back to the principle of triangulation. To get the right triangle from the equilateral triangle, fold the inscribed hexagon (p.344). It all goes back to three, to 3 diameters as the hexagon folds into the pentagon, square, and triangle. When compressed to a flat image, proportions change. The two-dimensional representations of these spatial sliding movements, plane *translation,* are rich in the development of flat design and folding into 3-D. This sliding movement is found within the symmetries of point *rotation* and line *reflection* (p.183).

Below. The triangle is divided into three isosceles triangles, which can be moved and opened holding parallel edges to explore different designs. In rearranging from left to right there is a change in division from a 2 to a 3-frequency equilateral triangle. This image represents the spiral opening of a center point as the triangles are move. The triangles slide back together in a right or left hand direction. In the last drawing the 3 corner isosceles triangles outside the original triangle boundary are equal in area to the space created in the center triangle. Notice the rearrangement of the center triangle.

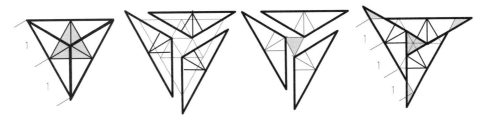

Below. The right triangle division of the equilateral triangle can be slid open and moved in multiple arrangements. This example shows ways to explore designing 2-dimensionally that can also open spatially replacing the flat shape with a polyhedral equivalent. The key is to be consistent to the divisions of the folds in the circle. The other approach is to play with an open-ended exploration, much like with Tangrams. Discoveries are made by what every ways we choose to play, if we are paying attention.

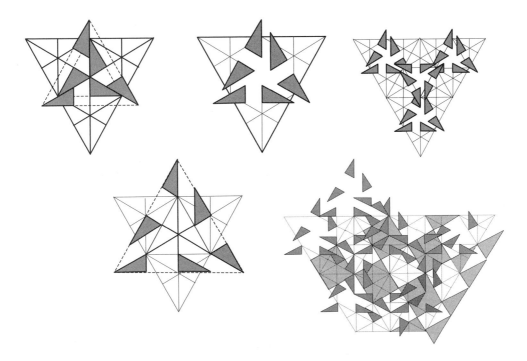

<u>Below</u>. This chart shows the importance of a center location of division to the potential of translation. Without division only the non-center triangle is structural. The movement stages of divisional change open from the center by sliding along the internal edges of division. The outer symmetry of the shapes are reflected in the opening movement. Moving shapes around on the floor is more instructive than drawing shapes, and is more accessible to young children. The exploration of the 2-dimensional designs inherent in these systems can be instructive to what we observe 3-dimensionally with polyhedra. The polyhedra are themselves only generalized models, but can also indicate something of what happens when a flower opens from a very small point in a cellular bud, or a seed expanding into space, becoming a large tree, or a human being, while at the same time expanding space within the growth of its own development.

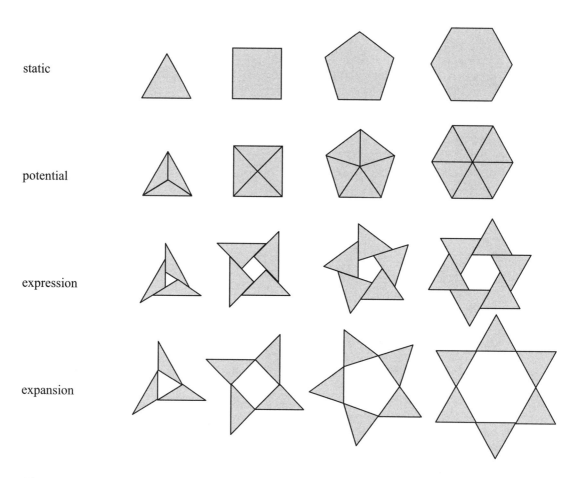

static

potential

expression

expansion

<u>Above</u>. Sliding individual parts of division, one half of the length of the other, is a systematic and consistent opening movement. That does not happen in static form or even through the concept of "potential" movement. Expression is interaction that happens only with movement. By continuing that movement to the full length of the adjacent sides expansion occurs and a point is formed joining the shapes forming an interval shape of the original in a different orientation. <u>The expanded hexagon is the only shape that generates an interval in the same orientation and size as the original hexagon.</u> The hexagon is the only division into equilateral triangles, all the rest are flat distortions of spatial division. To continue the flat movement outward would require separation based upon the inherent directives of organization that are found within the larger context that is not obvious in the individual separated polygons. Organization does not reside within the parts, it is a function of the contextual system of symmetrical order.

When using polyhedra and sliding the edges and faces against each other, things happen that cannot be anticipated in the image, even in a flat polygon as above. Fold circles into flat shapes and move them around on the floor, exploring what they will do (p. 70).

342

Below. Connect the outward points of the expanded figures. Continued sliding movement will necessitate separation as well as flipping the individual triangles (shown below). With separation there is the freedom to move rotationally, reflectively and transversally, and in combinations increasing the range of designs towards organizational expansion in symmetry to the first shape.

Below. Separation of connections between individual parts reflects the translation from established point connections to reveal a self-organizing quality, which is simply the reflection of a larger context of preformed structural pattern development.

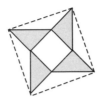
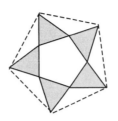
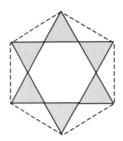

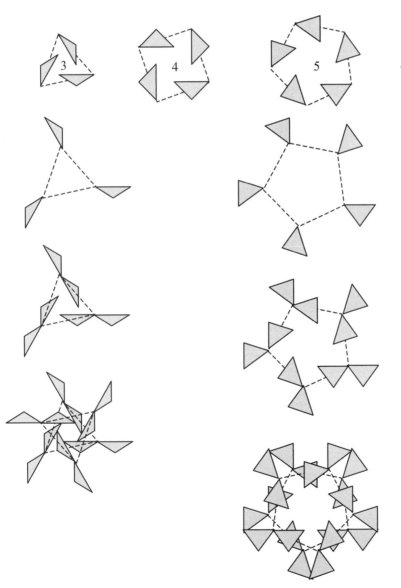

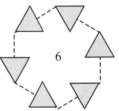

As the individual parts are translated to the outside of the new positions of points, and combined with the prior arrangement, we observe a dual forming of parts that changes the restraints. When reflected to the opposite position through rotation from the center, expanded forming of the initial order is generated. In the case of the triangle the hexagon context is revealed both into and out from the multiple arrangement of parts, constrained only by a directional growth. With the pentagon the full 5-10 symmetry is revealed. This kind of arranging is endlessly instructive.

INSCRIBED HEXAGON

The inscribed hexagon is the one exception in the book that extends beyond the 9 creases of the tetrahedron. It seems permissible using the diameters already there since the hexagon relationship is inherent, and adds much to the reforming of the circle.

An *inscribed* hexagon can be *infolded* by using the first 6 of the 9 lines creased into the circle. There are two basic approaches.

<u>Fold</u>   the tetrahedron open to the large triangle. Position the triangle with folded flaps facing up.

Fold the triangle in half using one of the bisecting diameters. Make sure the folded-over curved flap is tightly tucked inside when refolding on the already folded diameter. This is 1/3 of the circumference to be infolded.

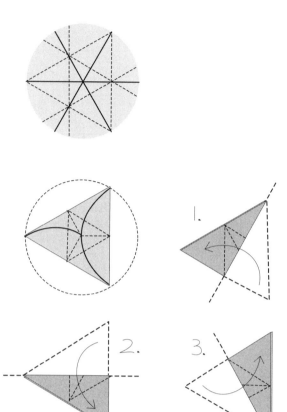

Recrease the same way on all 3 diameters.

recrease on 3 diameters

Open the circle. These new creases connect the 6 end points of the 3 diameters, forming an infolded hexagon where all points of the polygon lie on the circumference of the circle.

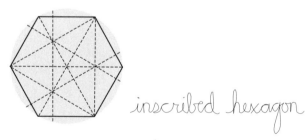

inscribed hexagon

Refold the triangle and flatten the new flaps out with the circumference folded into the center of the triangle. These newly-folded flaps form 3 small vesica shapes. The 360° of the circle has been folded into 180° of the triangle showing the relationship of the Pi function (p.39).

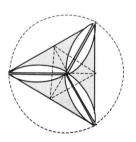

The second approach is more accurate when it is required. It takes a little longer and is less instructional to the corresponding grid lines and congruencies. Often doing the same thing in different ways brings up new information that is otherwise missed.

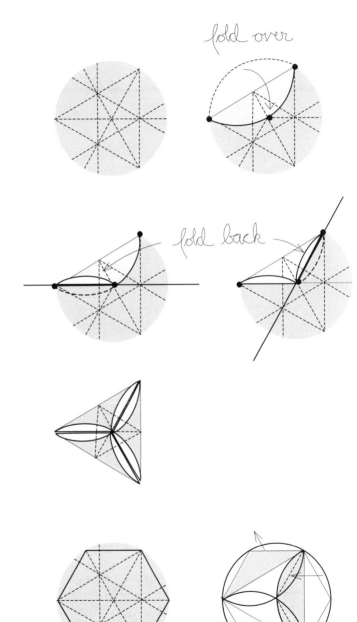

Fold   1/3 of the circumference to the center using already creased line. Locate the 2 end points of the folded edge and the center point to the circle.

Fold   a creased line between each end point and the center point. To make an accurate crease use a straight edge (a metal ruler works well) connecting the two points and fold over making a sharp crease. There are fewer layers of paper to fold, which should increase accuracy, but it takes longer to do.

Do the same thing to the remaining 2/3 of the circle. This will crease the next 4 sectors of the folded-over circumference.

The hexagon pattern inherent to the 3 diameters is now proportionally formed. Nothing has been added, taken away, constructed or regulated through measurement. There is a decrease in length from the circumference to the hexagon, to the triangle, and back to the circumference--a full circle. Three equal divisions of the circumference have been folded to the center showing the movement between the infolded hexagon and the triangle.

345

The infolded (inscribed) hexagon in the circle shows six radial-length lines have been added around the circumference. These lines show the distance between the end points of the 3 diameters, forming the six equilateral triangles to be the same in length as the radius. All six points are circles. The relationship between the straight-line hexagon and the circle is one of part to Whole.

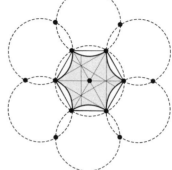

Above. Folding in the circumference around the inscribed hexagon shows the difference in perimeter and area between the circle and the hexagon. These lines give form to another size triangle of 30°, 60°, 90° relationship (p.332). Inherent in each arc is a complete circle. This reveals 6 more circles, 12 points around the center point, 13.

Below. This is another arrangement of 7 circles sharing radii. This image is a compressed viewing of spherical information (p.31).

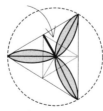
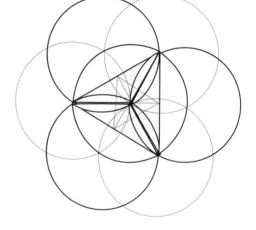

Fold the tetrahedron and then fold the inscribed hexagon. The folds reveal pre-formed relationships.

These 6 curved flaps group together in 3 pairs. Each pair forms a vesica or petal shape when folded flat.

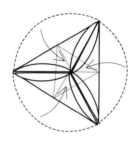

out from center   between points in   points to center

Each circle shares a radius with the center circle. Each circle is a center where every point of intersection is a circle (p.40). There are 3 sets of opposite points, tangent circles through the center. Three sets of 2 half vesicas are the difference between the hexagon and the circle which is represented by the number 3.1415... (p.39).

The circle moves out, forming the fullness of what it is. The circle moves into the hexagon, into the triangle, reforming into and generating the fullness of what it is.

346

REFORMING SPHERICAL VE

Infolding the hexagon opens options to reforming the vector equilibrium (VE) sphere (p.89). The hexagon flaps can now be folded in or out and in combinations with various reconfigurations.

<u>Fold</u>   2/3 of the triangle in with small curved flaps folded out. One third of the circle is left open.

The triangle point is folded to the opposite point on the same diameter line, folding the diameter in half. This is a similar forming as with the VE sphere (p.89). This forms 2 modified tetrahedra joined by an edge length. The edges are different than the VE while the angles are the same. Use a bobby pin to hold edges together.

Fold another circle exactly the same and join them together on the long edges using bobby pins to hold them. This is the same as the second step in forming the vector equilibrium, forming a quadrilateral from the 2 circles. There are also other ways to join these units.

Make another set of 2 circles in the same way and join them together on the short edge crossings, using bobby pins to hold them. This forms the spherical pattern of the VE 13 points reformed into a rectangular prism.

Reverse the joining and put the square openings together. Fold the circumference either in or out forming straight edges for joining. This forms 2 intersecting rhomboid prisms showing 14 points, the rhombidodecahedron. It is a stellated octahedron arrangement and a compound of two square prisms.

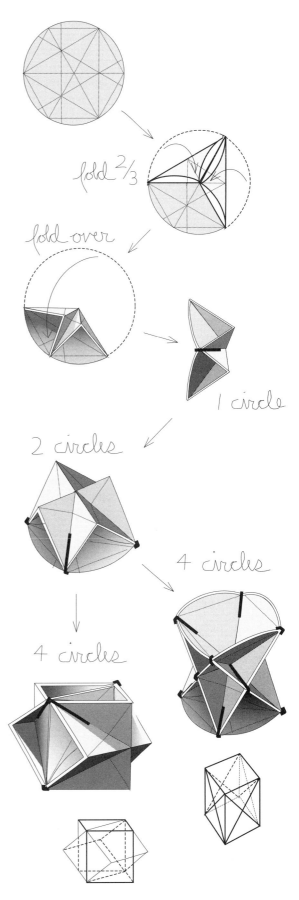

347

Both the rectangular prism and the VE sphere are the same pattern of 13 points of intersection, a function of 4 circles, of 3 diameters folded the same to different configurations. The prism shows 4 intersecting rectangles rather than 4 circles of the sphere. The difference is that of the circle and the inscribed hexagon. There would be no hexagon were there not a circle.

There are many combinations of curved and straight edges, in and out folded circumferences, with congruent open planes and surfaces. Spheres can be joined on the square and triangular faces by folding the circumference into a straight line edge.

Each modification of the sphere will determine how they can be joined and with what means for developing larger systems. There are many combinations of reconfigurations and variations in this spherical system.

Two spheres can be joined sharing rectangular faces, in the same way two bubbles join and share an internal plane. This type of joining can be done on one or more rectangular sides. This sharing of internal planes can be extended, generating endless clusters and chains of spheres.

Here are 3 spheres, 2 have one side folded in and the middle sphere has 2 opposite sides folded in. There are various degrees of infolding and many combinations where differently formed units are interchangeable.

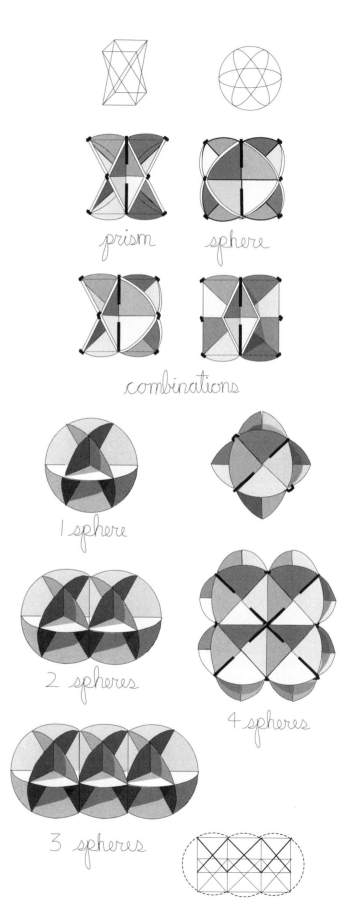

prism   sphere

combinations

1 sphere

2 spheres

4 spheres

3 spheres

348

Here we can see how much of the sphere gets folded in by changing the proportions of the 2 edge-joined tetrahedra on each circle. The 8 triangles and 6 squares show octahedra and tetrahedra configurations. The inside angles don't change. Each infolding reduces the volume of the sphere in the same way polygons reduce the area of the circle.

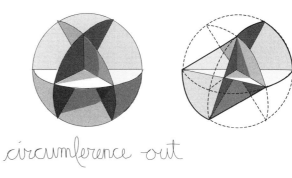

circumference out

The curved edges can be folded to form a straight line between the corner points of the top and bottom square spaces. The inscribed hexagon can be folded over to the outside or inside. This changes the look and methods of joining (pp.97, *Fig.4,5*).

folded in

octahedron

The square ends are congruent to the square intervals of the sphere as well as to the octahedron. They can be joined in linear, planer and spherical systems. In this way prisms and spheres can be used to form strut and hub systems. In attaching 6 prisms to the square openings on the sphere, the 3 axes of the octahedron are formed with a spherical center.

prisms end to end

349

The square ends of the prisms form 1/2 an octahedron cavity. The 6 square edges of the prisms are the center locations of the 3 axes of the octahedron/cube arrangements.

cube center

pattern formation

The 6 prisms joined together form a center inside cube by 12 edges from the 6 squares. The sides of the inverted pyramids, half octahedron at the square ends, form 12 rhomboids that make up the pattern of the rhombidodecahedron (p.232). The center joining is an intersection where these interrelated patterns form within a spherical system (p.97, *Fig.2b*).

expanded form

cubic system

Assembling 6 octahedron-patterned sets into a larger system consistent to the edge joining, shows the center cube relationship reflected on the outside as it is developed. The tetrahedron pattern of the cube is hidden and the cube form shows no face diagonals; the diagonals have been dropped out of the larger cubic system, they remain in the individual prisms that make up the system. The structural nature of this system is in the octahedral pattern of the sub-unit as they reform into the rhombidodecahedron.

The structural design can be seen in the diagonals of each unit. The rhombidodecahedra are formed by the diagonal edges of the rectangle prisms. This system is formed using tetrahedra/octahedra-patterned directives.

350

Forming this three-axial octahedral system using multiples will continue to repeat the octahedron arrangement in a fractal way, showing universal interconnectedness from the infinitely large to the infinitely small. Pattern not only replicates itself in as many forms as possible but towards both large and small extremes. Each multiple is a replica, even though we often call them patterns. This 2-D design is a compressed arrangement from a single point of viewing spatial organization. It is the same matrix viewed on page 253.

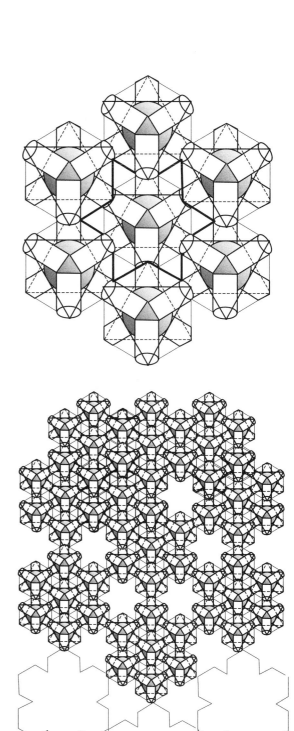

By using the drawing above and continuing the process of joining multiple units, a larger net, thus smaller division occurs. The pattern has not changed; the replicas are in greater number. *Scale* is a word we use to position relative size of parts within complexities of multiplication through a divisional process. This image shows hexagon star intervals that do not happen this way. The star areas would be spatial and not flat, they would be filled in with the density of layers in front and behind, making them impossible to locate. These star shapes represent simplified and generalized locations where the boundary stops and information has been dropped out. We can make wonderfully complex 2-dimensional designs simply by eliminating spatial information when compressed into a flat image. What is lost to product is the process and what gets left out we forget about; the forms distort, meaning has been diminished and origin is lost.

This drawing represents a spherical pattern that was started by folding a paper circle in half with consistent development of tetrahedra following directives from information generated in the folds of the circle.

351

12 POINT TETRAHEDRON STAR

<u>Fold</u>   tetrahedron with the infolded hexagon (p.344). Fold over 2 adjacent hexagon flaps that come together at one of the triangle points.

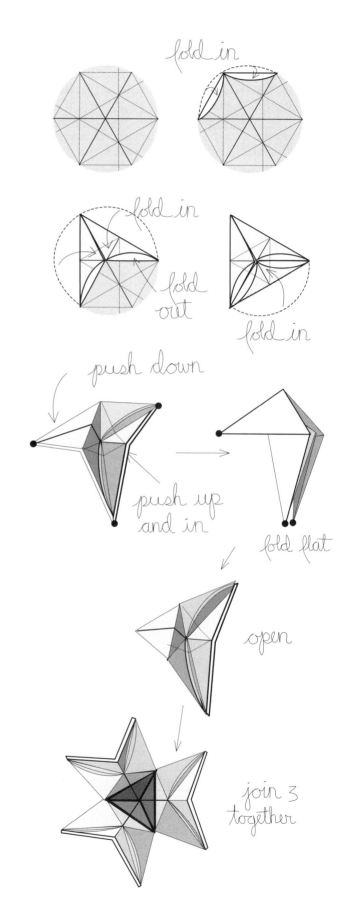

Fold in 2/3 of the equilateral triangle and bring the 2 folded straight edges together. Fold remaining circumference flaps out.

Fold in remaining 1/3 side to triangle, putting end of diameter to the center and folding the circumference flaps out, forming 2 vesicas.

Push down on the first folded triangle corner. Push up on the crease, from center out, opposite from point just folded down. The first triangle corner folded down will fold in the center along the diameter line and the entire folding will flatten out as shown. This will give a good memory to the folds in this position. If the points are a little ragged it will still work.

After it has been flattened open halfway, feel the movement of creases together.

Fold 2 more units the same. Flatten them and open as above. Tape the equilateral triangle folded ends together edge-to-edge, forming a tetrahedron with one open plane. Looking into the tetrahedron it will appear as a six-pointed star with the open tetrahedron center, an octahedron pattern.

Turn over the three-unit star and bring the adjacent edges at the 3 corners of the taped tetrahedron center together. Tape the edges so that the open face of the tetrahedron is to the outside

.

tape together edges

On the other side the tetrahedron surfaces will be inside the 3 points taped together with a slight opening to each point.

1 star unit

turn over

Make 3 more of these stars of 3 circles each.

Tape 2 stars together right triangle-face-to-triangle-face forming half a tetrahedron opening. Make 2 sets of 2 the same way.

join 2 star units together

Join the 2 sets, gluing right-triangle faces as above. This will complete the interior tetrahedron, forming an enclosed stellated tetrahedron. The 12 star points will form a vector equilibrium pattern.

4 star units

This diagram shows the 4 points of the tetrahedron centered in a star formation of 12 points.

Fig. 102 12 folded circles in 4 groups of 3 in a tetrahedron system. Each triangle side is formed by 3 stallions each within the open triangular arms. The 12 points form the 12 outside points of an irregular vector equilibrium (p.353).

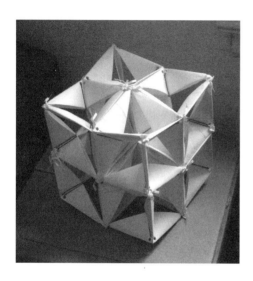

Fig. 103 32 circle reconfigured into 8 sets in a tetrahedron and assembled into a cubic system. The tetrahedral units are collapsible but when assembled as shown it becomes stable. This model is twisty tied together. For folding the units and process of assembly see page 218.

Fig. 104 This is the triangle prism star made with 6 folded circles and assembled as described on page 355-6.

Fig. 105 A square antiprism star made with 8 folded circles (p.*357*). It is the same folds and process but different symmetry as in *Fig.104*. Hair pins can be seen holding them together.

TRIANGLE PRISM STAR

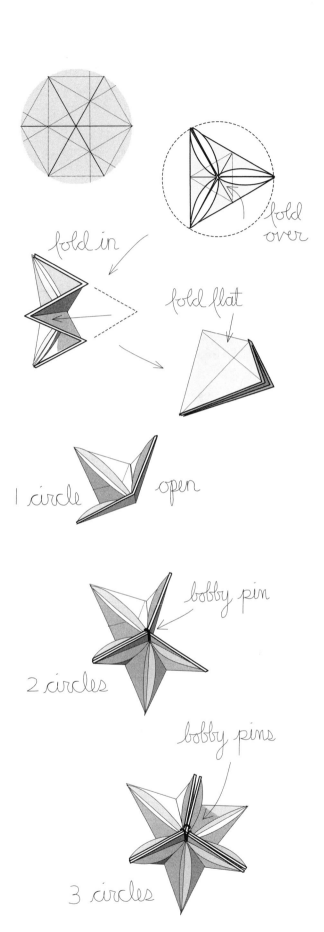

<u>Fold</u> the circle/tetrahedron with the infolded hexagon. This time fold a large triangle with all circumference flaps out, forming the 3 small vesicas.

Fold 1/3 of the triangle in along the long end of the bisector, from center to end point. As you do this, the other 2 points will begin to collapse towards each other. Make sure the circumference flaps are flat against the triangles as you continue to fold the triangle flat into a kite shape. Re-creasing gives memory to these folds for the next step.

Open about half way.

Fold 2 more units in the same way as above, making a total of 3. Attach 2, as shown to the right, using a bobby pin to hold them on the short side of the folded bisector.

Add the third in the same way. Three bobby pins will hold them together.

Crease another set of 3 circles, fold and join the same way with bobby pins

one set of 3

turn over

Turn the set over and look at the in-out design of the folds.

two sets of 3

join together

The 2 sets will fit one into the other, joining on the surfaces, with the end points going halfway in on the edge of the other. At that halfway point where they come together in a open "V" put a bobby pin to hold them together. There are 12 places to bobby pin. Or you can glue it if you do not want to take it apart. I would suggest bobby pins first, then glue.

This system of 6 reformed circles takes the arrangement of a slightly irregular triangle prism in an interesting 12-pointed star formation (p.354, *Fig.104*).

two views

356

SQUARE ANTI-PRISM STAR

<u>Fold</u>   the hexagon into the folded tetrahedron.

Reform as shown in making the single unit for the triangle prism star (p.355).

Bobbie pin 2 folded units together.

Make 2 sets of above and put them together in the same way as we did with 3. Adding one more unit makes 8 points instead of 6.

Make 2 sets of 4 in the same way. Again as we did with the triangle star, turn over and put the 4 in-out folds together (p.356). As they fit sliding half onto each other surface-to-surface, bobby pin at each of the eight overlapping "V"s. You might want to use bobby pins to hold the eight points closed where the vesicas come together, or you might want to glue the entire system.

The 2 right-angle crossings form 8 points of 2 squares twisted to each other makes an irregular square anti-prism arrangement (p.354, *Fig.105*).

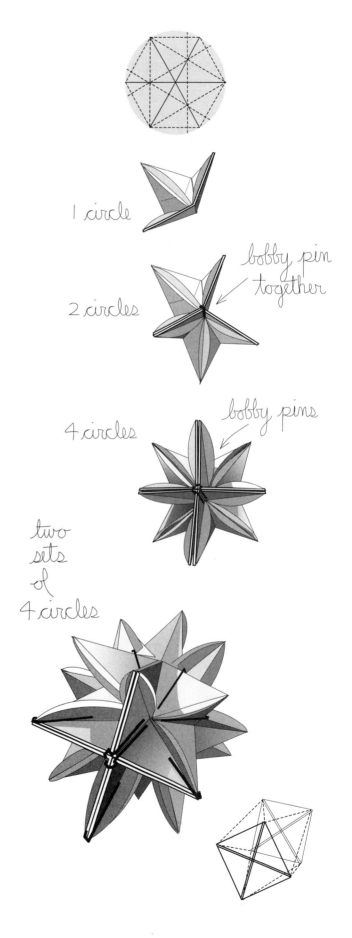

1 circle

2 circles

bobby pin together

4 circles

bobby pins

two sets of 4 circles

357

TWISTED TETRAHEDRON

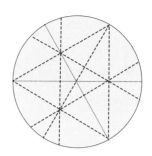

<u>Fold</u>   the tetrahedron and open the circle flat.

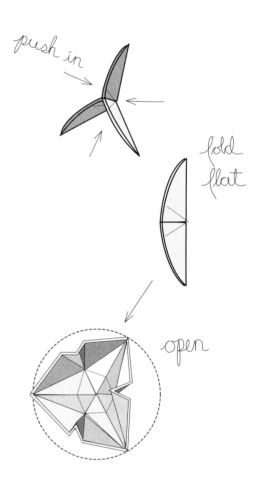

Fold the long parts of the 3 diameters up and the short parts to the back pushing in the midpoints of each side. This forms a three-pointed star with curved edges and straight edges (p.270).

Flatten the star. Fold 2 arms together and one by itself. Do this 3 times around so a memory is given to all the folds in 3 directions.

Open the star and at the mid-point on each edge of the large triangle find the 3 "V" tucks pointing in.

The open "V" tucks will fit into each other in multiples of 20 circles, that form an icosahedron pattern to a variation of the spherical icosidodecahedron (p.361, Fig.108).

Now close the triangle openings of the 3 "V"s using a bobby pin for each side to reform the circle and bring the large triangle to a flat plane.

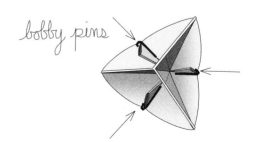

Fold one crease to meet the edge of the tuck (the tuck will be slightly longer than the crease). Put a bobby pin inside the tuck and over the creased edge, shown at right. This can be either a right or left hand direction.

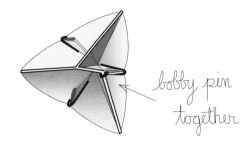

Do the same to the other two tucks. This will cause a twisting of the 3 ribs. Make sure all the bobby-pinned tucks are rotating in the same direction.

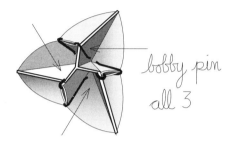

Fold 3 more circles in the same way, 4 in all.

Put them together into a tetrahedron pattern.
Join 3 units and bobby pin at the ends to hold them together. Join the fourth unit and bobby pin in the same way. This will leave an opening on each of the 6 edges forming a complex-looking tetrahedron open to the inside.

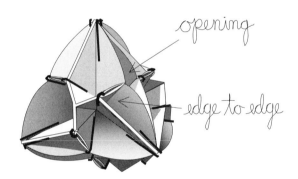

Variation: Close the openings by using bobby pins. Any time you use bobby pins you can also glue the edges and it will change the look when the pins are removed.

Leave edges open and slide the pinned tucks to the center of each side. This can also be done before the 4 units are assembled. Look for other variations.

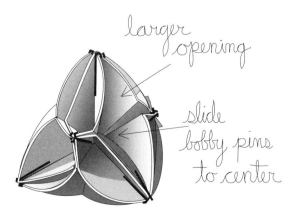

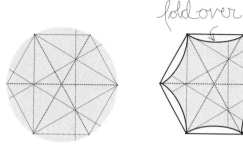

<u>Fold</u> the tetrahedron, infolding the hexagon (p.344). Fold the circumference flaps in.

Refold large triangle (p.358).

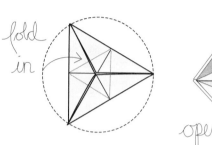

Open the triangle and fold "V" tucks behind, using tape to hold them flat to surface. This pulls the center point up in a stellated movement forming 3 radial ribs.

A.) Make 4 the same and put them together, joining the triangle planes in a tetrahedron configuration. Tape them on all 6 edges.

The joined triangles form a regular tetrahedron. The radial-formed ribs show the 12 edges of the cube. The corners are more than 90° angles. They deflect the square from the plane of the diagonals. This forms 12 congruent triangle planes rather than 6 square planes.

B.) Instead of taping the "V" tuck behind, bobby pin them on the top side where the ribs are. Open the "V" tuck by putting a pencil in and rounding a small cone.

In joining 4 it will take two pieces of tape to join each edge because of the rounded tucks sticking out. This triangle unit can be used to form other polyhedral systems.

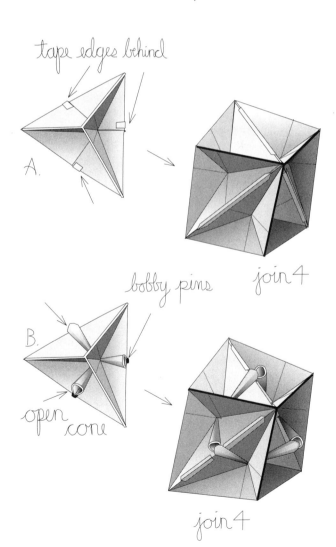

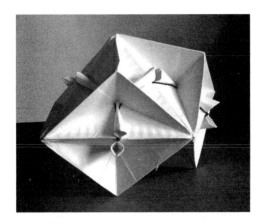

Fig.106   Two joined twisted cubes (p360). By removing one triangle from each unit and then joining them together on the open planes a bi-tetrahedron arrangement is formed looking like 2 interpenetrating cubes. This model is made using 6 folded tetrahedra from where the "V" tucks are opened and rounded. It is bobby pined on the tucks and and taped on the edges.

Fig. 107 Continuing this process by removing in the same way by adding open ended tetrahedron units. This forms a tetrahelix pattern with a very different look (p.202, *Fig.15*). The 3 twisting spines are still evident as well as 2 sets of 4 more non continious spines going in both directions. 18 paper plates are used with bobby pins and masking tape.

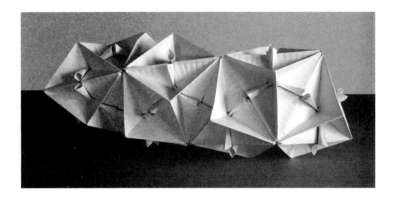

Fig. 108   The icosaheron is made from 20 paper plate units folded in the open position with the inscribed hexagon folded over before the "V" tucks are made (p358). The open tucks when one is put into the other atomatically form 12 open pentagons that give it a unique twisted look to this variation of the icosadodecahedron.

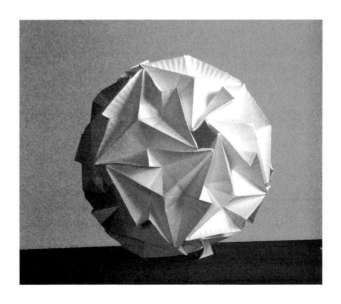

RHOMBOIDS

Start with the folded flat circle. Find a rhomboid along one of the three diameters.

<u>Fold</u>   steps 1-5 for the most even distribution of overlapping in folding the rhomboid shape.

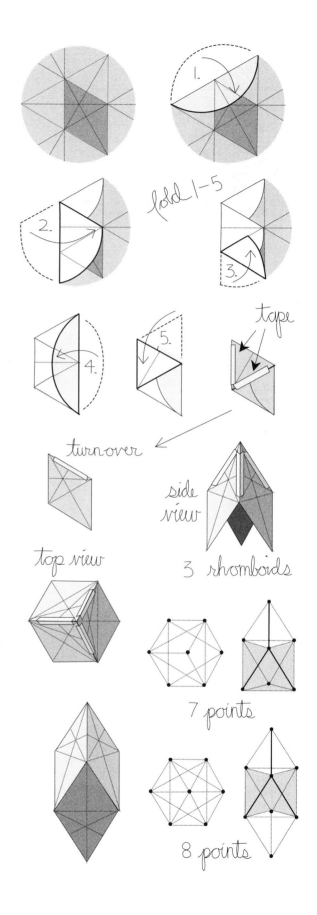

When folded, tape the edges down as shown.

Make 3 rhomboids as shown above. Put them together using the center circle triangles edge-to-edge forming an open-end tetrahedron. This arrangement shows 7 points in an octahedron and tetrahedron arrangement.

Make another set of 3 rhomboids joined in the same way. Join the 2 sets together edge-to-edge, similar to joining 2 open tetrahedra together forming an octahedron (p.129). Tape the edges. This forms one octahedron with a tetrahedra on each end, now an 8-point system of 6 rhomboid planes.   In the hexagon view 2 points are compressed into the center point.

362

With one set of 3 rhomboids, open one edge and add another rhomboid forming a square arrangement.

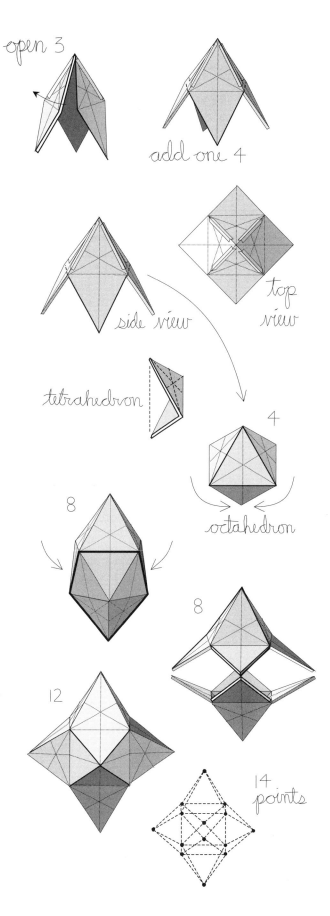

Each rhomboid is 2 equilateral triangles with 2 bisectors folded into it. Fold on the short bisector between the 3 triangles with each rhomboid. Bring the 4 triangles together edge-to-edge forming an octahedron.

Two open sets of 4 rhomboids, with the short bisectors folded slightly in and joined with another set of 4 fitting edge-to-edge, forms one of many *deltahedra* that can be made using individual equilateral triangles. These 2 sets of 4 rhomboids generate 8 interdependent irregular pentagons.

Open the 2 sets with triangle flaps touching each other point-to-point. This creates 4 spaces to be filled with 4 more rhomboids that have been slightly angled on the short bisector. Twelve rhomboids are used, similar to the rhombidodecahedron. Here the rhomboids are a different proportion and require a slight fold on the short bisector, changing one plane into 2 triangle planes, 24 triangles in a stellated cube relationship. This shows 14 points, which relate to the cube (p.93), and using 6 circles rather than 12 (p.169, *Fig.12*).

Open one side of the set of 4 rhomboids adding another rhomboid in the open space. This makes a set of 5 in a raised pentagon star.

(Adding a sixth rhomboid reduces the altitude to flat, making a hexagon star.)

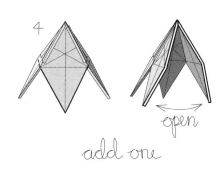

Tape the edges of the five rhomboids together.

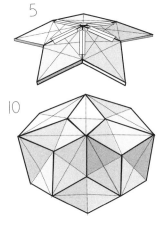

Add 5 more rhomboids around the pentagon star. Again they want to slightly fold on the short crease, making two triangles. The 120° angle of the rhomboid will not accommodate the openings for proper edge-to-edge connection without the planes changing. These 10 rhomboids form a larger pentagon pattern of 20 equilateral triangles in 5/10 symmetry.

With 10 rhomboids in a partial sphere add another row of 10 around the open edge. Notice the 120° angles again, causing the rhomboids to fold slightly making the triangles. This in/out of the triangles will form hexagons showing 3-6 point symmetry. The pentagon star points and the hexagon are all part of each other. The icosahedron and dodecahedron are beginning to emerge.

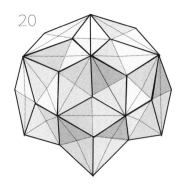

Continue to add rhomboids, forming pentagons until the sphere is completed. The last pentagon star is opposite the first star. Even though this was made with rhomboids, they are all slightly deformed to sets of equilateral triangles making a deltahedron of 60 faces.

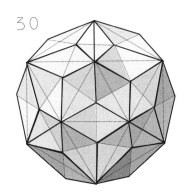

Five rhomboids of the pentagon plus 5 more makes a pentacap layer of 10 rhomboids. This is symmetrically reflected on the bottom. The band around the middle, "the belly band" has 10 rhomboids making a total of 30 rhomboids. Thirty gives reference to the number of edges for the icosahedron and dodecahedron.

By locating the pentagons, the dodecahedron becomes visible. By connecting the center points of the pentagon caps the icosahedron is revealed.

Looking at the vertex points where 3 edges come together shows the hexagon slightly folded in/out from the flat plane. There are 20 hexagons that are all intesecting each other as seen in the drawing of 2 intersecting circles. Here this happens in 5 rather than 6 which acounts for the folding of the rhomboids and spherical forming.

When assembling this deltahedron, if the triangle centers of each folded circle are consistently placed at the center of the pentagons, another set of relationships can be observed. Between the smaller 12 pentagons are 20 hexagon intervals. A truncated icosahedron is formed. Rather than cutting off the angles, we see a moving in and out, reforming the flat rhomboid surfaces. This is the traditional pattern of the soccer ball. Every child should know more about the design of the ball they play with.

There are 6-creased rhomboids in 3 sets of opposites at right angles to each other that reveal the 3 axes of the octahedron spherical division (pp.130, 181).

dodecahedron
and icosahedron

hexagons
and
pentagons

3 axes

As one of four points is lifted off the circle plane and folded over, a tetrahedron-patterned rhomboid is formed, two combined triangles. The rhomboid shape is a function of the triangulation of the circle and is folded on one of the 2 axial bisectors that form the tetrahedron interval (p.103).

By looking at the regular 60°/120° rhomboid we can get an idea of how other proportioned rhomboids function. The rhomboid appears in the circle with the folding of the 3 diameters. We also see it 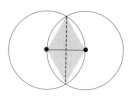 in the image of the dual circle radius. In the hexagon there are 6 interconnected rhomboid relationships (p.111). The square is a totally symmetrical rhomboid where 2 points on the circumference are exactly opposite each other so that the diameter and the curved line of movement bisect each other equally.

The rhomboid as a flat component to constructed polyhedra can be used in the same way we use equilateral triangles to construct various deltahedra. The rhomboid is the flat shape of 2 congruent triangles which, when folded on either bisector, forms an axial movement reflecting the first tetrahedral fold of the circle. There is a full tetrahedron range of 360° opening and closing. This is observed as one radius of the 3 diameters is folded out from the hexagon plane to form a pentagon. When the rhomboid/tetrahedral interval of the circle is completely collapsed by folding under, a square base pyramid is formed. The open base square is a rhomboid relationship of the 4 points. As soon as this square is out of square it becomes a tetrahedron that will collapse down into a pattern of the equilateral rhomboid (p.209). The rhomboid, as the square and pentagon, is a relationship of tetrahedra. The tetrahedron is in the pattern of spherical order where 4 is the expression of the spatial relationship of triangulation.

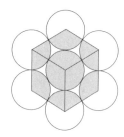 Here the rhomboid is shown as a relationship of centers of spheres in a hexagon layer. As a polygon, the rhomboid is only a small part of the information about the relationship of spheres. The spherical context shows space that breaks through the abstractness of surface plane and gives divisional information. The sphere is the context in which to understand the relationships of movement that occur in the space between the generalized and separated "solid shapes and forms".

Below. The 12 edge lines of the stellated octahedron/cube, when opened to 12 rhomboids, form an rhombidodecahedron. The 60°/120° proportioned rhomboid functions differently than those of the regular rhombidodecahedron. 1) Connections between the 3 and 4 symmetry vertex points form 12 quadrilateral intervals between the triangles of adjacent stellated arms. 2). Each interval of two triangle surfaces--4 points--form a tetrahedron interval. 3) There are 12 bisectors, 2 on each of 6 square faces, that define the 12 tetrahedral intervals between the arms, and correspond to 12 rhomboids. 4) The figure now changes, where the 12 spatial intervals, edge lines of the cube, are flattened to a single plane forming 12 rhomboids. This is another way to see the relationship between the form of the dual tetrahedron cube, stellated octahedron, and the rhombidodecahedron. They are all expressions of the three axis right-angle division of the sphere.

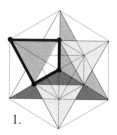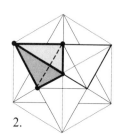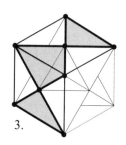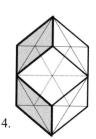

1.   2.   3.   4.

<u>Below</u>.   **5)** Three axes of symmetry through the cube, each perpendicular to the other.   **6)** Two opposite sides are stellated to the angle when the edges of the stellations are extended beyond the edges of the cube to the same length as the raised center points. Eight equilateral triangles of the octahedron are formed, the corners of the cube touching the midpoints of each formed edge. The cube has a pyramid on the top and bottom faces.   **7)** Truncating the 4 points around the middle of the octahedron reveals 4 rhombic faces plus 8 resultant rhomboids.   **8)** The 4-symmetry corresponds to the square diagonals, and the 3-symmetry to the cube corners. All 12 rhomboids are congruent.   **9)** The 8 corner points and the 6 points of the stellated faces of the cube are 14 points of the rhombidodecahedron; reflecting 14 spheres of the cubic arrangement (p.93). Euler's formula shows 14 points plus 12 planes minus 2 equals 24 edges (p.191). The numbers reveal relationships of transformational movement between different polyhedra.

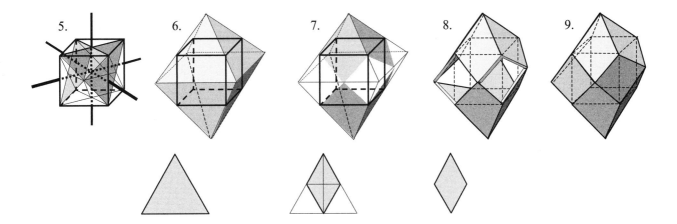

<u>Above</u>. The triangle shows the rhomboid as it lies in the two-frequency division. This indicates the changes from the cube to the rhombidodecahedron. Unlike this diagram where the short bisector is equal in length to the edge of the rhomboid, the regular rhombidodecahedron has irregular rhomboids where the square edge length, the short bisector, is longer than the edge length of the rhombus. This has to do with the altitude of the stellated square faces. One rhomboid is 1/2 of the triangle and 1/3 of the hexagon in the circle measure.

<u>Below</u>. Forming a stellated octahedron with the circumferences folded outside raises the center point of each 6 faces of the cube (p.232). The altitude of the movement out (stellating) is shorter than the length of the circumference arc. The difference between the regular rhomboid and the one found in the regular rhombidodecahedron is the circumference movement out through the square face of the cube

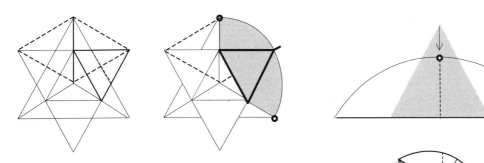

The vertical bisector of the triangle, altitude of tetrahedron, is less than 1/6 edge of the circle. When that is distributed over the circumference, the long bisector of the rhomboid, is extended, making the short bisector longer than the edge length of the triangle of the regular rhomboid.

367

<u>Below</u>. Opening the edge lines to vesicas shows the relationship of the twelve edges of the octahedron expanded to twenty-four edges of the rhombidodecahedron forming twelve rhomboids. **10)** Observe the octahedron with 8 triangle sides, 12 edges. **11)** Each side is a circle folded to a triangle where the 3 diameters intersect at the center location. **12)** Four of the 8 triangles (top half) are raised as the edges of each triangle are equally pulled in towards the raised center point. **13)** All 8 center points are raised and the circumference is pulled in towards the center of each point. The 12 edges of the octahedron open to 8 equidistance gravity centers in the triangles of the octahedron.

Each center point of the 8 triangles as an enlarged circle form 8 lesser circle divisions of a sphere that reveal 14 points of intersection forming the 12-vesica pedal shapes. On a flat plane two points is one line, on a sphere two points can also be an area of overlap or a line that has opened to become a plane (p.115).

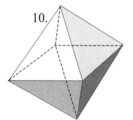 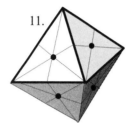 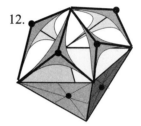 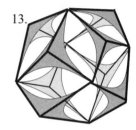

This is a stellations process that comes from the inside pushing outwards, pulling in the triangle edges towards the diameter lines which become the edge lengths of tetrahedral corners. This movement out from the surface moves the center of the circle 90° to the triangle plane that is moving inward. The openings between 12 triangle edges create the shape of 12 rhomboid spaces where the 4 points of each are all on the same plane. Both the vesica openings along with the rhomboid pattern appear simultaneously. The 8 vertexes of 3 planes coming together show the cube pattern, and the 6 vertexes of 4 planes coming together show the octahedron pattern in a position of balance. Here is a model where the octahedron, cube, and rhombidodecahedron, while individual, are aspects of each other.

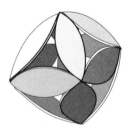

368

MOBIUS SURFACE

A paper strip joined end-to-end to itself is a loop with 2 continuous surfaces and 2 continuous edges. Traditionally the Mobius strip is a strip of paper that has been twisted 180° along its length and joined end-to-end, forming one continuous surface and one continuous edge. By twisting half of the 360° circle the number of discernable parts of the strip of paper has decreased by half; the length has increased twice.

By dividing the strip into a net of triangles, information is revealed that cannot be seen in the continuous flow of an undifferentiated surface. There are surface properties of the strip that can be extended to a surface of any shape.

<u>Fold</u>   3 tetrahedra and flatten to the triangle.

Fold one triangle onto the center triangle making a trapezoid shape. Tape the triangles down.

Attach the 3 trapezoid shapes in a line end-to-end and tape edge-to-edge with a hinge joint. There are 6 other natural hinges in the creases.

(There is a natural division of 5 triangles pointing up and 4 triangles pointing down; even numbers are in opposite orientation to the odd numbers. There is a nice symmetry of 3 sets of 3 numbers; 1+2+3=6, 4+5+6=15, 7+8+9=24. The difference between the numbers is 9. Each set of digits reduces to 6. 3x6=18. 1+8=9 or 3².) (p.123)

There are 4 ways to join the edges a, b, c, and d together forming a loop with an inside and outside surface. The combinations are (a to c) (a to d) (b to c) and (b to d). 1) The first 3 ways show the loop with 2 edges, one with a square interval and the other with a pentagon interval. 2) Folding b to d shows the hexagon with a tetrahedron interval.

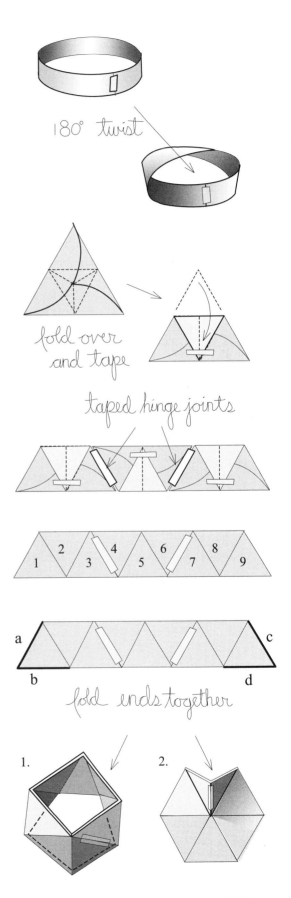

369

Tape each system together, joining on each side of each end. Play with them. Explore the different configurations as the pentagon collapses to a triangle, the square collapses to a triangle, to a line, an octahedron, tetrahedra, a helix formation. There is a lot here to be discovered.

By bringing the square and the pentagon together this system will collapse to a hexagon with two tetrahedra. The two tetrahedra are both open through to the opposite side like a hole. The hole allows connection from one side of the hexagon to the other.

Tape down the overlap triangle, forming the hexagon. Trace through the edge and the surface plane. It fits the description of the mobius strip but it is a hexagon with a hole. By closing part of one surface into itself, a flow from one side to the other is created. The edge starts from the center and returns to the center.

The Mobius strip can be a right handed or left-handed movement depending on which 2 triangles are lapped over, and on the direction of twisting. It will also take another form more open than the hexagon/tetrahedra form. This will show the square and the pentagon opened to each other.

As you play with these triangles it is not always a continuous surface, it moves out of the Mobius form into some open polyhedral forms. The square moves into the tetrahelix pattern of 4 tetrahedra, allowing for spin in either right or left direction. The 5 intervals go through various forms of pentagon, square and tetrahedral reformations.

One triangle doubled over another will keep the system of 8 from being able to move through its own center. It takes the ninth triangle for 8 to move through, much like a rolling set of 8 hinges with the ninth as fulcrum.

two surfaces

4 and 5 opening

3 and 6 opening

one surface hexagon with hole opened

hole closed

four in a tetrahelix

five

Go back to where we started with 9 triangles.

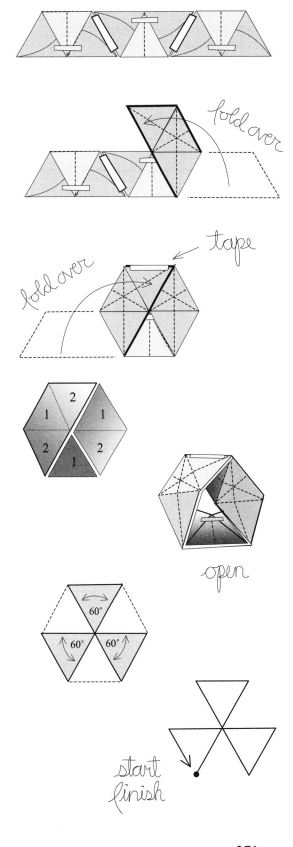

Fold over the first set of 3 triangles in the shape of a trapezoid.

Folding over the trapezoid on the left, leave the middle triangle exposed. Tape the outside edges where the 2 end triangles overlap. The strip has turned 180°.

This forms a hexagon with an overlap showing a triangle (1triangle), a rhomboid (2 triangles), and a trapezoid (3 triangles) with alternate overlap and no overlap. Ones are up, twos are down.

When the hexagon is open it forms a Mobius strip of 9 triangles. Each triangle will now successively move through the center space in a continuous movement around the ring, similar to the tetratorus (p.203).

Three sets of overlapping triangles are equivalent to turning the paper strip 180° to itself. (3 overlaps x 60° = 180° movement) The first move of the circle in half is 180°. When the 2 sides of 180° each are connected it becomes 360°, again reflecting the fundamental quality of 9.

The hexagon of 3 triangles can be drawn as a straight line with 6 continuous segments. This line from a starting point has 5 same-direction 60° turns, where every second distance is half the length of the first traveled. Six individual movements form 3 triangles revealing 3 more triangle relationships and one hexagon with seven points, all symmetrically balanced. Compare this to the regularity of drawing a circle (p.50).

This Mobius strip also functions as a torus, a single surface that rotates through an open center on a circular axis. By systematically pushing an outside point of this configuration to the center, one after the other in the same direction, it will rotate through the center space.

By closing 2 adjacent edges the unit changes form yet remains a single surface with one edge. It now has a tetrahedron bump with a concave interface and a bi-tetrahedron hole from one side to the other.

From the flat hexagon, pull open the top trapezoid and it will show a triangle opening on one side and a pentagon opening on the other, an inside/outside system. Open the remaining folded triangles and the two sides become one.

Opening the triangles.

Remove the three pieces of tape, which are holding down the triangles first folded over to make the trapezoids before taping the strip (p.369). Open the triangle flaps and flatten into a three-frequency triangle. As the triangle is opened it shows a Mobius surface of three slightly curved two-frequency triangles with folded over edges not found on a continuous and smooth surface.

With triangle flaps folded down, draw a centerline following the surface, marking a closed loop. Then open the triangle flaps and reform into a Mobius surface. The continuous line becomes discontinuous. The movement has not changed and the surface remains singular while the arrangement of lines has changed positions.

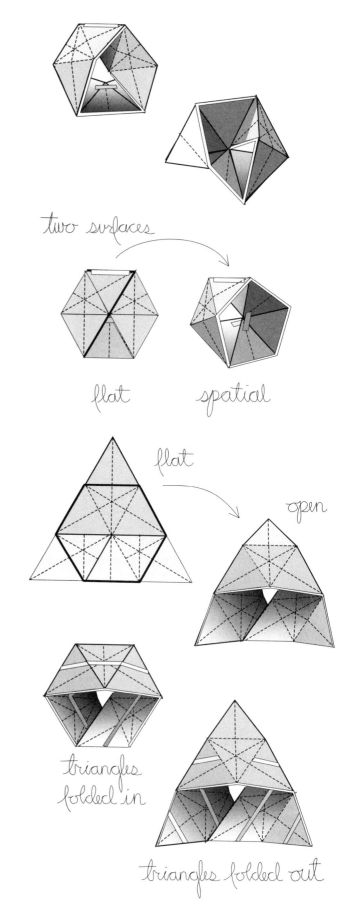

two surfaces

flat   spatial

flat

open

triangles
folded in

triangles folded out

This triangular form of the Mobius surface can be combined into multi-looped systems. The gray rhomboids to he right show areas of overlap.

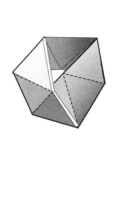

Being able to collapse the Mobius line of triangles back to the hexagon provides a way for connecting any number of loops in various arrangements, forming a single multi-formed surface of extended complexity.

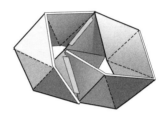 

One hexagon system paired with another system share a rhomboid by overlapping. This is similar to a one radius dual circle with shared vesica. Some overlapping will retain the single surface and some will not. When the matching is correct tape them together face-to-face.

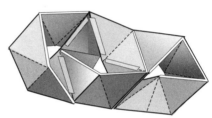 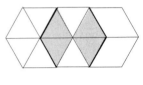

In exploring placement of multiple loops, there are different options of arrangement as well as orientation. Mobius loops can be joined in a circle, in lines or in polyhedral-patterned arrangements, in single continuous surfaces with many in and out combinations (p.376, *Fig.113*).

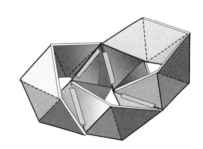 

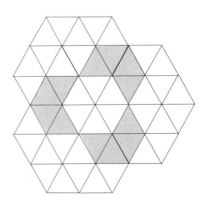

SOFT MOBIUS

We have seen examples of using the Mobius surface in soft folding in Section 8. Some of these are partial Mobius surfaces; the surface moves from front to back but also creates isolated surface areas. This does not fit a strict definition of a Mobius, but through 180° twisting and joining some interesting units can be developed into a variety of systems. Remember the circle is the exploration of all things possible and not just what we think.

<u>Fold</u> the circle into the triangle of 6 creases. You don't need the full tetrahedron folds. Tuck the flaps under each other.

Curl the left corner forward and inward. Curl the right side in and back. They both go in the same direction.

Continue to curve the left side towards the right, at the same time continue the twisting of the right corner over to meet the left corner so that the outside left overlaps the inside right in front of the triangle, rotating 180°.

It will look something like this. Tape the overlapping. Glue also works well for this joining.

Spread the opening. The third corner will curve to the back, up and around with the tip folded over the top to be glued or taped inside. The point will line up with the diameter crease on the inside. Even out and curve the edges with your finger, at the same time opening the spaces that move through each other.

This unit has been used for developing various systems (p.308, *Fig.95a,d*).

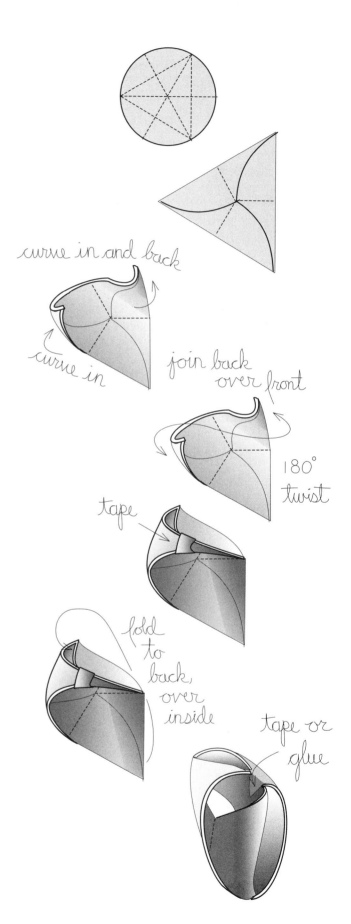

curve in and back

curve in

join back over front

180° twist

tape

fold to back, over inside

tape or glue

374

Another way to form a complex Mobius surface is to start with 2 large triangles.

Fold   each in half on one of the diameters.

Bring the 2 half-folded triangles together on the diameter edges. Tape them together from the center to the edge on the short segment of the diameter. Tape on both sides.

With the 2 long untaped parts of the diameters, twist and curve them towards each other, overlapping as much or as little as you want. (That will change the size of circle opening.) Tape or glue them together where they overlap. The front and backside of the triangle are now a continuous single surface and there is one edge that joins at the center.

Two options are to join on either of the outside curved edges. The edges can slide a little bit or all the way over to the center of each triangle, making a wider bridge between sides to the right hand or the left hand in direction.

There is a doubling over in the initial folding that will allow exploring multiple surfaces as well as the continuation of rearranging and increasing exposed surface.

With 2 sets of 2 ends there are 2 different combinations of a twisting that connects opposite surfaces into a continuous surface. (The same twisting shown on previous page.) This will form a three-open-looped surface. Curving the surface with your fingers is necessary.

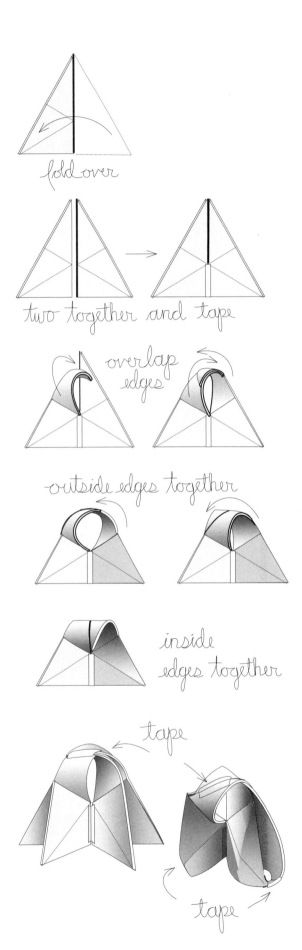

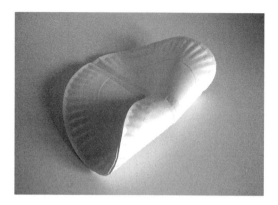

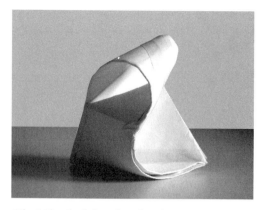

Fig.109 2 circles folded in half and joined on half the diameters edges. The untaped half of each circle is curved in opposite directions and joined to the center of the circle making a circle mobius surface. This fits the traditional one surface, one edge definition of a Mobius strip in circle form. The edge in this model goes through the surface from front to back. There is right hand and left hand folding of this surface. The triangulated circle is a variation (p.375).

Fig.110 2 right triangles (p375) joined and reconfigured to a tetrahedron mobius surface.

Fig.111 2 individually formed mobius surfaces using 2 triangles reconfigured differently.

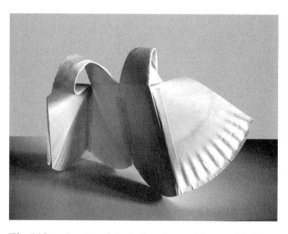

Fig.112 2 sets of 2 circles formed into a Mobius surface joined by a common edge (p.375).

Fig.113 3 Mobius strips joined forming 2 enclosed tetrahedra nodes of joining. The center line is seen as it traces the path of the single outside surface (p.373).

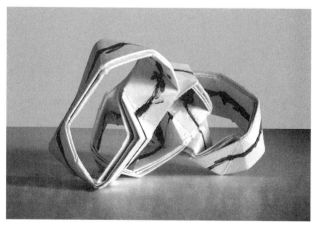

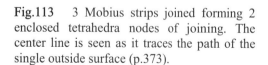

376

Eight seems to be the minimum number of equilateral triangles necessary to form a simple Mobius surface. In the same way 8 is the minimum number of regular tetrahedra that will form the rotating torus ring (p.203). When 9 triangles are used for the Mobius it can roll through its own center space. Using 18 triangles, 2 sets of 9, gives a closer approximation to the curve of the unfolded continuous strip. A 540° twist, one and a half turns (a 1:2 ratio), will double the Mobius twist keeping a single surface.

Joining the net of 8 equilateral triangles end-to-end in a circle, an open-end square anti-prism is formed with an inside and outside surface. When the strip is turned 180˚ and joined it will form 1/2 of a tetrahedron; one half of an octahedron and two tetrahedra. It still has an inside and an outside surface even though it has been turned 180° before joining.

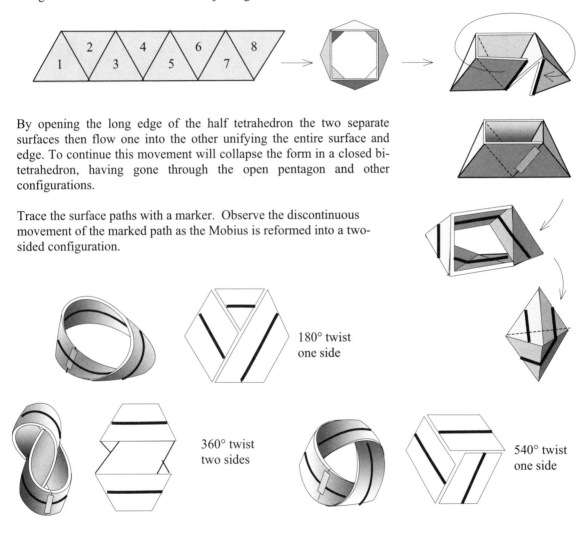

By opening the long edge of the half tetrahedron the two separate surfaces then flow one into the other unifying the entire surface and edge. To continue this movement will collapse the form in a closed bi-tetrahedron, having gone through the open pentagon and other configurations.

Trace the surface paths with a marker. Observe the discontinuous movement of the marked path as the Mobius is reformed into a two-sided configuration.

180° twist
one side

360° twist
two sides

540° twist
one side

The Mobius surface is not confined to a strip. It can take many forms of spatial complexity when considered as transitional movement between separated and continuous surfaces of a single system. The 180° movement makes sense because that is the first fold that reconfigures the circle appearing to be half of a circle. 360° is reformed into 180° by a right angle movement. Two half folded circles fully joined forms one Mobius circle (p.376, *Fig.109*). Ten tetrahedra can make a torus Mobius system (p.205).

THREE LOOPED MOBIUS

<u>Fold</u> 3 diameters and the two-frequency triangle.

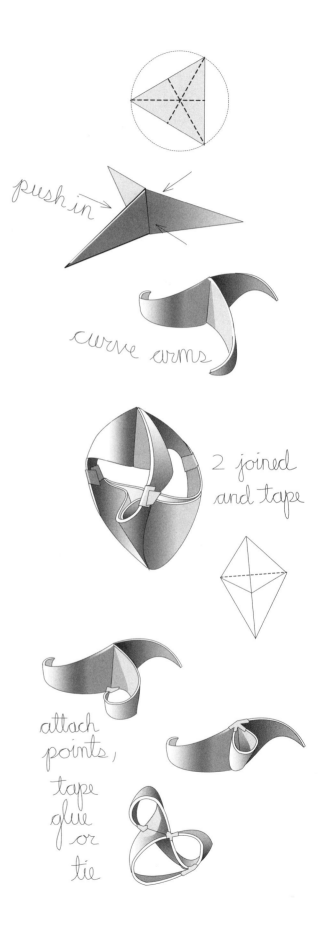

Fold in the short segments of the diameters making the long segments 3 ribs of the arms. Folding the hexagon will assure good creases with these folds (p.344). This is another direction of forming the basic 3–arm star (pp.75, 216, 218).

This time start by soft curving the arms going in the same direction. Use your fingers to form even curves. These are the units that are shown in the reformations on the next page.

Make another 3-arm star curving in the same direction. Put them together facing each other. Overlap the tips and tape together. The surface is now continuous. There are 3 edges that move through the 2 vertex points. This is a 3-looped Mobius surface using 2 circles (p.379, *Fig.117-8*). The inside of each half can be taped, glued, pinned, or left as they are. This is in a bi-tetrahedra pattern.

There are many variations to this basic 3-arm star in how the arms can be curved and attached. The end points can be attached at any point along the bottom edge, forming open spaces of various sizes. In the same way they can be attached to the top 3 spines. A few variations have been suggested (p.379, *Fig.116*).

These soft curved units are similar to those used for various helix and spirals systems (p.380, *Fig.123a,b*). This unit is the basis for the tangles (p.288, *Fig.61-62a,b*).

Looking at the simplicity of the form and folds of the 3-armed circle, the complexities of reconfigurations and combining into larger systems cannot be anticipated.

378

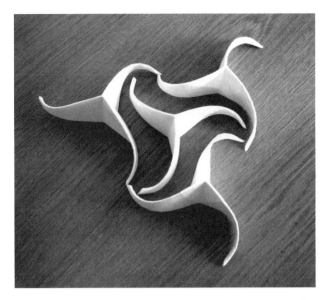

Fig. 115 The 4 stars in *Fig.114* have been rearranged into an open tetrahedron. Points are joined and held together with masking tape.

Fig. 114 4 three-arm stars in a triangle. The center circle shows the reverse orientation of the same configuration reflective of the 2-frequency triangle. This is the circular matrix (p.39).

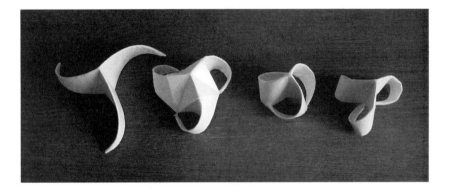

Fig. 116 3 of many possible reconfigurations of the star. Where the points are attached is makes the difference. The figure on the left next to star is from a fully folded tetrahedron with points attached to the bottom of the center axis. The figure on the right is from 6 creases with points joined to axis.

Fig. 117 The 3-looped Mobius with a single continuous line drawn on the surface. It is left hand to the model in *Fig.118*.

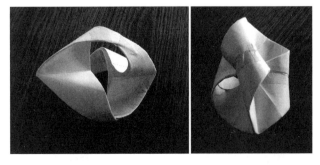

Fig. 119 Mobius where the 2 curved stars are joined edge-to-edge forming a tube closing in the open space *Fig.117*.

Fig. 118 A 3-looped Mobius using 2 circles. Joined on the points with open space between them (p.378). This is right hand to *Fig.117*.

379

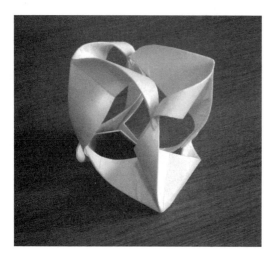

Fig.120   Another variation of 4 star soft formed circles in a tetrahedron arrangement. Unlike figure (p.379, *Fig115*) where 2 points are joined, 3 points are joined at each vertex making a relationship of 12 irregular triangles.

Fig.121a   The system from *Fig120* has been expanded by adding 4 more circles folded in the same way one onto the other.

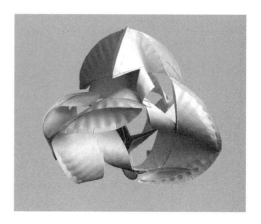

Fig. 121b   A view from the triangle plane of the tetrahedron arrangement from (*Fig121a*).

Fig.123a   10 units shown (p.379, *Fig116*) where only 2 points are attached to the center axis leaving one arm unattached. Each individual unit is partially slid into the next sequentially forming one complete revolution of a helix spin. This view is looking down through the central axis. (*Fig.122b*) shows the openness of the visual compression from this end view.

Fig.122b View from opposite end of (*Fig.123*). The individual units are more clearly visible.

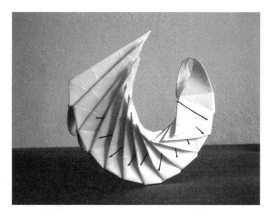

Fig.123a 12 paper plates folded to the unit on page 276, with 180° twisting forming ½ turn of a helix. Hair pins are holding it together.

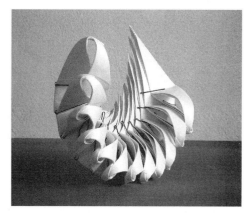

Fig.123b View of 12 from opposite side (*Fig.123a*) Here the mobius twisting is evident.

Fig.124 19 circles from (*Fig.123a-b*) in a different helix arrangement. The units have been changed slightly to give another angle relationship between them. The arms from one circle connect with the open spaces of other units on the developing helix.

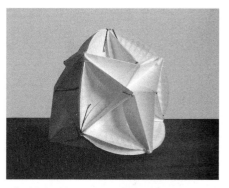

Fig.126a The same folding as (*Fig.125*) with 4 sets pushed in rather than still in a tetrahedron arrangement. The hexagon flaps are open in joining the 4 sets of 9" 3-armed star folded circles.

Fig.126b The same as above (*Fig.126a*) shows the edge, square symmetry (p,90). This is smaller scale using 2 ½" diameter circles and gluing together with the hexagon flaps folded over forming small vesicas.

Fig.125 2 scales of same tetrahedron arrangement using 12 circles. One with 9 inch diameter, the other using 2 ½ inch diameter circles. Sometimes it is problematic to assemble larger diameter circles and easier in a smaller scale. 3 folded 3-arm stars make a hexagon set with 4 sets in a tetrahedron joining.

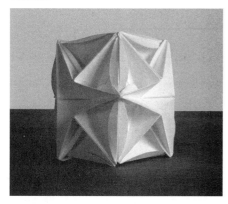

Fig.127 *5 circles in a 2 arm branching in one direction (p.54). Each diameter is reduced and folded in a 3-armed star.*

Fig.128 Fractal branching (*Fig.0*). in 3 directions. There are 46 circles in a 1, 3, 6, 12, 24 development.

Fig.129 Further development in one direction using the same 3 arm fold in fractal scaling (*Fig.128*).

Fig.130 This is the branching system from above (*Fig.128*) shifted from the flat plane to a semi-spherical arrangement.

Fig.131 This detail (*Fig.129*) shows how one arm fits into the folded slot of another. This allows for variations in position and angling along the length of the receiving arm. The circles around the 3-arms are 3 individual circles of a smaller diameter are reformed to fit into each of the 3 intervals holding the 3-arm folds rigid. They are similar to folded units in *Fig. 126-7* on page 381.

Fig132a

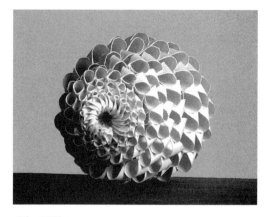

Fig.132b

Fig.132c

Fig.132d

Fig.132a-c. 4 views of a shell like spiral. Made from 162 circles starting from a 9" diameter paper plate down to a ¾" diameter circle. The sequence is in sets of 8 between each diminished diameter. This unit is seen on (p.296, 304, 277, *Fig.34*).

Fig.133a 19 paper plates in a helix arrangement. A slight variation of the mobius formed circle (p276).

Fig.133b Side view of 1 ½ turns of helix. Some of the circles have been crumpled and refolded for textural effect.

Fig.134 A number of curved 3-arm stars on a flat plane arrangement point-to-point show the relationship order of the circle matrix (p.39). The value of these reconfigurations are not in the forms themselves, though all are interesting, it is in the multiple associations and combinations that are possible through the changes inherent in the Whole circle/pattern.

Fig. 135a End point view of tetrahedron arrangement of 40 folded 3-arm star circles. The tetrahedral system (p.379, *Fig.115*) is shown in multiples forming a 6-frequency tetrahedron with only 3 layers of tetrahedral units.

Fig. 135b Edge view of the tetrahedral system in *Fig.121*. The units are taped together with bobby pins to hold the assembly of units. This can be understood as a spatially formation of the circle matrix (p.39).

Fig. 135c Detail of *Fig.121-122*. It is the point-to-point connections that hold the system in balance. This system is springy due to the curving of individual arms.

BIRD

We have modeled many forms that are recognizable as the more geometric parts of systems observable in our natural world. This last folding indicates one more direction using the same folds, which have not been explored in this book. As students become familiar with the basic folds they often come up with both imaginative towards realistic representational models of various kinds of creatures. All things formed with a square or other polygons can be folded within the circle; the circumference is the carrier of all information. It is the cell holding continuing generations of evolving forms.

<u>Fold</u>   2/3 of the circle into a triangle.

Bring the 2 short bisectors together folding the corner of the triangle up and into the space formed by the fold.

Use a bobby pin to hold edges together.

Pull out and down on the open vesica shape. At the same time bobby pinned edges will pull in a little bit.

Open the vesica with your finger and fold over one point to the opposite point. Curl over and curve end of the vesica up. This indicates 2 eyes, forehead, and a beak. You can give it specific personality by soft-forming different parts of the head.

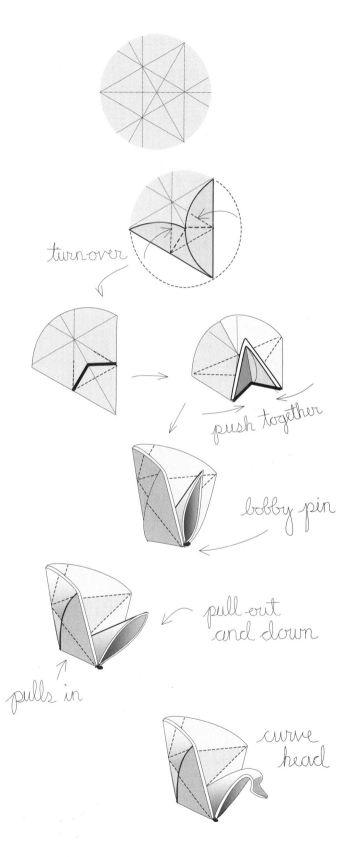

Curve the wings out with your fingers. You can do this to any angle desirable.

There are 2 creases on the bottom of the wings that can be used to open as the wings are curled to give more space and body to the bird. When using paper plates the fluting in the plate give a feather-like texture to the bird's wings.

Play around with combinations of the creases and soft folding. Curve wings to back or front, push up or pull down. Pull head out or pushed back in. The eyes and beak can be opened or push in. The beak can be rounded and folded down or formed long and sharp. Fold the head to one side, or the other. Spread bottom part to sit or close together to fly. Leave bobby pin in, take it out. This bird can be as individual as you like. Explore all the body changes and experiment with the subtle folding variations giving it different attitudes.

A streamlined bird with spread wings, similar to the one pictured to the right, will glide a bit with the right upward curves to the wings. Every fold and curve will reveal something new. Discoveries are made though exploration and informed through patterned interactions.

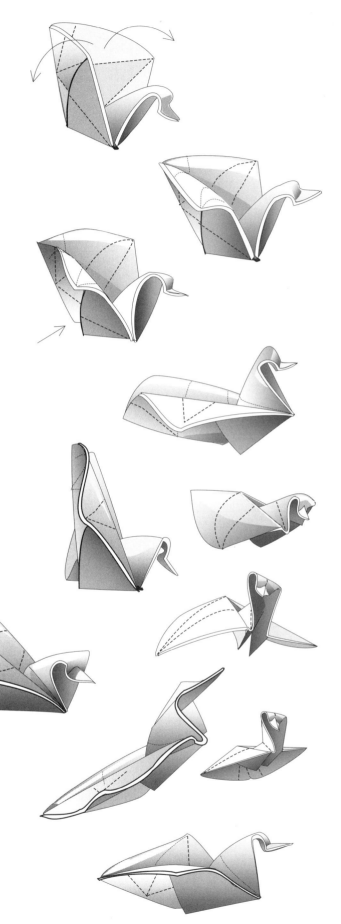